T0293547

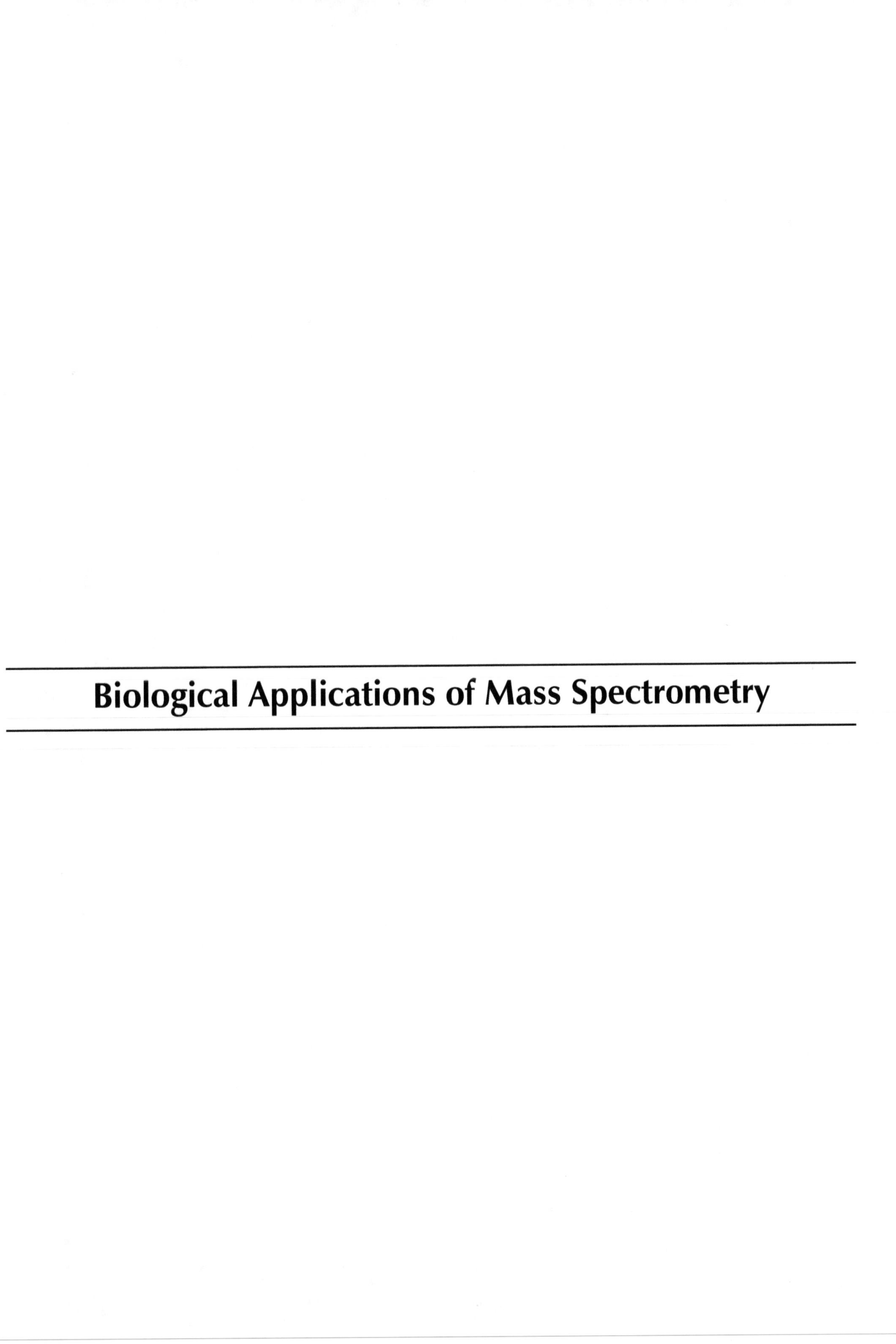

Biological Applications of Mass Spectrometry

Biological Applications of Mass Spectrometry

Editor: Max Corbyn

www.murphy-moorepublishing.com

MURPHY & MOORE

Cataloging-in-Publication Data

Biological applications of mass spectrometry / edited by Max Corbyn.
p. cm.
Includes bibliographical references and index.
ISBN 978-1-63987-736-2
1. Mass spectrometry. 2. Biochemistry--Technique. 3. Spectrum analysis.
4. Clinical chemistry. I. Corbyn, Max.
QP519.9.M3 B56 2023
572.36--dc23

Murphy & Moore Publishing
1 Rockefeller Plaza,
New York City,
NY 10020, USA

ISBN 978-1-63987-736-2

Contents

Preface

The world is advancing at a fast pace like never before. Therefore, the need is to keep up with the latest developments. This book was an idea that came to fruition when the specialists in the area realized the need to coordinate together and document essential themes in the subject. That's when I was requested to be the editor. Editing this book has been an honour as it brings together diverse authors researching on different streams of the field. The book collates essential materials contributed by veterans in the area which can be utilized by students and researchers alike.

Mass spectrometry (MS) refers to the measurement of mass-to-charge ratio of one or more molecules in a sample. These measurements are also utilized to determine the precise molecular weight of the sample constituents. It helps in the identification of unknown compounds through the determination of molecular weight, to determine chemical properties and structure of molecules, and measuring the known compounds. MS has grown to be an important analytical tool in biological research and is capable of characterizing a huge range of biomolecules, including oligonucleotides, sugars and proteins. It is useful in a variety of fields, such as metabolomics, biopharmaceutical and pharmaceutical research, forensic toxicology, clinical research, and proteomics. Food contamination detection, isotope ratio determination, carbon dating, pesticide residue analysis, protein identification, as well as drug discovery and testing are some of the specific uses of MS. This book explores all the important aspects of mass spectrometry and its biological applications. It aims to serve as a resource guide for students and experts alike and contribute to the growth of the knowledge on this topic.

Each chapter is a sole-standing publication that reflects each author's interpretation. Thus, the book displays a multi-facetted picture of our current understanding of application, resources and aspects of the field. I would like to thank the contributors of this book and my family for their endless support.

Editor

1

Determination and Pharmacokinetic Study of Gentiopicroside, Geniposide, Baicalin and Swertiamarin in Chinese Herbal Formulae after Oral Administration in Rats by LC-MS/MS

Chia-Ming Lu [1], Lie-Chwen Lin [1,2] and Tung-Hu Tsai [1,3,4,5,*]

1. Institute of Traditional Medicine, School of Medicine, National Yang-Ming University, No. 155, Sec. 2, Li-Nong St, Beitou District, Taipei 11221, Taiwan; E-Mails: a121060@gmail.com (C.-M.L.); lclin@nricm.edu.tw (L.-C.L.)
2. National Research Institute of Chinese Medicine, No. 155-1, Sec. 2, Li-Nong St., Beitou District, Taipei 11221, Taiwan
3. Graduate Institute of Acupuncture Science, China Medical University, No. 91, Hsueh-Shih Road, Taichung 404, Taiwan
4. School of Pharmacy, College of Pharmacy, Kaohsiung Medical University, No. 100, Shih-Chuan 1st Road, Kaohsiung 80708, Taiwan
5. Department of Education and Research, Taipei City Hospital, No.145, Zhengzhou Rd., Datong Dist., Taipei 103, Taiwan

* Author to whom correspondence should be addressed; E-Mail: thtsai@ym.edu.tw

---

Abstract: A sensitive and efficient liquid chromatography-tandem mass spectrometry (LC-MS/MS) method was developed for the simultaneous determination of gentiopicroside, geniposide, baicalin, and swertiamarin in rat plasma. To avoid the stress caused by restraint or anesthesia, a freely moving rat model was used to investigate the pharmacokinetics of herbal medicine after the administration of a traditional Chinese herbal prescription of Long-Dan-Xie-Gan-Tang (10 g/kg, p.o.). Analytes were separated by a C18 column with a gradient system of methanol–water containing 1 mM ammonium acetate with 0.1% formic acid. The linear ranges were 10–500 ng/mL for gentiopicroside, geniposide, and baicalin, and 5–250 ng/mL for swertiamarin in biological samples. The intra- and inter-day precision (relative standard deviation) ranged from 0.9% to 11.4% and 0.3% to 14.4%, respectively. The accuracy (relative error) was from −6.3% to 10.1% at all quality control levels. The analytical system provided adequate matrix effect and recovery with good precision and accuracy. The pharmacokinetic data demonstrated that the area under concentration-time curve (AUC) values of gentiopicroside, geniposide, baicalin, and swertiamarin were 1417 ± 83.8, 302 ± 25.8, 753 ± 86.2, and 2.5 ± 0.1 min μg/mL. The pharmacokinetic profiles provide constructive information for the dosage regimen of herbal medicine and also contribute to elucidate the absorption mechanism in herbal applications and pharmacological experiments.

Keywords: phytochemical analysis; LC-MS/MS; herbal medicine; pharmacokinetics; traditional Chinese medicine

---

1. Introduction

Long-Dan-Xie-Gan-Tang (LDXGT) is one of the best known traditional Chinese herbal prescriptions for the treatment of chronic hepatitis, jaundice, cystitis, and conjunctival congestion earache, as well as scrotum and extremitas inferior eczema [1,2]. According to the guidelines on Chinese herbal prescriptions (2013) from the Department of Chinese Medicine and Pharmacy, LDXGT consists of the following 10 herbal medicines: roots of *Gentiana scabra* (Chinese herbal name: long-dan-cao), roots of *Scutellaria baicalensis* (Chinese herbal name: huang-qin), fruits of *Gardenia jasminoides* (Chinese herbal name: zhi-zi), tubers of *Alisma orientalis* (Chinese herbal name: ze-xie), stems of *Clematis montana* (Chinese herbal name: mu-tong), seeds of *Plantago asiatica* (Chinese herbal name: che-qian-zi), roots of *Angelica sinensis* (Chinese herbal name: dang-gui), roots of *Rhemannia glutinosa* (Chinese herbal name: di-huang), roots of *Bupleurum chinense* (Chinese herbal name: chai-hu), and roots or rhizomes of *Glycyrrhiza uralensis* (Chinese herbal name: gan-cao), with a weight ratio of 4:2:2:4:2:2:2:2:4:2.

Although it has been reported that LDXGT has immunomodulatory effects on $CD4^+$, $CD25^+$ T cells and that it attenuates pathological signs in MRL/lpr mice [3], the bioactive components of the LDXGT compound have not yet been elucidated completely. Some of the components contained in the ingredient herbs, however, have been shown to have pharmacological properties. Several reports have indicated that gentiopicroside, geniposide, baicalin, and swertiamarin (the chemical structures are shown in Figure 1) have an important role in LDXGT. For example, gentiopicroside isolated from *Gentiana* root has been reported to promote the secretion of gastric juices and benefit the stomach [4]. Early pharmacological studies also have shown that gentiopicroside has antibacterial effects and free radical scavenging activity [5]. Furthermore, an animal study has documented that gentiopicroside provides protection against hepatitis induced by chemically and immunologically induced acute hepatic injuries [6]. Geniposide isolated from the fruits of *Gardenia* has shown anti-angiodenic and anti-inflammatory activities [7,8]. Baicalin, the flavonoids isolated from roots of *Scutellaria baicalensis* Georgi, have shown multi-functional efficacies, including antibacterial, antivirus, anti-inflammation and hepatoprotective activities [9,10]. Swertiamarin isolated from the *Gentiana* species has been reported to have important and extensive pharmacological activities, including antibacterial, hepatoprotective, antioxidant, anti-inflammatory, antinociceptive and antispastic properties, according to *in vitro* or *in vivo* pharmacodynamic experiments [11].

Figure 1. Structure formula of (**A**) gentiopicroside; (**B**) geniposide; (**C**) baicalin; (**D**) swertiamarin; and (**E**) carvedilol (IS).

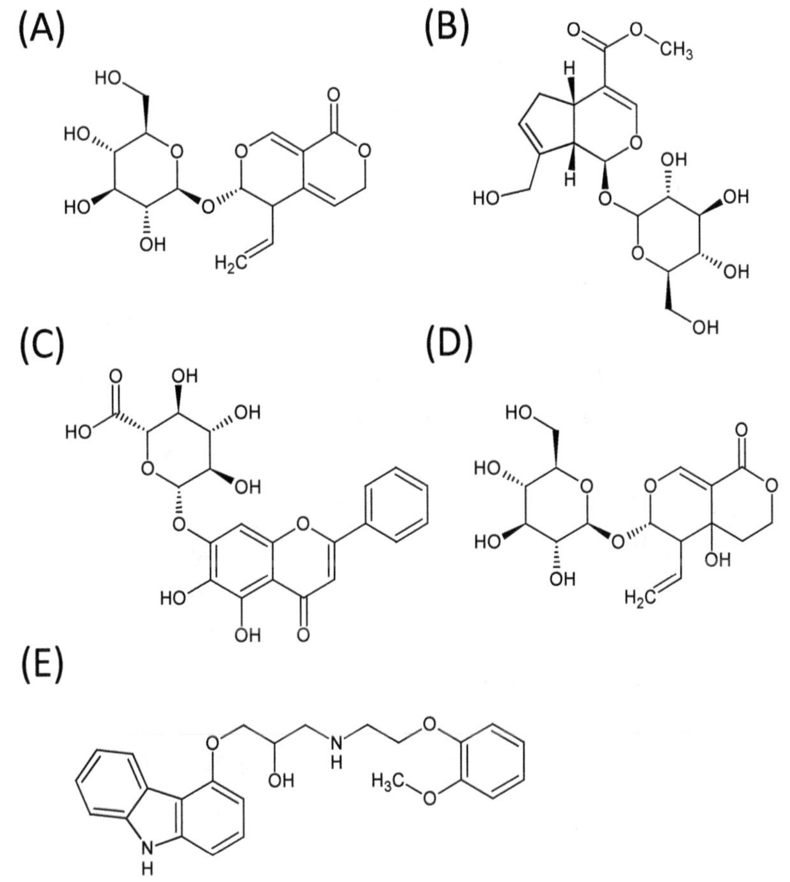

Recently, analytical methods have been established for analysis of the bioactive components of LDXGT by two-dimensional liquid chromatography with immobilized liposome chromatography column [12] and high-performance liquid chromatography coupled to photodiode array and electrospray ionization mass spectrometry (HPLC-DAD-ESI-MS) [13]. Although a previous study has investigated the plasma profiles and compared the pharmacokinetics of gentiopicroside in rats after oral administration of gentiopicroside alone, LDXGT, and a signal herb decoction of Radix Gentianae by HPLC [14], there is still limited information on simultaneous determination of the bioactive components of LDXGT by LC-MS/MS as applied to investigate its pharmacokinetics in freely moving rats.

The aim of this study was to develop and validate the LC-MS/MS method for the simultaneous determination of gentiopicroside, geniposide, baicalin, and swertiamarin in rat plasma after oral administration of LDXGT, and to investigate the absorption, distribution, and elimination of the multiple components in a traditional Chinese herbal prescription. It was expected that the four bioactive components of LDXGT would be detected in rat plasma and these herbal components were absorbed in multiple absorption sites or regulated by enterohepatic circulation. Furthermore, the pharmacokinetic study of the above components could help to elucidate the absorption mechanism of LDXGT for additional interpretation of traditional Chinese medicine.

2. Results and Discussion

2.1. Optimization of LC-MS/MS Conditions

To develop a sensitive and accurate analysis method for the determination of gentiopicroside, geniposide, baicalin, and swertiamarin, a triple quadruple mass spectrometer equipped with an electrospray ionization source is currently one of the most powerful tools for the simultaneous quantification of herbal components because of its high selectivity and sensitivity. Owing to investigation of the full scan and product ion scan mass spectra of analytes, the signal intensity in the positive mode was higher than that in the negative ion mode. Thus, all detection was carried out using the predominantly positive ion mode.

Qualification analysis of a triple-quadrupole mass spectrometer operated in the multiple reaction monitoring (MRM) mode consisting of two parts: selecting the precursor ion (MS 1), and selecting a specific fragment of product ion (MS 2). These two devices generated a very specific and sensitive response for the selected analyte, hence the integrated peak for the target component could be monitored in a sample after a simple one-dimensional chromatographic separation. The main mass fragments of the four active components are listed below: gentiopicroside: *m/z* 357.10 $[M+H]^+ \rightarrow 195.10$, collision energy: −10.0 eV; geniposide: *m/z* 406.10 $[M+NH_4]^+ \rightarrow 227.10$, collision energy: −10.0 eV; baicalin: *m/z* 447.00 $[M+H]^+ \rightarrow 271.05$, collision energy: −25.0 eV; swertiamarin: *m/z* 375.10 $[M+H]^+ \rightarrow 195.10$, collision energy: −10.0 eV; and carvedilol (IS): *m/z* 407.20 $[M+H]^+ \rightarrow 100.15$, collision energy: −35.0 eV, respectively (Table 1). The MS/MS parameters manually obtain the highest response for all of the precursor and product ion combinations.

Table 1. The analytical conditions of LC-MS/MS for the identification of the four components.

	Components	Molecular Weight	t_R (min)	Mass Fragments		Collision Energy (eV)
				Q1 Mass (amu)	Q3 Mass (amu)	
A	Gentiopicroside	356.32	4.0	357.10 $[M+H]^+$	195.10	−10.0
B	Geniposide	388.36	4.1	406.10 $[M+NH_4]^+$	227.10	−10.0
C	Baicalin	446.36	5.9	447.00 $[M+H]^+$	271.05	−25.0
D	Swertiamarin	374.12	3.6	375.10 $[M+H]^+$	195.10	−10.0
E	Carvedilol (IS)	406.48	5.8	407.20 $[M+H]^+$	100.15	−35.0

The retention time of gentiopicroside, geniposide, baicalin, swertiamarin, and carvedilol (IS) were 4.0, 4.1, 5.9, 3.6, and 5.8 min, respectively. To determine appropriate retention time, better resolution, and sensitivity, multiple chromatographic conditions were investigated. Several mobile phase systems in different ratios, such as acetonitrile–water and methanol–water, were examined in the course of analytical methods. The gradient elution, proper column, and flow rate were pivotal influences on separation for multi-components. Finally, a gradient system of methanol–water containing 1 mM ammonium acetate with 0.1% formic acid was chosen for the mobile phase in this study. The chromatographic conditions achieved are symmetric peak shape, good resolution, a short runtime (10 min), and appropriate ionization in the presence of endogenous species and co-elution (Figure 2).

Figure 2. Representative MRM chromatograms of gentiopicroside (channel 1), geniposide (channel 2), baicalin (channel 3), swertiamarin (channel 4), and carvedilol (IS) (channel 5) in rat plasma: (**A**) blank plasma samples; (**B**) blank plasma samples spiked with gentiopicroside, geniposide, baicalin, swertiamarin, and carvedilol (IS) at 500, 50, 100, 25, and 10 ng/mL, respectively; (**C**) diluted plasma sample (×10) of gentiopicroside, geniposide, and baicalin at 240 min, and plasma sample of swertiamarin at 90 min, after oral administration of Long-Dan-Xie-Gan-Tang (10 g/kg, p.o.).

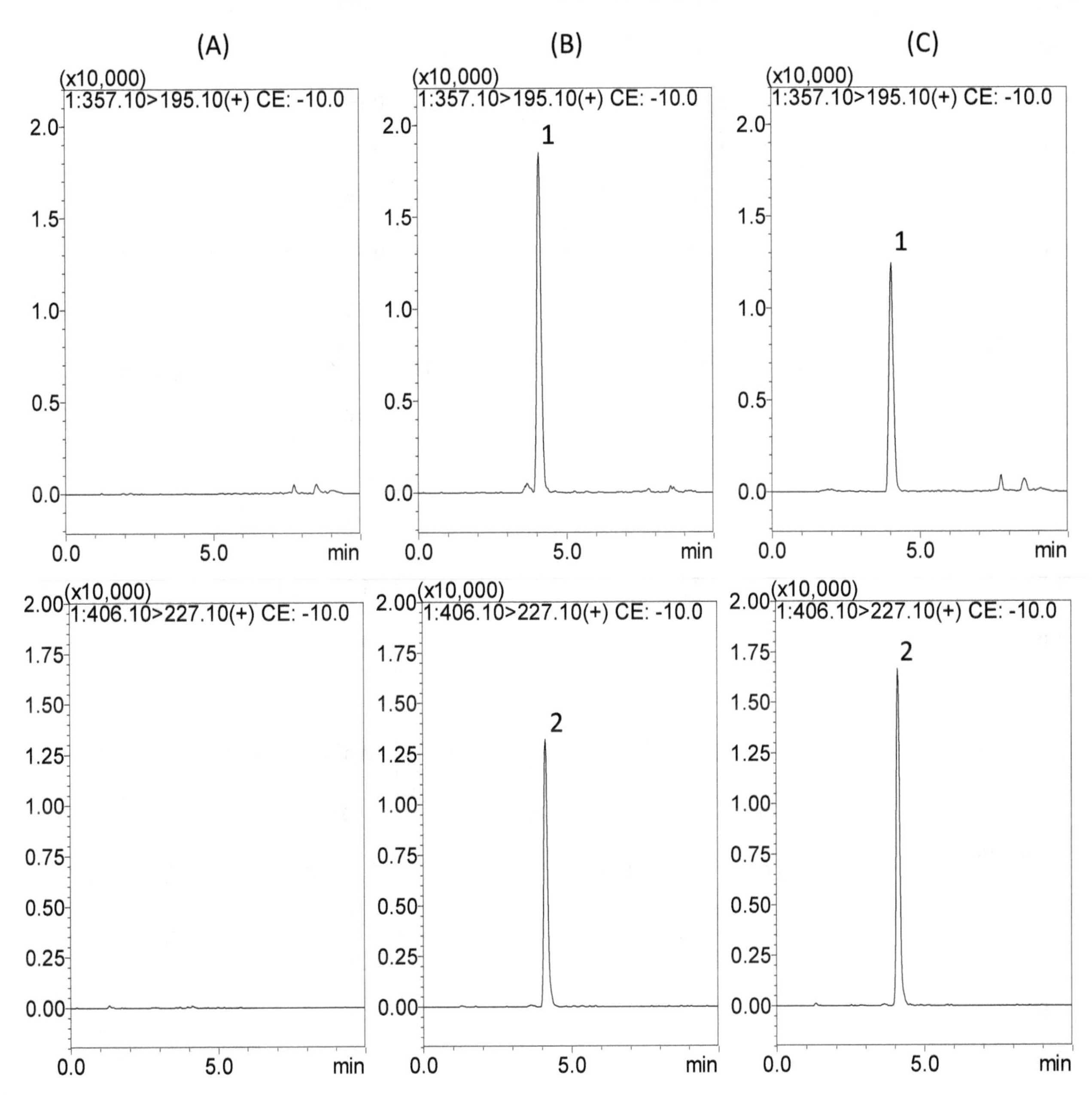

Figure 2. *Cont.*

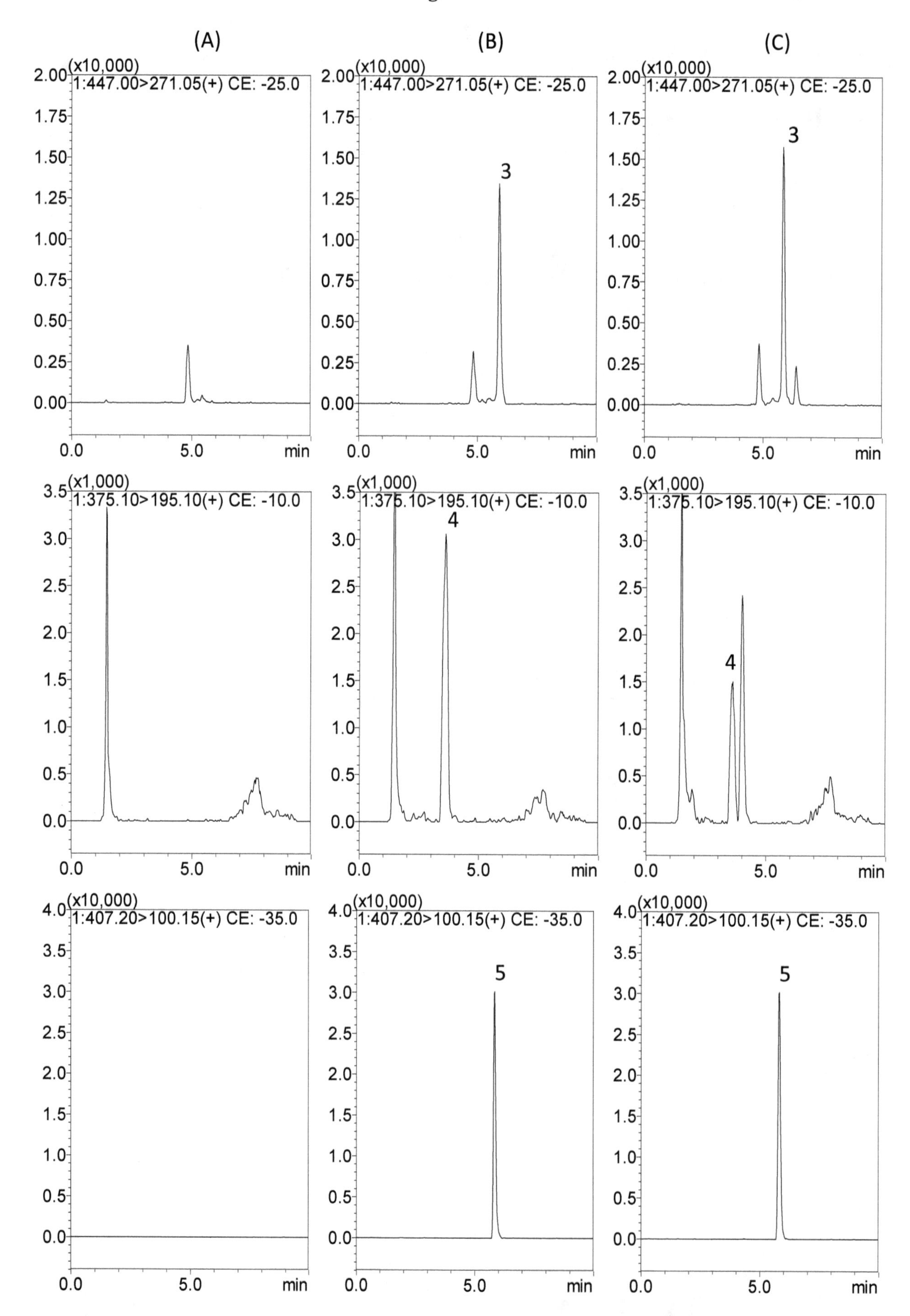

2.2. Protein Precipitation Methods for Sample Preparation

Since a great number of samples needed to be analyzed for pharmacokinetic analysis, having a simple, rapid, and economic sample preparation method is critical. Protein precipitation was more advisable and advantageous in the present work because it can not only ensure less endogenous interference, adequate recovery, and high sensitivity but also provides simple performance. The estimation of protein precipitation with some modifications was carried out following a previous report [15]. Methanol was more appropriate for reducing the matrix effect than acetonitrile. Satisfactory peak shape, matrix effect, and higher responses were obtained with the 5% formic acid addition. In the course of testing, methanol containing 5% formic acid solution (v/v) was chosen as the precipitation agent.

2.3. Method Validation

2.3.1. Selectivity

The selectivity was assessed by comparing the chromatograms of blank plasma samples obtained from six rats with corresponding spiked plasma samples. Figure 2 reveals that no interferences exist in the present analytical conditions.

2.3.2. Linearity, the Lower Limit of Determination (LLOD) and Lower Limit of Quantification (LLOQ)

For a standard calibration curve, the ratios of the chromatographic peak areas (analytes/internal standard) as ordinate variables were plotted *versus* the concentration of these drugs as abscissa. The linearity of calibration curves were demonstrated by the good determination of coefficients (r^2) obtained for the regression line. Good linearity was achieved over the calibration range, with all coefficients of correlation greater than 0.995. The mean values of regression equation of the analytes in rat plasma are listed as follows: $y = 0.0021x - 0.0013$ ($r^2 = 0.9998$, gentiopicroside); $y = 0.0069x + 0.0172$ ($r^2 = 0.9998$, geniposide), $y = 0.0050x - 0.0019$ ($r^2 = 0.9994$, baicalin), and $y = 0.0086x - 0.0099$ ($r^2 = 0.9998$, swertiamarin). The linear ranges were 10–500 ng/mL for gentiopicroside, geniposide, and baicalin, and 5–250 ng/mL for swertiamarin in biological samples.

The data showing the LLOD for the four components in rat plasma were gentiopicroside (3 ng/mL), geniposide (1 ng/mL), baicalin (3 ng/mL), and swertiamarin (1 ng/mL). Peak areas in chromatograms for the spiked plasma samples containing the above lowest concentrations were compared with the signal-to-noise ratio ≥ 3. Sensitivity is evaluated by the LLOQ determinations, which are defined as the lowest concentration that can be reliably and reproducibly measured in at least three replicates. The LLOQ values for the four components in rat plasma were gentiopicroside (10 ng/mL), geniposide (5 ng/mL), baicalin (10 ng/mL), and swertiamarin (5 ng/mL). Peak areas in the chromatograms for the spiked plasma samples containing the above lowest concentrations were compared with the signal-to noise ratio ≥ 10.

2.3.3. Precision and Accuracy

The intra- and inter-day precision and accuracy were determined by measuring six replicates of QC samples at six concentration levels. The performance data of the assay is summarized in Table 2. The intra- and inter-day precision (RSD, %) of four bioactive components ranged from 0.9% to 11.4% and from 0.3% to 14.4%, respectively. The accuracy (relative error, RE) was from −6.3% to 10.1% at all QC levels. These results are within the acceptable criteria for the FDA Bioanalytical Method Validation guidelines and show that the LC-MS/MS method was accurate, reliable, and reproducible for the quantitative analysis of gentiopicroside, geniposide, baicalin, and swertiamarin in rat plasma.

Table 2. Intra-day and inter-day precision and accuracy for the determination of four components.

Nominal conc.(ng/mL)	Intra-day			Inter-day		
	Observed conc. (ng/mL)	Precision, RSD (%)	Accuracy, Bias (%)	Observed conc. (ng/mL)	Precision, RSD (%)	Accuracy, Bias (%)
Gentiopicroside						
10	9.82 ± 0.98	10.0	−1.8	10.1 ± 1.45	14.4	0.9
25	23.5 ± 2.68	11.4	−6.0	25.7 ± 2.11	8.2	2.9
50	49.9 ± 4.29	8.6	−0.3	49.2 ± 1.26	2.6	−1.6
100	98.1 ± 2.52	2.6	−1.9	99.6 ± 2.91	2.9	−0.4
250	259 ± 6.67	2.6	3.5	251 ± 5.72	2.3	0.3
500	494 ± 5.09	1.0	−1.1	500 ± 2.82	0.6	−0.1
Geniposide						
10	9.37 ± 0.88	9.4	−6.3	9.77 ± 1.16	11.9	−2.3
25	25.4 ± 1.19	4.7	1.5	26.0 ± 1.96	7.5	3.9
50	49.5 ± 1.01	2.0	−0.9	49.8 ± 3.06	6.1	−0.4
100	98.2 ± 4.04	4.1	−1.8	100 ± 3.02	3.0	0.1
250	256 ± 5.66	2.2	2.3	249 ± 3.50	1.4	−0.4
500	497 ± 7.04	1.4	−0.5	500 ± 1.48	0.3	0.1
Baicalin						
10	11.0 ± 1.01	9.1	10.1	10.2 ± 1.51	14.8	2.1
25	24.9 ± 1.65	6.6	−0.3	25.8 ± 2.70	10.5	3.3
50	50.9 ± 1.92	3.8	1.9	47.3 ± 1.12	2.4	−5.4
100	98.4 ± 4.51	4.6	−1.6	99.3 ± 3.04	3.1	−0.7
250	245 ± 8.61	3.5	−1.9	250 ± 4.71	1.9	−0.1
500	491 ± 19.6	4.0	−1.7	505 ± 9.20	1.8	1.0
Swertiamarin						
5	5.35 ± 0.33	6.2	7.0	4.74 ± 0.25	5.3	−5.1
10	9.91 ± 0.66	6.7	−0.9	10.1 ± 0.53	5.2	1.0
25	24.8 ± 0.37	1.5	−0.9	24.9 ± 0.75	3.0	−0.4
50	50.0 ± 1.06	2.1	−0.1	49.8 ± 1.16	2.3	−0.4
100	99.6 ± 1.28	1.3	−0.4	101 ± 1.35	1.3	0.9
250	251 ± 2.20	0.9	0.5	253 ± 7.21	2.9	1.1

Data are expressed as mean ± S.D. ($n = 6$).

2.3.4. Matrix Effect and Recovery

To measure the matrix effect, it was determined at three different concentrations for all analytes. An absolute matrix effect was observed for some of the analytes at some of the low, middle, and high concentrations. Nevertheless, no relative matrix effects were seen, since the coefficients of variation at each concentration level were within ±20% for all components except geniposide (116.9% ± 6.9%) and baicalin (85.0% ± 13.6%), which were slightly out of range at 10 ng/mL. The ion suppression/enhancement in signal ranged substantially between 80% and 120% for the three different levels, indicating that the matrix effect on the ionization of analytes is not obvious under those conditions.

The extraction recoveries were also determined for three replicates from rat plasma spiked with low, middle, and high concentrations of the four components and internal standard. The mean recoveries of most samples were within 80%–120%, except geniposide. The recovery of geniposide was 119.4% ± 7.9% at a low concentration of 10 ng/mL. Even though it was slightly out of range, the extraction recovery of geniposide was stable and acceptable. The matrix effect and extraction recoveries from rat plasma are shown in Table 3.

Table 3. Matrix effect and extraction recovery of the four components and internal standard in rat plasma after sample preparation.

Nominal conc. (ng/mL)	Peak Area Set 1	Set 2	Set 3	Matrix Effect (%) [a]	Recovery (%) [b]
Gentiopicroside					
10	3911 ± 321	4001 ± 595	4312 ± 285	102.0 ± 7.4	108.7 ± 9.2
50	17829 ± 774	20372 ± 283	21498 ± 1749	114.4 ± 3.6	105.6 ± 9.9
250	89698 ± 3141	93932 ± 2505	104970 ± 4074	104.8 ± 4.1	111.7 ± 2.5
Mean ± S.D.				107.0 ± 7.3	108.7 ± 7.3
Geniposide					
10	12233 ± 190	14306 ± 1062	17028 ± 644	116.9 ± 6.9	119.4 ± 7.9
50	58115 ± 1045	67062 ± 1062	71407 ± 1764	115.4 ± 1.6	106.5 ± 1.1
250	295991 ± 3955	316422 ± 6847	346662 ± 8807	106.9 ± 3.7	109.6 ± 0.6
Mean ± S.D.				113.1 ± 6.1	111.8 ± 7.1
Baicalin					
10	9173 ± 232	7795 ± 1246	7825 ± 1054	85.0 ± 13.6	101.1 ± 10.7
50	45932 ± 1992	39518 ± 2684	37717 ± 460	86.0 ± 3.1	95.7 ± 5.6
250	224485 ± 4019	198185 ± 9541	186908 ± 6196	88.3 ± 3.2	94.4 ± 2.8
Mean ± S.D.				86.4 ± 7.3	97.1 ± 6.9
Swertiamarin					
10	16564 ± 490	16634 ± 397	17422 ± 660	100.5 ± 4.5	104.7 ± 1.7
50	76343 ± 1289	82585 ± 5095	88245 ± 4264	108.1 ± 4.8	107.0 ± 5.2
250	404293 ± 4777	432683 ± 3140	467239 ± 7470	107.0 ± 0.5	108.0 ± 1.0
Mean ± S.D.				105.2 ± 4.9	106.6 ± 3.1
Carvedilol (IS)					
10	76558 ± 6622	87040 ± 5353	83146 ± 3942	113.9 ± 2.9	95.7 ± 4.7

Data are expressed as mean ± S.D. ($n = 3$); [a] Matrix effect expressed as the ratio of the mean peak area of an analyte spiked post extraction (set 2) to the mean peak area of the same analyte standard (set 1) multiplied by 100. A value of >100% indicates ionization enhancement, and a value of <100% indicates ionization suppression; [b] Recovery calculated as the ratio of the mean peak area of an analyte spiked before extraction (set 3) to the mean peak area of an analyte spiked post extraction (set 2) multiplied by 100.

2.3.5. Stability

The stability of gentiopicroside, geniposide, baicalin, and swertiamarin in rat plasma was evaluated by using low, medium, and high concentrations of analytes. The four components were generally stable under the storage and analytical process conditions, including three freeze-thaw cycles, short-term (room temperature for 3 h), long-term (−20 °C for 7 days) and autosampler stability (8 °C for 12 h) in biological samples, and the deviation of the mean measured three different concentrations (10, 50, and 250 ng/mL) of the samples from the nominal concentration within ±15% of the initial values (Table 4). These stability results show no significant differences between the initial concentration and these QC samples.

Table 4. Stability of gentiopicroside, geniposide, baicalin, and swertiamarin in rat plasma QC samples.

Analytes/	Freeze-thaw Stability		Short-term Stability		Long-term Stability		Autosampler Stability	
Spiked Concentration (ng/mL)	Measured Concentration (ng/mL)	Accuracy (%)	Measured Concentration (ng/mL)	Accuracy (%)	Measured Concentration (ng/mL)	Accuracy (%)	Measured Concentration (ng/mL)	Accuracy (%)
Gentiopicroside								
10	11.3 ± 0.28	113.0	11.0 ± 0.40	110.2	9.82 ± 0.21	98.2	11.0 ± 1.09	109.8
50	53.7 ± 1.44	107.4	55.5 ± 1.68	111.0	51.1 ± 2.71	102.2	49.0 ± 2.47	97.9
250	265 ± 6.60	106.0	263 ± 16.2	105.1	240 ± 6.72	95.9	253 ± 10.7	101.1
Geniposide								
10	10.0 ± 0.94	100.3	11.4 ± 0.80	114.1	9.10 ± 0.77	91.1	10.3 ± 0.65	102.8
50	56.5 ± 1.00	113.0	52.9 ± 1.12	105.8	46.4 ± 4.16	92.7	48.9 ± 1.86	97.7
250	265 ± 10.9	106.0	260 ± 6.85	103.9	235 ± 15.4	94.1	248 ± 5.47	99.4
Baicalin								
10	10.3 ± 1.01	103.0	11.2 ± 1.14	111.8	9.54 ± 0.97	95.4	9.28 ± 0.72	92.8
50	49.5 ± 0.75	99.0	48.4 ± 1.18	96.7	49.7 ± 6.84	99.5	50.5 ± 4.33	101.0
250	231 ± 12.0	92.2	231 ± 10.8	92.6	238 ± 33.2	95.3	231 ± 1.57	92.4
Swertiamarin								
10	10.6 ± 0.76	105.7	11.3 ± 0.99	112.7	9.50 ± 0.50	95.0	10.6 ± 0.78	106.1
50	49.8 ± 0.66	99.6	48.5 ± 0.23	96.9	45.7 ± 3.21	91.3	49.9 ± 1.08	99.7
250	247 ± 2.74	98.7	252 ± 8.03	100.9	224 ± 10.7	89.8	240 ± 3.52	96.1

Data are expressed as mean ± S.D. (n = 3). The accuracy (%) was calculated as follows: Accuracy (%) = C_{obs}/C_{nom} × 100%; Freeze-thaw stability: three freeze-thaw cycles; Short-term stability: room temperature for 3 h; Long-term stability: kept at −20 °C for 7 days; Autosampler stability: 8 °C for 12 h at the autosampler.

2.4. Application of the Analytical System in Pharmacokinetic Study

The validated LC-MS/MS method was successfully applied for the pharmacokinetic study of the four active components after LDXGT administration (10 g/kg, p.o.) in freely moving rats. The mean plasma concentration-time profiles are illustrated in Figure 3 and the pharmacokinetic parameters are presented in Table 5.

Gentiopicroside showed a single absorption phase in the concentration-time curves and was rapidly absorbed with detectable levels of gentiopicroside at 5 min in rat plasma after LDXGT oral administration [15]. A previous study also indicated that gentiopicroside could be absorbed immediately in mice [16]. After rapidly achieving maximal levels of plasma concentrations, gentiopicroside declined sharply and was followed by a slower phase of decrease until 4 h. Moreover, the pharmacokinetic parameters of gentiopicroside showed the largest C_{max} at 5767 ± 412 ng/mL ($p < 0.05$) and AUC at 1417 ± 83.8 min μg/mL ($p < 0.05$) among the four bioactive components after LDXGT administration.

Figure 3. Mean plasma concentration-time profile of (**A**) gentiopicroside; (**B**) geniposide; (**C**) baicalin; and (**D**) swertiamarin in rat plasma after oral administration of Long-Dan-Xie-Gan-Tang (LDXGT; 10 g/kg). The herbal contents of LDXGT are gentiopicroside 17.04 mg/g, geniposide 25.08 mg/g, baicalin 9.94 mg/g, and swertiamarin 0.22 mg/g. Data are expressed as mean ± S.E.M. ($n = 6$).

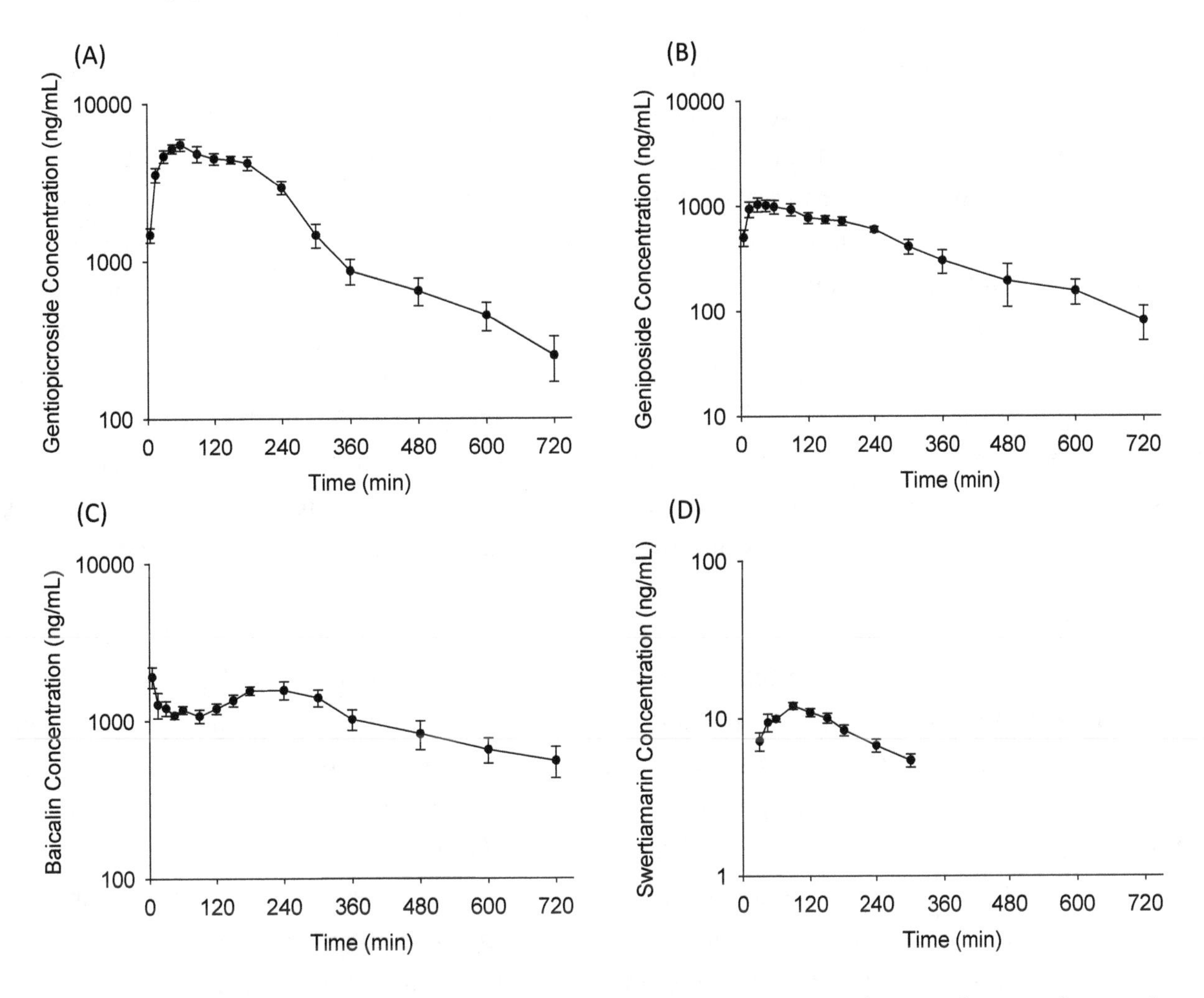

Table 5. Pharmacokinetic parameters of the four components in rat plasma after oral administration of LDXGT (10 g/kg, p.o.).

Parameter	Gentiopicroside	Geniposide	Baicalin	Swertiamarin
C_{max} (ng/mL)	5767 ± 412 [a]	1164 ± 144	2008 ± 265	13.0 ± 0.5
T_{max} (min)	60.0 ± 6.7	55.0 ± 12.6	15.8 ± 9.0	97.5 ± 14.4
$t_{1/2}$ (min)	127 ± 7.7	168 ± 27.4	314 ± 56.3	188 ± 27.6
AUC (min µg/mL)	1417 ± 83.8 [b]	302 ± 25.8	753 ± 86.2	2.5 ± 0.1
Cl (mL/min/kg)	118 ± 6.5	784 ± 55.5	112 ± 20.7	581 ± 58.5
MRT (min)	193 ± 7.9	213 ± 18.2	298 ± 11.6	144 ± 3.1

Data expressed as mean ± S.E.M. ($n = 6$); C_{max}: the maximum plasma concentration, T_{max}: time to reach the maximum concentrations, $t_{1/2}$: half-life, AUC: area under the concentration-time curve, CL: clearance, MRT: mean resident time; [a] Significantly different ($p < 0.05$) from the other active components in C_{max}; [b] Significantly different ($p < 0.05$) from the other active components in AUC.

Geniposide displayed a plateau absorption phase in the concentration-time curves and showed a slower phase of elimination in our pharmacokinetic profiles. Although the content of geniposide (25.08 mg/g) was higher than gentiopicroside (17.04 mg/g) in LDXGT, geniposide did not show the largest C_{max} (1164 ± 144 ng/mL) and AUC (302 ± 25.8 min μg/mL) in pharmacokinetics. There are many possible explanations, and one of the influencing factors might be that the relative bioavailability of gentiopicroside (10.3%) was higher than geniposide (4.2%) [15,17]. In addition, the previous results of Sun *et al.* [18] illustrated that the oral bioavailability of geniposide was dramatically enhanced in different combinations of its constituent herbs. It can be deduced that the absorption and oral bioavailability of geniposide in rats may significantly vary from herb–herb interaction [17,19].

Swertiamarin showed the lowest level of AUC (2.5 ± 0.1 min μg/mL) and fell below the lower limit of quantification (LLOQ) at 300 min in the concentration-time curves after LDXGT administration. The data of Li *et al.* [20] also demonstrated that swertiamarin showed rapid absorption and elimination, and high concentrations were found in the liver and kidneys, indicating that swertiamarin was rapidly metabolized in the liver and eliminated by the kidneys. Furthermore, Wang *et al.* [21] previously reported that swertiamarin was biotransformed to erythrocentaurin and 3,4-dihydro-5-(hydroxymethyl) isochroman-1-one by human intestinal bacteria. It was assumed that this low level in rat plasma was due to either the low content in LDXGT (0.22 mg/g) or the first pass effect. The possible reasons should be confirmed in our further work and we will seek to identify and detect the major metabolites of swertiamarin *in vivo*.

Baicalin displayed rapid and sustained absorption (T_{max} = 15.8 ± 9.0 min, $t_{1/2}$ = 314 ± 56.3 min). The concentration-time curves of baicalin presented a double-peak absorption phase in the plasma profile. There were many possible explanations for the phenomenon of multiple-peak behavior. Shaw *et al.* [22] indicated that several mechanisms have been proposed for the phenomenon: (1) enterohepatic recycling; (2) the presence of absorption sites along the stomach and different gastrointestinal segments; and (3) variable gastric emptying. Furthermore, the pharmacokinetic results of Zhang *et al.* [23] also demonstrated that baicalin presented a bimodal phenomenon in the plasma profile, and it is generally assumed that baicalin is poorly absorbed from the gastrointestinal tract in its native form and must be hydrolyzed by microflora enzymes (bacterial glucuronidase) in the gut into aglycone-baicalein, and then reconverted back to baicalin. Our pharmacokinetic results indicate that there would be multiple absorption sites or regulation of enterohepatic circulation of baicalin in rats.

In this study, we first focused our investigation on the pharmacokinetic profile of major bioactive components in *Gentiana scabra* (gentiopicroside and swertiamarin), *Gardenia jasminoides* (geniposide), and *Scutellaria baicalensis* (baicalin). The validated LC-MS/MS method was successfully applied to a pharmacokinetic study in freely moving rats after administration of a Chinese herbal prescription of Long-Dan-Xie-Gan-Tang.

3. Experimental

3.1. Chemicals and Reagents

The chemicals gentiopicroside and swertiamarin were purchased from Nacalai Tesque, Inc. (Kyoto, Japan). Geniposide, baicalin, carvedilol (internal standard, IS), pentobarbital, and heparin were obtained from Sigma-Aldrich Chemicals (St. Louis, MO, USA). Ammonium acetate, formic acid, methanol of HPLC grade, and other reagents were purchased from E. Merck (Darmstadt, Germany). Triply deionized water was prepared by Millipore (Milford, MA, USA) and used for all preparations in this study.

3.2. Herbal Preparation of Long-Dan-Xie-Gan-Tang

The 10 herbs of LDXGT were purchased from a Chinese traditional herbal medicine store in Taipei and prepared in the National Research Institute of Chinese Medicine, Taipei, Taiwan. According to a previous study on the preparation of herbal formulas [22], the crushed herbs of *Gentiana scabra* (69.3 g), *Scutellaria baicalensis* (34.6 g), *Gardenia jasminoides* (34.6 g), *Alisma orientalis* (69.3 g), *Clematis montana* (34.6 g), *Plantago asiatica* (34.6 g), *Angelica sinensis* (34.6 g), *Rhemannia glutinosa* (34.6 g), *Bupleurum chinense* (69.3 g), and *Glycyrrhiza uralensis* (34.6 g) were co-boiled with 9000 mL water in a water bath at 70 °C for 9 h and filtered. This process was repeated once and then the filtrates were combined and concentrated to 500 mL by rotary evaporator at 60 °C. The extracted solution was evaporated under vacuum and partitioned. Water was removed by freeze-drying. The above crushed herbs were extracted to 144.65 g, so the extraction yield of the decoction was about 32.14% (144.65:450.00, w/w). The lyophilized powder of LDXGT was used for the following experiment.

To calculate the administration dosage, the contents of gentiopicroside, geniposide, baicalin, and swertiamarin in LDXGT extract were determined by LC-MS/MS with the same analytical conditions used in Section 3.3. The results demonstrated that the contents of the four components of LDXGT were quantitated as follows (mg/g): gentiopicroside 17.04 mg/g, geniposide 25.08 mg/g, baicalin 9.94 mg/g, and swertiamarin 0.22 mg/g in LDXGT.

3.3. LC-MS/MS and Analytical Conditions

The LC-MS/MS analysis was performed using a Shimadzu LCMS-8030 triple-quadrupole mass spectrometer equipped with an electrospray ionization (ESI) interface and coupled to the UFLC system, equipped with two chromatographic pumps (LC-20AD XR), an autosampler (SIL-20AC XR), a DGU-20A5 degasser, a forced air-circulation-type column oven (CTO-20A), and a photo-diode array detector (SPD-M20A) (Shimadzu, Kyoto, Japan). The optimized instrument settings were as follows: interface voltage, 4.5 kV; desolvation line temperature, 250 °C; heat block temperature, 400 °C; desolvation gas, nitrogen; desolvation gas flow rate, 3 L/min; drying gas, nitrogen; drying gas flow rate, 17 L/min; collision gas, argon; and collision gas pressure, 230 kPa.

The MS/MS measurements operating parameters were performed in the multiple reaction monitoring (MRM) mode. The chromatographic separation was achieved using a Acquity UPLC BEH C18 column (2.1 mm × 100 mm, 1.7 μm, Waters Corporation) with a KrudKatcher Ultra In-Line Filter (AF0-8497, 0.5 μm Depth Filter × 0.004 in, Phenomenex Corporation), and the column temperature was maintained at 35 °C. The mobile phase consisted of A (1 mM ammonium acetate and 0.1% formic acid in water) and B (0.1% formic acid in methanol) with a linear gradient elution of 20%–95% (v/v) B at 0–5 min, 95% B at 5–7 min, and 20% B at 7–10 min. The flow rate was 0.2 mL/min and the injection volume was 5 μL.

3.4. Preparation of Calibration Standards and Quality Controls

The stock solutions were prepared by dissolving 1.0 mg each of gentiopicroside, geniposide, baicalin, and swertiamarin into 1.0 mL of 100% methanol to a final concentration of 1.0 mg/mL. All stock solutions were stored at −20 °C before use. A series of working standard solutions was freshly prepared by spiking aliquots of the stock solutions into drug-free plasma samples to obtain the following concentrations: 1, 5, 10, 25, 50, 100, 250, 500, and 1000 ng/mL. Working solutions for QC samples with low, medium, and high concentrations were prepared in the same manner.

3.5. Method Validation

Full validation of the analytical method estimated in this study was according to the US Food and Drug Administration guidelines for validation of bioanalytical methods [24]. The analytical method was considered valid according to assays of its selectivity, linearity, accuracy, precision, lower limit of quantification (LLOQ), matrix effect, extraction recovery, and stability.

3.5.1. Selectivity

The selectivity was assessed by comparing the chromatograms of blank plasma samples obtained from six rats with corresponding spiked plasma samples.

3.5.2. Linearity, the Lower Limits of Detection (LLOD) and Lower Limits of Quantification (LLOQ)

The sample preparation for calibration curves involved creating freshly spiked plasma (45 μL) samples with stock working solution (5 μL) of analytes at concentration ranges of 5–1000 ng/mL and extracted as described in Section 3.6.4. The concentration of each sample was derived from the calibration curve and corrected by the respective dilution volume. For a standard curve, the ratios of the chromatographic peaks area (analytes /internal standard) as ordinate variables were plotted *versus* the concentration of these drugs as abscissa. All linear curves were required to have a coefficient of estimation of at least > 0.995. The lower limits of detection (LLOD) and lower limits of quantification (LLOQ) were determined at a signal-to-noise ratio of about 3 and 10 by analyzing the diluted standard solution.

3.5.3. Precision and Accuracy

The intra- and inter-day variability for the four analytes were determined by six replicates at concentrations of 5, 10, 25, 50, 100, 250, 500, and 1000 ng/mL using the LC-MS/MS method described above on the same day (intra-day) and six different days (inter-day), respectively. The accuracy (bias, %) was calculated from the mean value of observed concentration (C_{obs}) and nominal concentration (C_{nom}) using the relationship accuracy (Bias, %) = $[(C_{obs} - C_{nom})/C_{nom}] \times 100\%$. The relative standard deviation (RSD, %) was calculated from the observed concentrations as precision (RSD, %) = [standard deviation (SD)/C_{obs}] × 100%. Accuracy and precision values within ±15% were considered acceptable in the experimental concentration range, and the LLOQ values were all less than ±20%.

3.5.4. Matrix Effect and Recovery

Following the previous report of Hou *et al.* [25], three sets of extraction methods were prepared to evaluate the matrix effect and recovery in the quantitative bioanalytical method.

Set 1. Standard solutions were constructed using neat standard solutions in the mobile phase. The samples were prepared by placing 5 μL of the appropriate concentrations of standard solutions and 145 μL of the mobile phase (150 μL, total volume) into 1.5-mL centrifuge tubes. After mixing, the solutions were transferred to autosampler vials, and 5 μL was injected directly into LC-MS/MS system.

Set 2. Standard solutions spiked after extraction were constructed in three different lots of blank plasma by placing 50 μL of plasma in 1.5-mL centrifuge tubes, followed by the addition of 100 μL methanol (containing 5% formic acid, v/v) for protein precipitation (vortex 5 min). After centrifugation (13,100× *g* for 10 min, at 4 °C), the supernatant was filtered using a 0.22-μm mini syringe filter. The supernatant (145 μL) was supplemented with 5 μL of appropriate concentrations of standard solutions. After mixing, the solutions were transferred to autosampler vials, and 5 μL was injected into LC-MS/MS for analysis. In *set 2*, the standard solutions were spiked after extraction into different lots of plasma, whereas in *set 3*, the standard solutions were spiked into different lots of plasma before extraction.

Set 3. The standard solutions spiked before extraction were constructed in three different lots of plasma by placing 45 μL of plasma in 1.5-mL centrifuge tubes to which 5 μL of appropriate concentrations of standard solutions were added before extraction, followed by the addition of 100 μL of methanol (containing 5% formic acid, v/v) for protein precipitation (vortex 5 min). After centrifugation (13,100× *g* for 10 min, at 4 °C), the supernatant was filtered using a 0.22-μm mini syringe filter. Finally, the solutions were transferred to autosampler vials, and 5 μL was injected into LC-MS/MS for analysis.

By comparing the peak areas of the standard solutions, standard solutions spiked before and after extraction into different lots of plasma at low, medium, and high concentrations using three replicates, the recovery and ion suppression or enhancement associated with a given lot of plasma were assessed.

Results obtained in this manner were used to determine the matrix effect (ME) and recovery (RE) of the extraction. The peak areas obtained in neat standard solutions in *set 1* were indicated as A, the corresponding peak areas for standard spiked after extraction into plasma or feces homogenate extracts as B (*set 2*), and the areas for standard spiked before extraction as C (*set 3*); the ME and RE values could be calculated as follows: ME (matrix effect, %) = (B/A) × 100%; RE (recovery, %) = (C/B) × 100%.

3.5.5. Stability

To estimate the stability of analytes in rat plasma during the experiments, storage, and analysis processes, spiked samples with low, median, and high concentrations of analytes were designed and conducted under different conditions, including freeze-thaw cycle analysis, short-term stability, long-term stability, and autosampler stability. Freeze-thaw stability was assessed over three freeze and thaw cycles. Short-term stability was determined by keeping the samples at room temperature for 3 h. Long-term stability was evaluated by analyzing samples kept at −20 °C for 7 days. Autosampler stability was determined for samples kept at the autosampler temperature (8 °C) for 12 h. All stabilities were calculated as the ratio of average concentration and freshly prepared samples (n = 3). The stability (accuracy, %) was calculated as follows: stability (%) = (C_{obs}/C_{nom}) × 100%.

3.6. Pharmacokinetic Application

3.6.1. Experimental Animals

All experimental protocols involving animals were reviewed and approved by the Institutional Animal Care and Use Committee (IACUC) of National Yang-Ming University, Taipei, Taiwan. (IACUC Approval No: 1021003). Male Sprague Dawley rats (200–260 g, 6–7 weeks of age) were obtained from the Laboratory Animal Center at National Yang-Ming University, Taipei, Taiwan. The animals were specifically pathogen-free and housed in standard cages kept in a temperature-controlled room with a 12-h light/dark cycle, and free access to food (Laboratory Rodent Diet 5001, PMI Feeds, Richmond, IN, USA) and water *ad libitum*.

3.6.2. Freely Moving Rat Model

A freely moving rat model was used in this experiment to avoid the stress caused by restraint or anesthesia. Sprague Dawley rats were initially anesthetized with pentobarbital (50 mg/kg, i.p.), and polyethylene tubing (PE-50) was implanted in the right carotid artery for blood sampling. The PE-50 tube was exteriorized, fixed in the dorsal region of the neck. Patency of the tube was maintained by flushing with heparinized saline (20 IU/mL). During the period of surgery, the body temperature of the rats was maintained at 37 °C with a heating pad. After surgery, the rats were kept in cages individually and allowed to recover for one day.

3.6.3. LDXGT Administration and Sample Collection

The dose of LDXGT (10 g/kg, p.o.) was appropriate for oral administration in rats. About 200 μL of blood samples were withdrawn from the cannula implanted in the carotid artery and placed into a heparin-rinsed vial at 0, 5, 15, 30, 45, 60, 90, 120, 150, 180, 240, 300, 360, 480, 600, and 720 min in 6 healthy rats after a single administration of LDXGT (10 g/kg, p.o.). Then the samples were centrifuged at 3000× *g* for 10 min at 4 °C and the separated plasma samples were frozen in polypropylene tubes at −20 °C until analysis.

3.6.4. Sample Preparation

Plasma samples were prepared with protein precipitation. An aliquot of 50 μL plasma sample and 100 μL methanol (containing 5% formic acid solution and 10 ng/mL of internal standard, v/v) were combined and vortexed for 5 min. The mixture was centrifuged at 13,100× *g* for 10 min at 4 °C. The supernatants were collected, filtered using a 0.22-μm mini syringe filter, and transferred to autosampler vials. Finally, 5 μL was injected into the LC-MS/MS system. If the concentration of analyte was not within the linearity range, the plasma sample was diluted to the appropriate concentration with drug-free plasma and the concentration of analyte was obtained using back-calculation.

3.7. Statistical Analysis

Pharmacokinetic calculations were performed on each individual set of data using the pharmacokinetic software WinNonlin Standard Edition, version 1.1 (Scientific Consulting Inc., Apex,

NC, USA). A non-compartmental analysis was applied to obtain blood pharmacokinetic parameters, including maximum plasma concentration (C_{max}), time to reach the maximum concentrations (T_{max}), half-life ($t_{1/2}$), area under concentration-time curve (AUC), and clearance (CL). Statistics and graphics were performed with version 10.0 (SPSS, Chicago, IL, USA) and SigmaPlot 8.0 software. Data were expressed as the mean ± S.D. or mean ± S.E.M. Comparisons were performed using one-way analysis of variance (ANOVA) followed by Dunnett's test, and the difference was considered statistically significant if $p < 0.05$.

4. Conclusions

In this study, a rapid, sensitive, and selective LC-ESI-MS/MS method has been developed and validated for the quantitation of gentiopicroside, geniposide, baicalin, and swertiamarin in rat plasma after oral administration of LDXGT. These pharmacokinetic studies were performed through an experimental model of freely moving rats. The assay provided adequate matrix effect and recovery with good precision and accuracy. These pharmacokinetic results reveal a constructive contribution to comprehending the multiple components of the extensive action mechanism of absorption, distribution, and excretion of a traditional Chinese herbal prescription. Furthermore, the pharmacokinetic profile provides a firm basis for the design of dosing regimens that can contribute to preclinical studies of herbal applications and pharmacological experiments.

Acknowledgments

Funding for this study was provided in part by research grants from the National Science Council (NSC102-2113-M-010-001-MY3), Taiwan and TCH 103-02; 10301-62-021 from Taipei City Hospital, Taipei, Taiwan.

Author Contributions

Chia-Ming Lu made substantial contributions to conception and design, acquisition of data, and analysis and interpretation of data. Lie-Chwen Lin gave suggestions for writing. Chia-Ming Lu and Tung-Hu Tsai participated in drafting the article and revising it critically for important intellectual content; and gave final approval of the version to be submitted and any revised version.

References

1. Wang, Y.; Kong, L.; Hu, L.; Lei, X.; Yang, L.; Chou, G.; Zou, H.; Wang, C.; Annie Bligh, S.W.; Wang, Z. Biological fingerprinting analysis of the traditional Chinese prescription Longdan Xiegan Decoction by on/off-line comprehensive two-dimensional biochromatography. *J. Chromatogr. B Anal. Technol. Biomed. Life Sci.* **2007**, *860*, 185–194.

2. Chen, F.P.; Kung, Y.Y.; Chen, Y.C.; Jong, M.S.; Chen, T.J.; Chen, F.J.; Hwang, S.J. Frequency and pattern of Chinese herbal medicine prescriptions for chronic hepatitis in Taiwan. *J. Ethnopharmacol.* **2008**, *117*, 84–91.
3. Lee, T.Y.; Chang, H.H. Longdan Xiegan Tang has immunomodulatory effects on CD4+CD25+ T cells and attenuates pathological signs in MRL/lpr mice. *Int. J. Mol. Med.* **2010**, *25*, 677–685.
4. Rojas, A.; Bah, M.; Rojas, J.I.; Gutierrez, D.M. Smooth muscle relaxing activity of gentiopicroside isolated from Gentiana spathacea. *Planta Med.* **2000**, *66*, 765–767.
5. Kumarasamy, Y.; Nahar, L.; Sarker, S.D. Bioactivity of gentiopicroside from the aerial parts of Centaurium erythraea. *Fitoterapia* **2003**, *74*, 151–154.
6. Kondo, Y.; Takano, F.; Hojo, H. Suppression of chemically and immunologically induced hepatic injuries by gentiopicroside in mice. *Planta Med.* **1994**, *60*, 414–416.
7. Koo, H.J.; Lee, S.; Shin, K.H.; Kim, B.C.; Lim, C.J.; Park, E.H. Geniposide, an anti-angiogenic compound from the fruits of Gardenia jasminoides. *Planta Med.* **2004**, *70*, 467–469.
8. Koo, H.J.; Lim, K.H.; Jung, H.J.; Park, E.H. Anti-inflammatory evaluation of gardenia extract, geniposide and genipin. *J. Ethnopharmacol.* **2006**, *103*, 496–500.
9. Li, H.Y.; Hu, J.; Zhao, S.; Yuan, Z.Y.; Wan, H.J.; Lei, F.; Ding, Y.; Xing, D.M.; Du, L.J. Comparative study of the effect of baicalin and its natural analogs on neurons with oxygen and glucose deprivation involving innate immune reaction of TLR2/TNFalpha. *J. Biomed. Biotechnol.* **2012**, *2012*, 267890.
10. Trinh, H.T.; Joh, E.H.; Kwak, H.Y.; Baek, N.I.; Kim, D.H. Anti-pruritic effect of baicalin and its metabolites, baicalein and oroxylin A, in mice. *Acta Pharmacol. Sin.* **2010**, *31*, 718–724.
11. Xu, G.L.; Li, H.L.; He, J.C.; Feng, E.F.; Shi, P.P.; Liu, Y.Q.; Liu, C.X. Comparative pharmacokinetics of swertiamarin in rats after oral administration of swertiamarin alone, Qing Ye Dan tablets and co-administration of swertiamarin and oleanolic acid. *J. Ethnopharmacol.* **2013**, *149*, 49–54.
12. Wang, Y.; Kong, L.; Lei, X.; Hu, L.; Zou, H.; Welbeck, E.; Bligh, S.W.; Wang, Z. Comprehensive two-dimensional high-performance liquid chromatography system with immobilized liposome chromatography column and reversed-phase column for separation of complex traditional Chinese medicine Longdan Xiegan Decoction. *J. Chromatogr. A* **2009**, *1216*, 2185–2191.
13. Wang, Y.; Yang, L.; He, Y. Q.; Wang, C. H.; Welbeck, E. W.; Bligh, S. W.; Wang, Z.T. Characterization of fifty-one flavonoids in a Chinese herbal prescription Longdan Xiegan Decoction by high-performance liquid chromatography coupled to electrospray ionization tandem mass spectrometry and photodiode array detection. *Rapid Commun. Mass Spectrom.* **2008**, *22*, 1767–1778.
14. Wang, C.H.; Cheng, X.M.; Bligh, S.W.; White, K.N.; Branford-White, C.J.; Wang, Z.T. Pharmacokinetics and bioavailability of gentiopicroside from decoctions of Gentianae and Longdan Xiegan Tang after oral administration in rats--comparison with gentiopicroside alone. *J. Pharm. Biomed. Anal.* **2007**, *44*, 1113–1117.
15. Chang-Liao, W.L.; Chien, C.F.; Lin, L.C.; Tsai, T.H. Isolation of gentiopicroside from Gentianae Radix and its pharmacokinetics on liver ischemia/reperfusion rats. *J. Ethnopharmacol.* **2012**, *141*, 668–673.

16. Wang, C.H.; Wang, Z.T.; Bligh, S.W.; White, K.N.; White, C.J. Pharmacokinetics and tissue distribution of gentiopicroside following oral and intravenous administration in mice. *Eur. J. Drug Metab. Pharmacokinet.* **2004**, *29*, 199–203.
17. Cheng, S.; Lin, L.C.; Lin, C.H.; Tsai, T.H. Comparative oral bioavailability of geniposide following oral administration of geniposide, Gardenia jasminoides Ellis fruits extracts and Gardenia herbal formulation in rats. *J. Pharm. Pharmacol.* **2014**, *66*, 705–712.
18. Sun, Y.; Feng, F.; Yu, X. Pharmacokinetics of geniposide in Zhi-Zi-Hou-Pu decoction and in different combinations of its constituent herbs. *Phytother. Res.* **2012**, *26*, 67–72.
19. Long, Z.M.; Bi, K.S.; Huo, Y.S.; Yan, X.Y.; Zhao, X.; Chen, X.H. Simultaneous quantification of three iridoids in rat plasma after oral administration of Zhi-zi-chi decoction using LC-MS. *J. Sep. Sci.* **2011**, *34*, 2854–2860.
20. Li, H.L.; He, J.C.; Bai, M.; Song, Q.Y.; Feng, E.F.; Rao, G.X.; Xu, G.L. Determination of the plasma pharmacokinetic and tissue distributions of swertiamarin in rats by liquid chromatography with tandem mass spectrometry. *Arzneimittelforschung* **2012**, *62*, 138–144.
21. Wang, S.; Tang, S.; Sun, Y.; Wang, H.; Wang, X.; Zhang, H.; Wang, Z. Highly sensitive determination of new metabolite in rat plasma after oral administration of swertiamarin by liquid chromatography/time of flight mass spectrometry following picolinoyl derivatization. *Biomed. Chromatogr.* **2014**, *28*, 939–946.
22. Shaw, L.H.; Lin, L.C.; Tsai, T.H. HPLC-MS/MS analysis of a traditional Chinese medical formulation of Bu-Yang-Huan-Wu-Tang and its pharmacokinetics after oral administration to rats. *PLoS One* **2012**, *7*, e43848.
23. Zhang, Z.Q.; Liua, W.; Zhuang, L.; Wang, J.; Zhang, S. Comparative pharmacokinetics of baicalin, wogonoside, baicalein and wogonin in plasma after oral administration of pure baicalin, radix scutellariae and scutellariae-paeoniae couple extracts in normal and ulcerative colitis rats. *Iran. J. Pharm. Res.* **2013**, *12*, 399–409.
24. Nowatzke, W.; Woolf, E. Best practices during bioanalytical method validation for the characterization of assay reagents and the evaluation of analyte stability in assay standards, quality controls, and study samples. *AAPS J.* **2007**, *9*, E117–E122.
25. Hou, M.L.; Chang, L.W.; Chiang, C.J.; Tsuang, Y.H.; Lin, C.H.; Tsai, T.H. Pharmacokinetics of di-isononyl phthalate in freely moving rats by UPLC-MS/MS. *Int. J. Pharm.* **2013**, *450*, 36–43.

2

Micropreconcentrator in LTCC Technology with Mass Spectrometry for the Detection of Acetone in Healthy and Type-1 Diabetes Mellitus Patient Breath

Artur Rydosz

Department of Electronics, AGH University of Science and Technology, Av. Mickiewicza 30, Krakow 30-059, Poland; E-Mail: artur.rydosz@agh.edu.pl

---

Abstract: Breath analysis has long been recognized as a potentially attractive method for the diagnosis of several diseases. The main advantage over other diagnostic methods such as blood or urine analysis is that breath analysis is fully non-invasive, comfortable for patients and breath samples can be easily obtained. One possible future application of breath analysis may be the diagnosing and monitoring of diabetes. It is, therefore, essential, to firstly determine a relationship between exhaled biomarker concentration and glucose in blood as well as to compare the results with the results obtained from non-diabetic subjects. Concentrations of molecules which are biomarkers of diseases' states, or early indicators of disease should be well documented, *i.e.*, the variations of abnormal concentrations of breath biomarkers with age, gender and ethnic issues need to be verified. Furthermore, based on performed measurements it is rather obvious that analysis of exhaled acetone as a single biomarker of diabetes is unrealistic. In this paper, the author presents results of his research conducted on samples of breath gas from eleven healthy volunteers (HV) and fourteen type-1 diabetic patients (T1DM) which were collected in 1-l SKC breath bags. The exhaled acetone concentration was measured using mass spectrometry (HPR-20 QIC, Hiden Analytical, Warrington, UK) coupled with a micropreconcentrator in LTCC (Low Temperature Cofired Ceramic). However, as according to recent studies the level of acetone varies to a significant extent for each blood glucose concentration of single individuals, a direct and absolute relationship between blood glucose and acetone has not been proved.

Nevertheless, basing on the research results acetone in diabetic breath was found to be higher than 1.11 ppmv, while its average concentration in normal breath was lower than 0.83 ppmv.

Keywords: breath analysis; low temperature cofired ceramics (LTCC) technology; micropreconcentrators; exhaled acetone measurements

---

1. Introduction

Breath analysis has been developing for many years and there have been many instances of research into its potential for diagnosing diseases, e.g., diabetes [1–5]. In general, diabetes is diagnosed on a basis of glucose concentration in blood. A non-invasive and painless method for diagnostic, preventive and monitoring diabetes is still needed. Breath analysis could be faster, cheaper, more flexible and comfortable than blood analysis. Many researchers around the world have been searching for an one-to-one correspondence between a single exhaled compound and a given plasma metabolite [6,7]. Recently, breath analysis has concentrated on acetone as a biomarker of diabetes [8–10]. However, the results obtained have often been inconclusive [11]. An overview of the studies of breath analysis in diabetes, focusing on the breath metabolites and their potential utilization in clinical applications, is widely discussed and presented [12]. Careful studies using an appropriate trace gas analytical methods are necessary to the same extent as certified sampling procedures to reduce different outcomes for nominally similar analytical methodologies. The sampling guidance for the breath research community is one of the challenge of the International Association of Breath Research (IABR).

The exhaled acetone is usually in the range of 0.2–1.8 ppmv for healthy people, and in the range of 1.25–2.4 ppmv for people with diabetes [13]. Based on literature review, exhaled breath acetone measurement in its developmental stage is currently characterized by intensive studies and creation of experimental prototype devices [14,15]. Unfortunately, portable devices for exhaled acetone measurements remain currently unavailable. Acetone and other volatiles in breath are present in nanomolar quantities. In order to measure such low concentrations the laboratory systems are applied, *i.e.*, proton transfer reaction mass spectrometry (PTR-MS) [16], atmospheric pressure chemical ionization mass spectrometry (APCI-MS) [17], gas chromatography-mass spectrometry (GC-MS) [18], selected ion flow tube mass spectrometry (SIFT-MS) [19,20]. All mentioned methods are in fact very expensive and require well-qualified personnel, therefore their use is exclusively restricted to laboratories.

Breath analysis as a supplementary tool for diagnosing and monitoring diabetes makes sense only in case of utilization of portable analyzers. However, commercially available gas sensors are developed for measuring samples at several tens part per million (ppm). One of the cheap and very effective methods to increase the limit of detection, in order to measure such low amounts of volatiles, is the use of a micropreconcentrator structures. During a study on T1DM subjects and healthy volunteers, the author aimed to establish the accuracy, repeatability and selectivity response for measurement of exhaled acetone using the micropreconcentrator in low temperature cofired ceramics (LTCC) technology with HPR-20 QIC MS. Based on a literature review, the proposed micropreconcentrator structure in LTCC technology is a novel solution and it have not yet been investigated. It could be an alternative to the commonly presented and discussed preconcentrators (micropreconcentrators) in silicon and silicon-glass technology. In the present study, the exhaled acetone concentrations were measured with micropreconcentrator coupled with mass spectrometry (MS) as a very sensitive detector. Furthermore, the proposed micropreconcentrator can be used with commercially available gas sensors.

2. Experimental Section

All experiments involving human subjects were performed according to the "Declaration of Helsinki" and in accordance with Polish law. All patients and volunteers declared a written consent to participate in the investigation.

2.1. Reference Gas and Breath Samples

As a reference gas the certified acetone in the concentrations of 80 ppmv, 8 ppmv and 0.8 ppmv (Air Products, Warszawa, Poland) was used. As the exhaled human breath (37 °C) is almost saturated with water, the reference acetone samples were humidified using the Drechsel's bottle filled with water. The humidity level was controlled by homemade software (LabView), a humidity sensor (SHT25, Sensirion, Stäfa, Switzerland) and mass flow controllers (MKS Instrument, Munchen, Germany). The breath samples were collected with a FlexFoil Plus SKC® (Chicago, IL, USA) Breath-gas Analysis Bag of 1.0 L volume. The bags were stored at room temperature (25 °C), filled with a single exhalation and warmed up to 45 °C for 3 min in order to avoid condensation. Prior to use, the new bags were cleaned three times by using pure (99.9999%) nitrogen. Additionally, to remove any contaminations after each measurement the bags were purified with pure N_2.

2.1.1. Breath Measurements with Volunteers

Eleven healthy volunteers with no history of any respiratory disease participated in the research, including one (male) heavy smoker, and one pregnant woman. There were five women with an average age of 31 years old and six men with average of 33 years old. Volunteers attended ten visits within two weeks, all visits occurred in the Department of Electronics AGH, Krakow, Poland. The analysis of exhaled acetone was performed between 8:00 am and 3:00 pm for nine volunteers, and between 3:00 pm and 8:00 pm for another two. Subjects of the experiment refrained from eating or drinking for at least two hours before being tested. After five minutes of rest, the subjects were asked to inhale moderately and then to exhale in the 1-l breath bags and again after 2 h following the consumption of 75 g of glucose. A sample of ambient air was collected at the same time. After collection, gas samples in the breath bags were analyzed using the MS with micropreconcentrator structure. The glucose in the blood was measured using a commercial glucometer Acuu Check Active (Roche Diagnostics®, Basel, Switzerland). The blood glucose measurement as well as breath acetone measurement were performed at least two times to avoid any incidental results.

2.1.2. Breath Measurements with Diabetes Patients

Fourteen patients diagnosed with type-1 diabetes were asked to breathe into breath bags and measured glucose in blood using their own glucometers. There were eight women with an average age of 31 years and six men with an average age of 25 years. Ten patients used Acuu Check Active (Roche Diagnostics®,

Basel, Switzerland), four patients used One Touch Mini Select (LifeScan, Inc.©, Milpitas, CA, USA), Contour TS (Bayer GmbH®, Leverkusen, Germany), Acuu Check Go and Acuu Check Performa (Roche Diagnostics®, Basel, Switzerland), respectively. The breath sample procedure was the same as for the healthy volunteers.

2.2. Apparatus

The HPR-20 QIC real time gas analyzer (Hiden Analytical, Warrington UK) with a 6-ports Valco Injector (VICI Instruments Co. Inc., Schenkon, Switzerland) was used for all the measurements. The micropreconcentrator design and fabrication process was precisely reported in [21]. A schematic view of the measurement setup is presented in Figure 1. The instrument was calibrated using a calibration gas mixture as well as the certified acetone concentrations. As the carrier gas, pure N_2 was employed and the injector was maintained at 180 °C. MS interface temperature and the ionization voltage were set at 250 °C and 70 eV, respectively. The acquisition range in SEM detector was set from 10^{-7} Tr to 10^{-13} Tr. The mass spectrometry was operated with a multiple ion detection (MID) mode set at *m/z* of 43 (acetone) and 58 (acetone), see Figure 2a. A scan rate of 1 scan/s was applied. The data collection was made with MASsoft Professional Software (Hiden Analytical, Warrington, UK).

Figure 1. Schematic view of the measurement system based on micropreconcentrator in low temperature cofired ceramics (LTCC) technology and HPR-20 QIC mass spectrometry for exhaled acetone analysis.

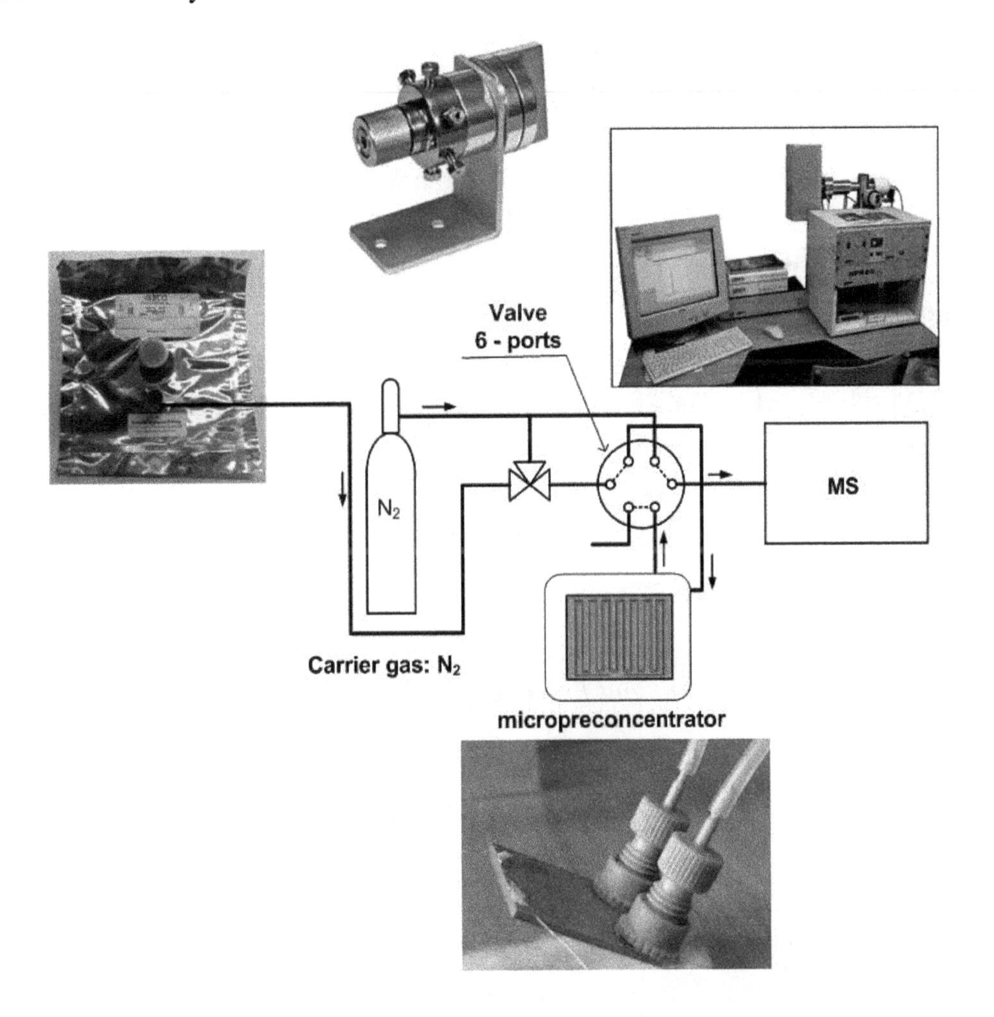

Figure 2. MS spectrum of acetone (**a**), a result of multiple ion detection (MID) analysis of mass spectrograph for the exhaled acetone and acetone after preconcentration for three different adsorption times: 5 min, 10 min, 30 min (**b**).

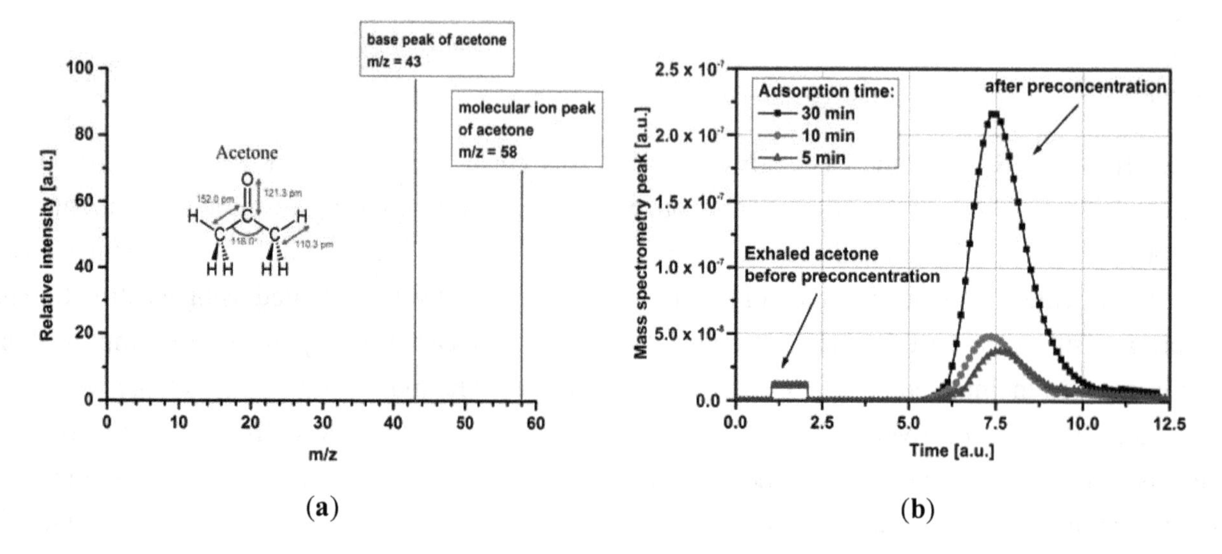

(a) (b)

2.3. Experimental Methods and Conditions

The experimental procedure consisted of few stages. Firstly, the micropreconcentrator was filled with Carboxen-1018. It is a hydrophobic adsorbent material with a large surface area and grain diameters suitable to channel dimensions. It is also recommended by the company as the best adsorbent material for breath analysis. Before using the micropreconcentrators, the adsorbent material have to be activated. The author used typical time-temperature profiles for adsorbent activation. The activation temperature changed in the range of 100 °C–400 °C and the total activation time equals approximately 3 h. After preparing the micropreconcentrator the measurement setup was calibrated. The preconcentration process is precisely described in [22]. Based on the previously obtained results, gas flow of 25 mL/min was passed through the active material during 5, 10 and 30 min. At the end of each measurement the gas lines were purged with N_2. Then, an electrical pulse was applied to the heater of the device in order to reach a desorption temperature of 220 °C. Figure 2b shows mass spectrometry peak for the acetone taken from an one T1DM subject before and after preconcentration. In the present study, a concentration factor (CF) is defined as the ratio of gas preconcentration after and before preconcentration process (*i.e.*, the ratio of peak area before and after desorption). The obtained concentration factor is around 3.1, 5.85, 16.35 at 5 min, 10 min, and 30 min adsorption time, respectively.

3. Results and Discussion

3.1. Detection of Acetone in Breath without Preconcentration

Acetone concentrations in 14 diabetic patients and 11 healthy volunteers were summarized in Table 1. Both groups were examined 10 times. However, Table 1 only shows the results for the lowest and the highest blood glucose concentration obtained during the measurements. The median acetone in the healthy group was 0.63 ± 0.12 ppmv, while in diabetic patients breath was 2.08 ± 0.47 ppmv. The acetone concentration in normal breath was ranged from 0.49 to 0.83 ppmv. The measurement results confirmed the statement that the exhaled acetone concentration for non-diabetic is less than 1 ppmv. Breath acetone concentrations from the diabetes patients were from 1.11 to 3.11 ppmv. The obtained results are in accordance with the results reported in [12,23,24].

Table 1. The information of diabetic patients and the breath acetone level.

Subject number	Group	Sex [a]	Age	Years of diabetes	Breath Acetone (ppmv)	Blood glucose (mg/dl)
1 [b]	Healthy volunteer	M	43	X	0.49	91
					0.56	111
2 [c]		F	30		0.57	86
					0.47	132
3		M	28		0.63	101
					0.83	137
4		M	41		0.74	74
					0.81	123
5		F	32		0.65	103
					0.73	150
6		M	28		0.48	86
					0.69	146
7		F	30		0.59	84
					0.71	136
8		M	27		0.63	102
					0.72	145
9		F	39		0.45	86
					0.80	135
10		F	26		0.48	90
					0.79	140
11		M	29		0.45	86
					0.73	151
12	Patients with diabetes type I	F	33	15	1.52	60
					1.70	168
13		M	23	2	1.62	70
					3.12	196
14		M	22	2	1.80	100
					2.75	231
15 [d]		F	29	15	2.16	64
					1.30	170
16		M	23	13	1.51	174
					3.12	367
17		F	25	4	1.11	65
					1.79	372
18		F	21	5	1.55	82
					3.09	324
19		F	40	10	1.63	70
					2.70	350

Table 1. *Cont.*

Subject number	Group	Sex [a]	Age	Years of diabetes	Breath Acetone (ppmv)	Blood glucose (mg/dl)
20		F	23	3	1.70	75
					1.85	127
21		M	20	1	1.87	181
					3.12	270
22		M	24	8	1.78	63
					1.87	165
23		F	22	2	1.86	90
					2.41	451
24		F	55	15	1.61	70
					2.98	257
25		M	35	21	1.65	62
					3.09	389

[a] M: male and F: female; [b] heavy smoker; [c] in the 15th week of pregnancy;[d] patient with a reverse trend.

Usually, we expect a linear relationship between breath acetone and blood glucose concentration for T1DM (Figure 3a) as well as for healthy volunteers (Figure 3b). However, these studies do not definitively prove a linear and reliable correlation. Therefore, it is clear that as the levels of acetone vary so widely for each blood glucose concentration in each individual, a direct and absolute relationship between blood glucose and acetone does not exist [11]. Furthermore, a reverse trend was observed for one female patient (15 years with diagnosed T1DM) and one healthy volunteer. In both cases, the measurement were performed at the same time of day: 5:00 pm–5:30 pm, the subjects had the same individual diet and resisted of taken any drugs, except insulin by T1DM (NovoRapid, Warsaw, Poland). The correlation coefficient in linear regression was 0.9654 and 0.7312 for T1DM subject and HV, respectively. The obtained results with linear fitting are presented in Figure 4.

Figure 3. Relations between breath acetone as measured by mass spectrometry and blood glucose for T1DM subjects (**a**) and healthy volunteers (**b**)

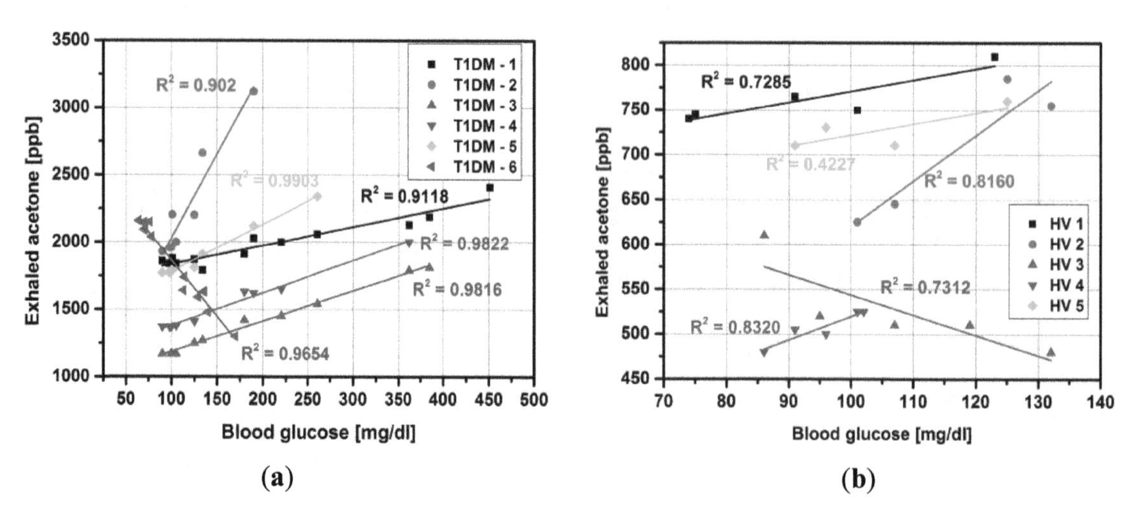

Figure 4. Correlation between breath acetone as measured by mass spectrometry and blood glucose for T1DM subject (**a**) and healthy volunteer (**b**).

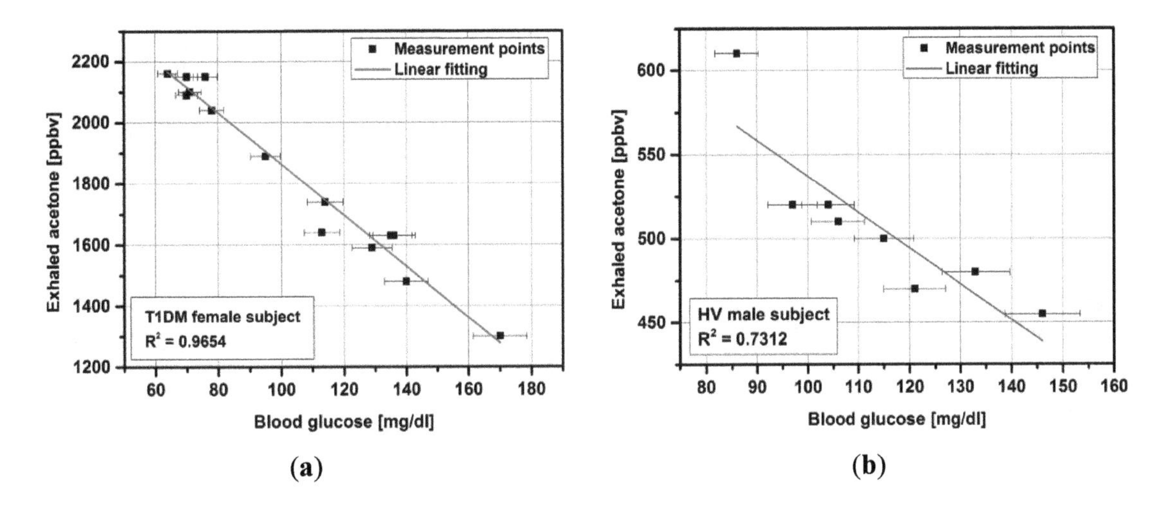

3.2. Impact of Micropreconcentrator on Breath Acetone Detection

Based on the successful acetone preconcentration with the micropreconcentrator structure in LTCC technology [21] as well as in MEMS technology [22], breath samples taken from T1DM patients and healthy volunteers were determined. As was already stated, breath analysis only makes sense in the case of utilization of portable analyzers. Semiconductor gas sensors are a good alternative to laboratory systems with their low cost, ease of control and compatibility with microelectronic technology, including LTCC technology. The main drawbacks are low selectivity, sensitivity at ppm level and response drift. However, in recent years, research in semiconductor sensors has focused on nanomaterials, which resulted in improved parameters such as: sensitivity, selectivity and stability. The exhaled acetone concentration was in 0.52–3.12 ppmv range in both investigated groups. Unfortunately, it is below a limit of detection (LOD) for commercially available semiconductor acetone sensors. One of the cheapest and very effective method to increase LOD to measure such low amounts of acetone is using the micropreconcentrator structures. It accumulates and concentrates the acetone over an adsorption time period and desorbs mainly based on thermal desorption. Such solutions are common in breath analysis [25–28]. To start with, the exhaled acetone concentration was measured directly from the breath bags. Then, the preconcentration process started with various adsorption times, *i.e.*, 5 min, 10 min, and 30 min. During the preconcentration process, the breath sample was taken from breath bag with flow rate set to 25 mL/min. The acetone concentration measured after preconcentration was in the range of 1.61–51.01 ppmv. The concentration of the initial gas was enhanced to the ppm range, placing it within the sensitivity range of the available gas sensors. Figure 5 shows the concentration factor with respect to adsorption time for acetone obtained from breath samples: four healthy volunteers and four T1DM subjects. It is obvious that the best results are obtained with a longer preconcentration time. It was previously proved that after 90 min adsorption time, the concentration factor reaches the constant value. However, after 30 minutes adsorption, the acetone concentration measured from diabetic patients breath samples increases approximately 17 times and reaches the LOD for commercially available acetone sensors [29].

Figure 5. Concentration factor with respect to adsorption time for breath samples taken from T1DM and healthy subjects.

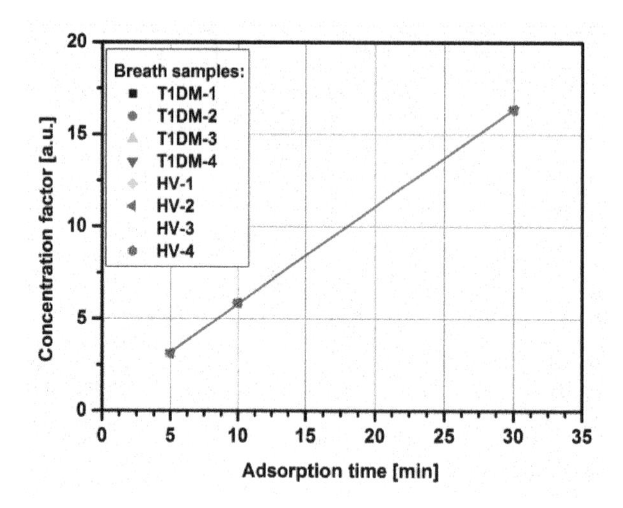

4. Conclusions and Perspectives

The aim of the presented investigation was the detection of exhaled acetone in healthy and type-1 diabetic patients. The exhaled acetone concentration was in 0.52–3.12 ppmv range in both investigated groups. The breath samples were further determined using micropreconcentrator structure filled with Carboxen-1018 adsorbent material. The adsorption time was reduced to 5, 10 and 30 min. The effect of the concentration factor changes for acetone samples at different initial concentration as well as various flow rates, desorption temperature, *etc.* was investigated in previous work [21]. Therefore, it was essential to investigate how the concentration factor changes for acetone from diabetic and non-diabetic breath (Figure 5). The concentration of the initial gas was enhanced to the ppm range, placing it within the sensitivity range of the available gas sensors. The exhaled acetone higher than 1.11 ppmv was found in diabetic breath and less than 0.83 ppmv in healthy controls. The obtained results proved that type-1 diabetic patients have higher exhaled acetone concentration than healthy people. However, the real potential of breath acetone will rely on its ability to monitor the small changes in breath acetone that parallel changes in blood glucose. It seems that such correlation never exist.

Acknowledgments

The author acknowledges to Heike Bartsch de Torres and Jens Mueller from IMN MacroNano, Tehnische Universitaet Ilmenau for helping with LTCC fabrication. The work was financial supported by the National Science Center, decision number DEC-2013/09/N/ST7/01232

References

1. Tassopoulos, C.; Barnett, D.; Russell, F.T. Breath-acetone and blood-sugar measurements in diabetes. *Lancet* **1969**, *239*, 1282–1286.

2. Guo, D.; Zhang, D.; Zhang L.; Lu, G. Non-invasive blood glucose monitoring for diabetics by means of breath signal analysis. *Sens. Actuators* **2012**, *173*, 106–113.
3. Miekisch, W.; Schubert, J.K.; Noeldge-Schomburg, G.F.E. Diagnostic potential of breath analysis: Focus on volatile organic compounds. *Clin. Chim. Acta* **2004**, *347*, 25–39.
4. Fleischer, M.; Simon, E.; Rumpel, E.; Ulmer, H.; Harbeck, M.; Wandel, M.; Fietzek, C.; Weimar, U.; Meixner, H. Detection of volatile compounds correlated to human diseases through breath analysis with chemical sensors. *Sens. Actuators* **2002**, *83*, 245–249.
5. Zhang, Q.; Wang, P.; Li, J.; Gao, X. Diagnosis of diabetes by image detection of breath using gas-sensitive laps. *Biosens. Bioelectron.* **2000**, *15*, 249–256.
6. Amann, A.; Smith, D. *Breath Analysis for Clinical Diagnosis and Therapeutic Monitoring*; World Scientific: Singapore, Singapore, 2005; pp. 305–316.
7. Wang, C.J.; Mbi, A.; Shepherd, M. A study on breath acetone in diabetic patients using a cavity ringdown breath analyzer: Exploring correlations of breath acetone with blood glucose and glycohemoglobin A1C. *IEEE Sens. J.* **2010**, *10*, 43–63.
8. Kinoyama, M.; Nitta, H.; Watanabe, A.; Ueda, H. Acetone and isoprene concentrations in exhaled breath in healthy subjects. *J. Health Sci.* **2008**, *54*, 471–477.
9. Worrall, A.D.; Bernstein, J.A.; Angelopoulos, A.P. Portable method of measuring gaseous acetone concentrations. *Talanta* **2013**, *112*, 26–30.
10. Righettoni, M.; Tricoli, A.; Gass, S.; Schmid, A.; Amann, A.; Pratsinis, S.E. Breath acetone monitoring by portable Si:WO3 gas sensors. *Anal. Chim. Acta* **2012**, *738*, 69–75.
11. Smith, D.; Spanel, P.; Fryer, A.A.; Hanna, F.; Ferns, G.A. Can volatile compounds in exhaled breath be used to monitor control in diabetes mellitus? *J. Breath Res.* **2001**, doi:10.1088/1752-7155/5/2/022001.
12. Minh, T.D.C.; Blake, D.R.; Galassetti, P.R. The clinical potential of exhaled breath analysis for diabetes mellitus. *Diabetes Res. Clin. Pract.* **2012**, *97*, 195–205.
13. Mohamed, E.; Linder, R.; Perriello, G.; di Daniele, N.; Poeppl, S.; de Lorenzo, A. Predicting type-2 diabetes using an electronic nose-based artificial neural network analysis. *Diabetes Nutr. Metabol.* **2002**, *15*, 215–221.
14. Phillips, M.; Herrera, J.; Krishnan, S.; Zain, M.; Greenberg, J.; Cataneo, R.N. Variation in volatile organic compounds in the breath of normal humans. *J. Chromatogr. B* **1999**, *729*, 75–88.
15. Musa-veloso, K.; Likhodii, S.S.; Rarama, E.; Benoit, S.; Liu, Y.M.C.; Chartrand, D.; Curits, R.; Carmant, L.; Lortie, A.; Comeau, F.J.E.; *et al*. Breath acetone predicts plasma ketone bodies in children with epilepsy on a ketogenic diet. *Nutrition* **2006**, *22*, 1–8.
16. Massick, S.M.; Vakhtin, A.B. Breath acetone detection optical methods in the life science. *Proc. SPIE* **2006**, doi:10.1117/12.685276.
17. Moser, B.; Bodrogi, F.; Eibl, G.; Lechner, M.; Rieder, J.; Lirk, P. Mass spectrometric profile of exhaled breath-field study by PTR-MS. *Respir. Physiol. Neurobiol.* **2005**, *145*, 295–300.
18. Koyanagi, G.K.; Kapishon, V.; Blagojevic, V.; Boheme, D.K. Monitoring hydrogen sulfide in simulated breath f anesthetized subjects. *Int. J. Mass Spectrom.* **2013**, *354–355*, 139–143.
19. Fan, G.-T.; Yang, C.-L.; Lin, C.-H.; Chen, C.-C.; Shih, C.-H. Applications of Hadamard transform-gas chromatography/mass spectrometry to the detection of acetone in healthy human and diabetes mellitus patient breath. *Talata* **2014**, *120*, 386–390.

20. Diskin, A.M.; Spanel, P.; Smith, D. Time variation of ammonia, acetone, isoprene and ethanol in breath: A quantitative SIFT-MS study over 30 days. *Physiol. Meas.* **2003**, *24*, 107–120.
21. Rydosz, A.; Maziarz, W.; Pisarkiewicz, T.; de Torres, H.B.; Mueller, J. A Micropreconcentrator Design Using Low Temperature Cofired Ceramics Technology for Acetone Detection Applications. *IEEE Sens. J.* **2013**, *13*, 1889–1896.
22. Rydosz, A.; Maziarz, W.; Pisakiewicz, T.; Domański, K.; Grabiec, P. A gas micropreconcentrator for low level acetone measurements. *Microelectron. Reliability* **2012**, *52*, 2640–2646.
23. Storer, M.; Dummer, J.; Lunt, H.; Scotter, J.; McCartin, F.; Cook, J.; Swanney, M.; Kendall, D.; Logan, F.; Epton, M. Measurement of breath acetone concentrations by selected ion flow tube mass spectrometry in type 2 Diebetes. *J. Breath Res.* **2011**, doi:10.1088/1752-7155/5/4/046011.
24. Kundu, S.K.; Bruzek, J.A.; Nair, R.; Judilla, A.M. Breath acetone analyzer: Diagnostic tool to monitor dietary fat loss. *Clin. Chem.* **1993**, *39*, 87–92.
25. Ueta, I.; Saito, Y.; Hosoe, M.; Okamoto, M.; Ohkita, H.; Shirai, S.; Tamura, H.; Jinno, K. Breath acetone analysis with miniaturized sample preparation device: In-needle preconcentration and subsequent determination by gas chromatography-mass spectroscopy. *J. Chromatogr. B* **2009**, *877*, 2551–2556.
26. Groves, W.; Zellers, E.; Frye, G. Analyzing organic vapors in exhaled breath using a surface acoustic wave sensor array with preconcentration: Selection and characterization of the preconcentrator adsorbent. *Anal. Chim. Acta* **1998**, *371*, 131–143.
27. Dow, A.; Lang, W. Design and fabrication of a micropreconcentrator focuser for sensitivity enhancement of chemical sensing systems. *IEEE Sens. J.* **2012**, *12*, 2528–2534.
28. Ihira, S.-I.; Toda, K. Micro gas analyzers for environmental and medical applications. *Anal. Chim. Acta* **2008**, *619*, 143–156.
29. Figaro Product Information. TGS 822-for the detection of Organic Solvent Vapors. Available online: http://www.figarosensor.com/products/822pdf.pdf (accessed on 22 July 2014).

3

Pharmacokinetic Comparisons of Benzoylmesaconine in Rats Using Ultra-Performance Liquid Chromatography-Tandem Mass Spectrometry after Administration of Pure Benzoylmesaconine and Wutou Decoction

Pei-Min Dai [1,2], Ying Wang [1,2], Ling Ye [1], Shan Zeng [1], Zhi-Jie Zheng [1], Qiang Li [1], Lin-Liu Lu [1,2] and Zhong-Qiu Liu [1,2,*]

1 Department of Pharmaceutics, School of Pharmaceutical Sciences, Southern Medical University, Guangzhou 510515, Guangdong, China; E-Mails: daipeimin1989@163.com (P.-M.D.); IITCM_Wangy@126.com (Y.W.); lotus623@126.com (L.Y.); zengshan_jessica@126.com (S.Z.); zhijiezheng@126.com (Z.-J.Z.); liqiangsmu@163.com (Q.L.); iitcm_lu@126.com (L.-L.L.)

2 International Institute for Translational Chinese Medicine, Guangzhou University of Chinese Medicine, Guangzhou 510006, Guangdong, China

* Author to whom correspondence should be addressed; E-Mail: liuzq@gzucm.edu.cn

---

Abstract: Wutou decoction is widely used in China because of its therapeutic effect on rheumatoid arthritis. Benzoylmesaconine (BMA), the most abundant component of Wutou decoction, was used as the marker compound for the pharmacokinetic study of Wutou decoction. The aim of the present study was to compare the pharmacokinetics of BMA in rats after oral administration of pure BMA and Wutou decoction. Pure BMA (5 mg/kg) and Wutou decoction (0.54 g/kg, equivalent to 5 mg/kg BMA) were orally administered to rats with blood samples collected over 10 h. Quantification of BMA in rat plasma was achieved using sensitive and validated ultra-performance liquid chromatography-tandem mass spectrometry (UPLC-MS/MS). Specifically, the half-life ($T_{1/2}$) and mean residence time values of pure BMA were 228.3 ± 117.0 min and 155.0 ± 33.2 min, respectively, whereas those of BMA in Wutou decoction were decreased to 61.8 ± 35.1 min and 55.8 ± 16.4 min, respectively. The area under the curve (AUC) of BMA after administration of Wutou decoction was significantly decreased (five-fold) compared with that of pure BMA. The results indicate that the elimination of BMA in rats after the administration of Wutou decoction was significantly faster compared with that of pure BMA.

Keywords: benzoylmesaconine; Wutou decoction; pharmacokinetics; UPLC-MS/MS

---

1. Introduction

Herbal medicine is mostly administered in combination based on the principle of traditional Chinese Medicine to achieve effect optimization or toxicity reduction. It is generally recognized that the combined use of prescriptions have a greater impact on the therapeutic effects. Pharmacokinetic study, a useful method to predict the *in vivo* process, has been used to elucidate the different processes between pure chemicals and multiple-ingredient prescriptions; therefore, it is important to conduct pharmacokinetic studies to evaluate the rationality of the traditional prescriptions.

Wutou decoction, a well-known traditional prescription, comprises Radix *Aconiti* (derived from the dried root of *Aconitum carmichaeli* Debx.), Herba *Ephedrae* (derived from the stem of *Ephedra sinica* Stapf.), Radix *Paeoniae Alba* (derived from the dried root of *Paeonia lactiflora* Pall.), Radix *Astragali* (derived from the dried root of *Astragalus membranaceus*) and Radix *Glycyrrhizae* (derived from the dried root of *Glycyrrhiza uralensis* Fisch). The decoction has been used for hundreds of years in China for rheumatoid arthritis (RA) treatment.

To date, many studies have investigated the effects of Wutou decoction in patients with RA. Radix *Aconiti*, a naturally occurring product, is considered a prominent component of Wutou decoction because of its therapeutic effects against RA [1,2]. However, Radix *Aconiti* is a toxic herb that can cause serious cardiac poisoning [3–6]. Previous studies revealed a 58% incidence rate of toxic reactions among 188 patients who had taken aconitum [7–9]. Aconitum alkaloids are the effective chemicals of Radix *Aconiti*. Notably, the high levels of toxicity of Radix *Aconiti* are derived from diester aconitum alkaloids, including aconitine, mesaconitine (MA) and hypaconitine, which can be hydrolyzed to benzoylaconine, benzoylmesaconine (BMA) and benzoylhypaconine, respectively. The MA hydrolysate BMA demonstrates lower toxicity [7], higher abundance and a better pharmacological effect than the other two monoester diterpenoid alkaloids in Wutou decoction [10]. This study selected BMA as a marker compound for the pharmacokinetic study of Wutou decoction.

BMA exhibits biological effects, including analgesic [11,12], antiviral and antifungal activities [13]. It can also stimulate cytokine secretion [14]. Although BMA is less toxic than diester diterpene alkaloids [10], excessive intake of BMA still causes toxic reactions [15]. Hence, several pharmacokinetic studies have been performed to understand the *in vivo* process of BMA by using the LC-MS method [16–20]. For instance, some researchers [16] found that BMA demonstrates fast absorption and elimination *in vivo* after the administration of pure BMA ($T_{1/2\,\beta}$ = 407 ± 180 min). Others [17] verified that BMA has short T_{max} (36.17 ± 1.72 min) compared with the other two monoester aconitum alkaloids after intravenous drop infusion of "SHEN-FU" injectable powder. Recently, research on the pharmacokinetics of BMA was conducted in combination with other herbs or components, such as Dahuang-Fuzi decoction and Radix *et Rhizoma Rhei* [18,19]. The pharmacokinetic parameters (C_{max}, AUC) of BMA were remarkably reduced; $T_{1/2}$ and mean residence time (MRT) were delayed with oral administration of Dahuang-Fuzi decoction [18]. The additional studies demonstrated that co-administration with Rhizoma *Zingiberis* could reduce the C_{max} of BMA with the increased $T_{1/2}$, T_{max} and AUC [19]. The co-administration suggests that Rhizoma *Zingiberis* or other ingredients in Dahuang-Fuzi decoction could not accelerate the elimination of BMA, therefore having little influence on the toxicity reduction.

Although the above studies focused on the pharmacokinetics of BMA [16–20], little is known about the pharmacokinetics after the administration of Wutou decoction and pure BMA. Since Wutou decoction has complex ingredients, herb-herb interactions may exert an influence on the efficacy of Chinese medicine; therefore, this study aims to compare the pharmacokinetics of BMA after oral administration of pure BMA and Wutou decoction to rats using a rapid, sensitive and reliable UPLC-MS/MS method. The information obtained might be useful for understanding different herb-herb interactions between BMA and other ingredients, which can be used as a reference for the clinical administration of Wutou decoction.

2. Results and Discussion

2.1. Validation of UPLC-MS/MS Methods

An Acquity UPLC BEH C18 column was selected for the chromatographic separation of BMA (Figure 1A) and testosterone (Figure 1B), because of its good peak shapes and acceptable retention times. A 2-mM ammonium acetate and acetonitrile solution led to better peak shapes compared to other mobile phases. Testosterone was selected as an internal standard, because of its similar chromatographic behavior and complete separation with BMA.

Figure 1. Chemical structures of benzoylmesaconine (BMA) (**A**) and the internal standard, testosterone (**B**).

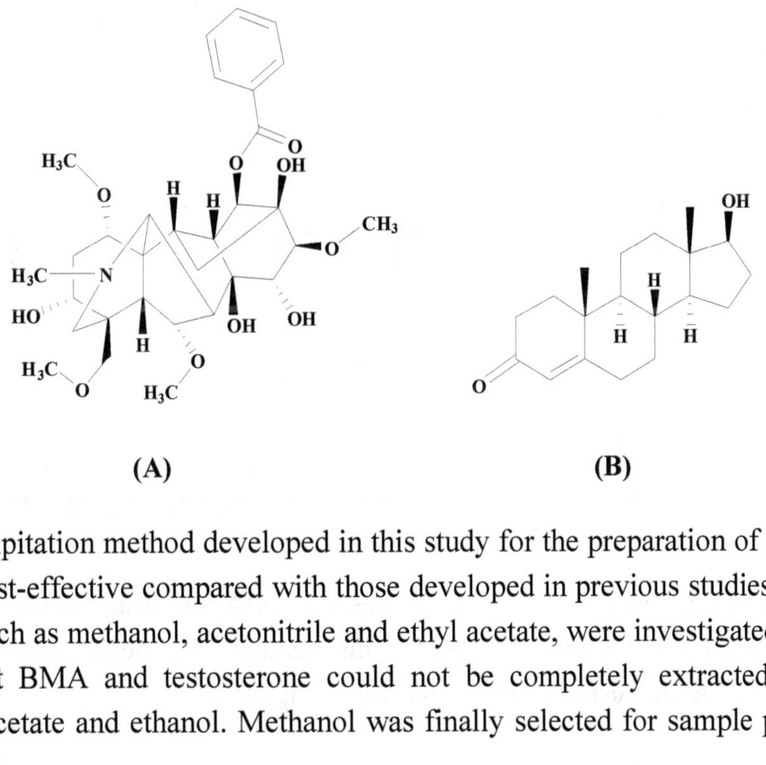

(A) **(B)**

The protein precipitation method developed in this study for the preparation of plasma samples was more simple and cost-effective compared with those developed in previous studies [17,21,22]. Several organic solvents, such as methanol, acetonitrile and ethyl acetate, were investigated for extraction. The results showed that BMA and testosterone could not be completely extracted simultaneously by acetonitrile, ethyl acetate and ethanol. Methanol was finally selected for sample preparation, because of its relatively high extraction recovery. Methanol-water (1:1), instead of the mobile phase, has been used to reconstitute the residue, because of the better peak shape. Therefore, a one-step process of protein precipitation with methanol that exhibited acceptable recovery was selected. This process was time saving and helpful to the stability of plasma samples.

The concentrations of BMA in the plasma samples were too low to be determined by UPLC-UV. Thus, an UPLC-MS/MS method with higher sensitivity than UPLC-UV was developed in this study. The MS and MS/MS daughter scan spectrograms of BMA and testosterone in the ESI source are shown in Figures 2 and 3, respectively.

Figure 2. Full-scan (**A**) and MS/MS daughter scan (**B**) spectrograms of BMA.

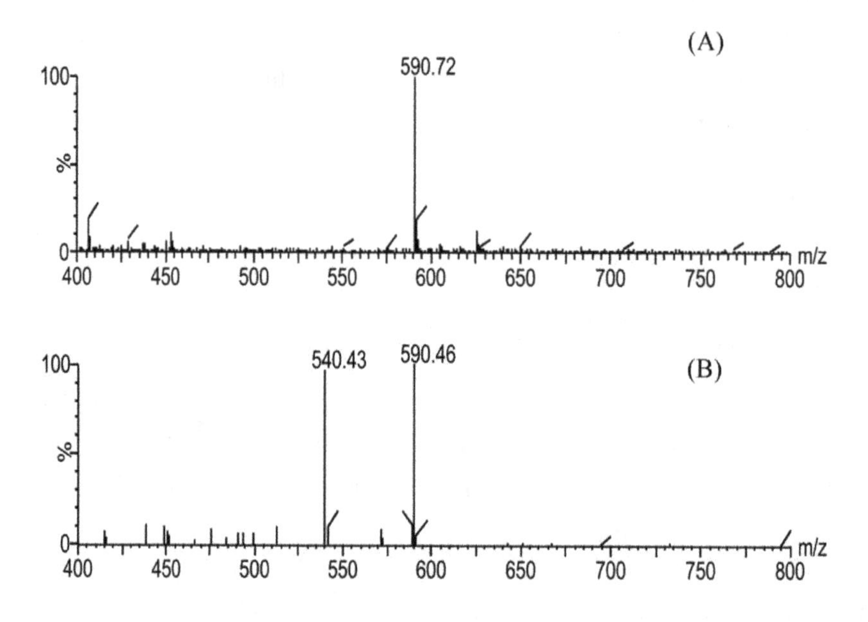

Figure 3. Full-scan (**A**) and MS/MS daughter scan (**B**) spectrograms of testosterone.

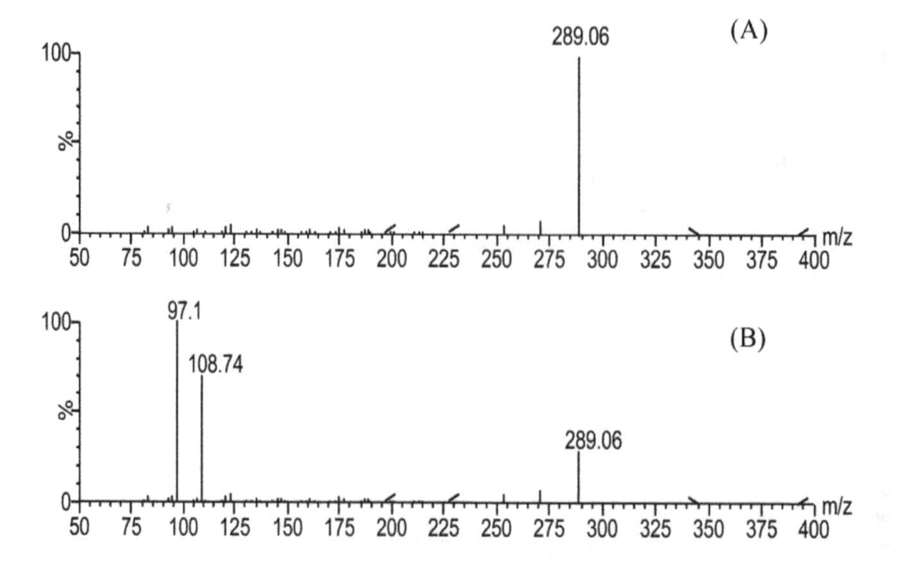

The chromatograms showed excellent peak shapes for BMA and testosterone at retention times of 2.8 and 3.6 min, respectively, with no endogenous interference. The peak responses of blank plasma samples spiked with BMA and testosterone are shown in Figure 4.

Figure 4. Chromatograms for BMA and testosterone in blank plasma (**A**); blank plasma spiked with BMA at 6 ng/mL and testosterone at 100 nM (**B**); plasma sample of BMA and testosterone after administration of pure BMA (**C**); and Wutou decoction (**D**). The retention time of BMA and testosterone was 2.8 and 3.3 min, which could be separated completely.

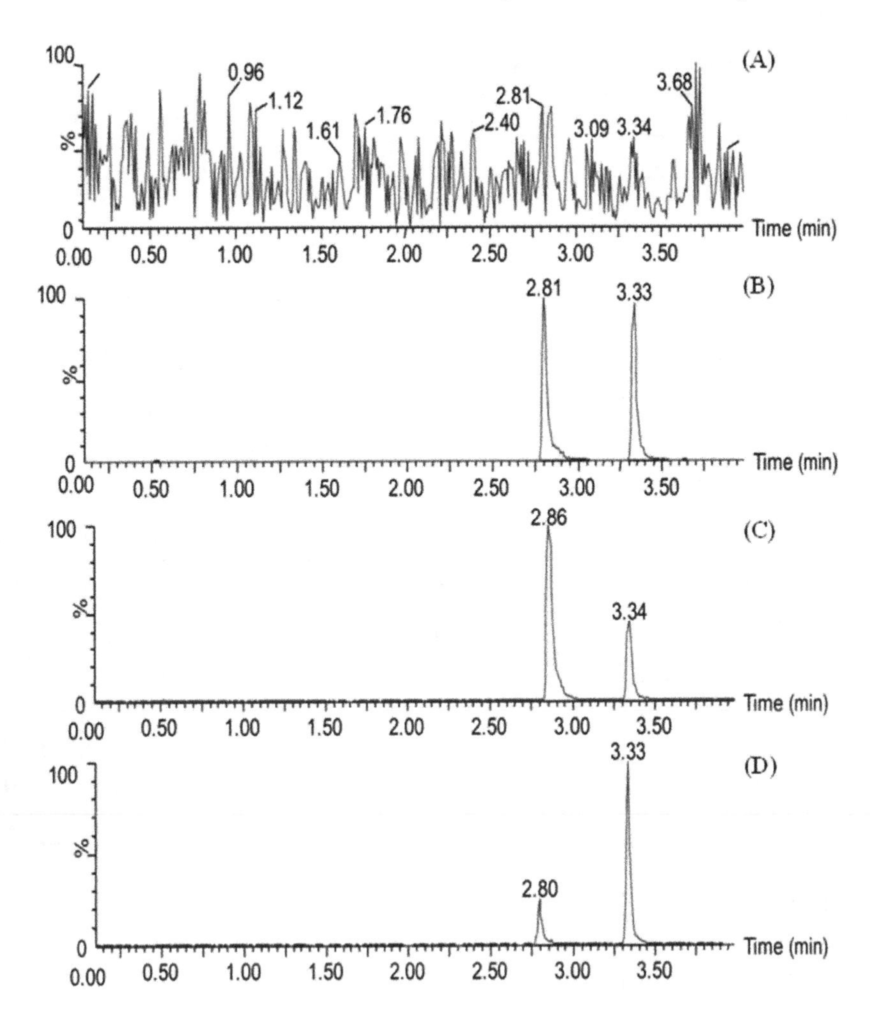

The calibration curve was investigated in the range of 0.3 ng/mL to 60 ng/mL. All curves had correlation coefficients of >0.990. The lower limit of quantification (LLOQ) of BMA under the developed UPLC-MS/MS method was 0.3 ng/mL, with a signal-to-noise ratio of >10.

The precision and accuracy of the method for detecting BMA were evaluated at three concentrations. The RSD% values for the intra-day and inter-day precision were smaller than 6.1% and 13.9%, respectively. The RE% values for the lowest concentration were 17.6% and smaller than 15.0% for the other two concentrations (see in Table 1).

Table 1. Accuracy and precision of BMA in rat plasma (n = 5).

Concentrations (ng/mL)	Accuracy (RE%)	Precision (RSD%)	
		Intra-day	Inter-day
1.2	17.6	6.1	13.9
6	12.7	6.1	7.9
30	10.3	3.7	4.7

The extraction recovery values of BMA were >85% at three concentrations. The matrix effects were over the range of 126.1% to 139.0% at low, middle and high concentrations, respectively, suggesting that significant ion suppression or enhancement did not occur at the expected retention times of the targeted ions (see in Table 2).

Table 2. Extraction recovery and matrix effect of BMA in rat plasma (n = 5).

Concentrations (ng/mL)	Extraction Recovery		Matrix Effect	
	Mean (%)	RSD (%)	Mean (%)	RSD (%)
1.2	92.0	5.4	126.1	5.9
6	92.9	9.3	139.0	2.3
30	85.2	6.4	134.1	6.5

The accuracy biases of BMA in the rat plasma samples ranged from 94.7% to 109.8%, 99.2% to 114.6%, 90.2% to 101.7% and 92.6% to 100.0% for the stability evaluation of short-term storage, long-term storage, three freeze-thaw cycles and auto-sampler storage, respectively (see in Table 3).

Table 3. Stability evaluation of BMA in rat plasma (n = 5).

Concentrations (ng/mL)	Short-Term Storage		Long-Term Storage		Three Freeze-Thaw Cycles		Auto-Sampler Stability	
	Accuracy (%)	RSD%	Accuracy (%)	RSD%	Accuracy (%)	RSD%	Accuracy (%)	RSD%
1.2	94.7	13.3	114.6	2.4	90.2	10.2	100.0	3.9
6	109.8	5.1	110.3	3.3	100.8	4.0	106.4	8.3
30	109.4	3.3	99.2	8.2	101.7	5.0	92.6	4.8

2.2. Content Determination of BMA in the Lyophilized Powder

The content of BMA in the lyophilized powder was determined as follows. Lyophilized powder was dissolved in pure water to obtain the decoction solution. The solution was vortexed for 3 min and then centrifuged at 13,000 rpm for 30 min. The supernatant was injected into the UPLC-MS/MS system for determination. The result showed that 0.92% of BMA was present in the lyophilized powder.

2.3. Pharmacokinetic Study

Although previous studies already investigated the pharmacokinetics of BMA [21,22], the present study is the first to compare the pharmacokinetic behavior of pure BMA and BMA as a marker compound of Wutou decoction in rats.

The developed UPLC-MS/MS method was used for the quantitation of plasma samples after oral administration of pure BMA (5 mg/kg) and Wutou decoction (0.54 g/kg, equivalent to BMA 5 mg/kg). As the predominant component of Wutou decoction, BMA exhibited the highest content in rat plasma after oral administration, which is in accordance with a previous study [21]. The plasma concentration *versus* time profile of BMA in the two rat groups is shown in Figure 5.

Figure 5. Plasma concentration *versus* time profile of BMA after oral administration of BMA and Wutou decoction at a dosage of 5 mg/kg (BMA). Each point represents the mean and standard deviation (mean ± SD) of five rats. BMA plasma concentrations at 240 min were approximately 0 after oral administration of Wutou decoction, and time points after that were deleted.

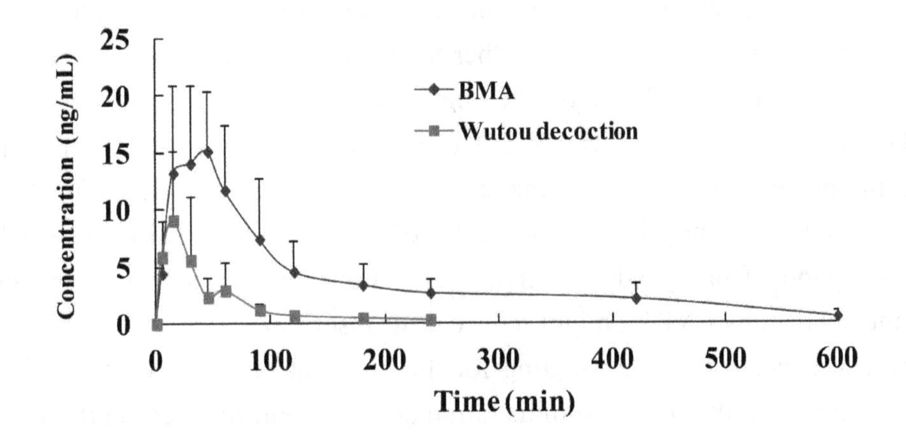

The pharmacokinetic parameters were expressed as the mean ± SD. All parameters are shown in Table 4. After the oral administration of pure BMA and Wutou decoction, the plasma concentration-time curve produced a fast rising trend followed by a sharp decline with $T_{1/2}$ of 228.3 ± 117.0 min and 61.8 ± 35.1 min, respectively, until the levels fell below the detection limits. BMA was absorbed at a fast rate and reached a maximum concentration at C_{max} of 16.2 ± 6.7 ng/mL and 10.0 ± 5.8 ng/mL within 35.0 ± 11.2 min and 13.0 ± 4.5 min after pure BMA and Wutou decoction administration, respectively. The mean residence time (MRT) values for pure BMA were 155.0 ± 33.2 min, whereas that for Wutou decoction were 55.8 ± 16.4 min. The $AUC_{(0\text{-}t)}$ values of BMA for pure BMA and Wutou decoction were 2247.4 ± 1171.9 and 447.8 ± 292.2 ng·min/mL, respectively, and the relative bioavailability of BMA after oral administration of Wutou decoction was 19.9%. Furthermore, the results of the pharmacokinetic study indicated that the absorption and elimination of BMA after the administration of pure BMA and Wutou decoction were very fast, within 13.0 min to 35.0 min and 61.8 min to 228.3 min, respectively.

Table 4. Pharmacokinetic parameters of BMA in Sprague-Dawley rat plasma (n = 5) after oral administration of BMA and Wutou decoction at 5 mg/kg (BMA). $AUC_{(0\text{-}t)}$, $T_{1/2}$, C_{max}, T_{max}, mean residence time (MRT) and relative bioavailability (RF) are shown in the table. Each value represents the mean and standard deviation (mean ± SD) of five rats.

Parameters	Unit	BMA	Wutou Decoction
$AUC_{(0\text{-}t)}$	ng·min/mL	2,247.4 ± 1,171.9	447.8 ± 292.2 *
$T_{1/2}$	min	228.3 ± 117.0	61.8 ± 35.1
C_{max}	ng/mL	16.2 ± 6.7	10.0 ± 5.8
T_{max}	min	35.0 ± 11.2	13.0 ± 4.5 *
MRT	min	155.0 ± 33.2	55.8 ± 16.4 *
RF	%	-	19.9

* $p < 0.05$ between two groups.

Notably, the pharmacokinetic process of BMA in Wutou decoction was significantly different compared with that in pure BMA (ANOVA, $p < 0.05$). $AUC_{(0\text{-}t)}$, $T_{1/2}$ and MRT decreased, suggesting that other complex ingredients in Wutou decoction might have an influence on the pharmacokinetics of BMA, which possibly demonstrated shorter retention time *in vivo* and lower distribution in the target organs, leading potentially to toxicity attenuation.

Numerous comparative pharmacokinetic studies have been reported on the absorption of effective components enhanced after combination with other herbs, some of which results from the competitive inhibition between components in herbs on the elimination via efflux transporters or the metabolism via drug metabolizing enzymes, such as CYP3A (microsomal cytochrome P450) [18]. The possible mechanism for the decrease in pharmacokinetic parameters is that the drug efflux transporters and CYP3A are distributed abundantly in the enterocytes of the gastrointestinal tract, which is the crucial place for the absorption of orally administered drugs. Therefore, the excretion by intestinal efflux transporters or metabolism by CYP3A might reduce drug absorption.

Although the exact mechanisms accounting for the different pharmacokinetic behaviors of BMA after pure BMA and Wutou decoction administration are not clear, different herb-herb interactions in Wutou decoction might be one possible explanation. The compounds in Wutou decoction might play an important role in affecting the elimination of BMA, influencing several links of ADME (absorption, distribution, metabolism and elimination) of BMA in rats, which leads to the decreased AUC and $T_{1/2}$. For example, the efflux transporter, MRP2 (multi-drug resistance associate protein 2), was demonstrated to be involved in the efflux of BMA [6]. The diminished $AUC_{(0\text{-}t)}$ might suggest that other complex ingredients in Wutou decoction reduced the absorption and bioavailability of BMA via induction of MRP2 efflux.

Furthermore, it was also reported that drug metabolizing enzymes, including CYP3A (microsomal cytochrome P450), played a significant role in the elimination of some herbs *in vitro* [23,24]. Meanwhile, BMA could be also metabolized by CYP3A [15]. Accelerating the metabolic elimination of BMA would reduce the oral bioavailability of BMA. Thus, another factor that may contribute to the faster elimination of BMA in Wutou decoction is metabolic induction by the other coexisting constituents. $T_{1/2}$ and MRT decreased, which indicated that some components in Wutou decoction might induce the metabolism of CYP3A, thereby shortening the retention time of BMA *in vivo*, which might contributed to the toxicity reduction. Among all of the components of Wutou decoction, diammonium glycyrrhizinate reportedly has a positive effect on the toxicity reduction of aconitum alkaloids [25,26].

Thus, the faster elimination of BMA in rats after Wutou decoction administration is likely to be attributed to efflux through MRP2 or metabolism by CYP3A4/5 [6]. Some *in vivo* herb-herb interactions between the complex ingredients of Wutou decoction might exist to decrease the AUC and MRT of BMA via the enhanced expression of MRP2 or CYP3A. However, direct research on the induced expression of CYP3A or MRP2 by the complex ingredients in Wutou decoction is still limited, leading to more studies on a deeper level.

3. Experimental Section

3.1. Chemicals and Reagents

BMA (≥98% purity) was obtained from the National Institute for the Control of Pharmaceutical and Biological Products (Beijing, China). Testosterone (≥98% purity) was purchased from Nacalai Tesque (Kyoto, Japan). Radix *Aconiti*, Herba *Ephedrae*, Radix *Paeoniae Alba*, Radix *Astragali* and Radix *Glycyrrhizae* (identified by the Sichuan Institute for Food and Drug Control) were purchased from Dongguan China Herbal Medicine Co., Ltd. (Dongguan, China). A voucher specimen was deposited at the Laboratory of Pharmaceutics, School of Pharmaceutical Sciences, Southern Medical University (Guangzhou, China). All other chemicals were of analytical grade.

3.2. Animals

Male Sprague-Dawley rats weighing between 230 and 280 g were supplied by the Laboratory Animal Center of Southern Medicine University (license: SCXK, Guangdong, 2006–0015) (Guangzhou, China). The rats were housed four per cage in a unidirectional airflow room under controlled temperature (20 °C to 24 °C), relative humidity (40% to 70%) and a 12-h light/dark cycle. The animal experimental protocol was approved by the ethics committee of Southern Medicine University (No: 2011–0015, Date: 24 October 2012). All animal studies were carried out according to the guide for care and use of laboratory animals. The rats were fasted, but allowed free access to water for at least 12 h before the experiment.

3.3. Instruments and Conditions

The UPLC-MS/MS system (Waters, Milford, MA, USA) consisted of a binary solvent manager, a column compartment, an auto-sampler manager and a single quadrupole mass spectrometer. All data acquisition was performed with Waters Masslynx V4.1 software.

The liquid chromatography separation of plasma samples was carried out on an Acquity UPLC BEH C18 column (1.7 μm; 2.1 mm × 50 mm) at 40 °C. Two micromolar ammonium acetate (A) and 100% acetonitrile (B) were used as the mobile phase. The flow rate was set at 0.3 mL/min. The gradients used for the elution were 0 min, 95% A; 0 min to 1.2 min, 95% to 88% A; 1.2 min to 3 min, 88% to 20% A; and 3 min to 4 min, 20% to 95% A. The injection volume was 5 μL.

Electro spray ionization in the positive mode was used to detect BMA in the plasma samples. The MS tune parameters were as follows: capillary voltage, 3.0 kV; cone voltage, 30 V; ion source temperature, 120 °C; desolvation temperature, 350 °C; desolvation gas flow, 650 L/h; and cone gas flow, 50 L/h. The mass spectrometer was operated in the multiple reaction monitoring (MRM) mode.

3.4. Preparation of Wutou Decoction Lyophilized Powder

Radix *Aconiti* (6 g) was added to 600 mL of boiling pure water and then boiled for 30 min to extract its active constituents. A mixture of *Herba Ephedrae* (9 g), Radix *Paeoniae Alba* (9 g), Radix *Astragali* (9 g) and Radix *Glycyrrhizae* (9 g) was added and boiled for 30 min. The solution was filtered and concentrated to obtain Wutou decoction. The decoction was concentrated to 100 mL using a rotary evaporator and then stored at −20 °C overnight for freezing. The frozen decoctions were freeze dried by a lyophilizer to obtain the lyophilized powder form of the decoction. The content of BMA in the lyophilized powder was quantified by UPLC-MS/MS the day before the pharmacokinetic experiment.

3.5. Preparation of Stock Solutions

A stock solution of BMA was prepared in methanol at a concentration of 5 mM, and the working solution of BMA was diluted with methanol. The testosterone stock solution was prepared in acetonitrile at a concentration of 20 mM, and the working solution of testosterone was diluted to 1000 nM with acetonitrile. All stock solutions were stored at −20 °C prior to use.

3.6. Biosample Collection

Pure BMA and Wutou decoction lyophilized powder was dissolved in 0.9% saline to obtain a concentration of 0.25 mg/mL (BMA). Ten rats were randomly divided into two groups. Pure BMA (5 mg/kg) and Wutou decoction (0.54 g/kg, equivalent to 5 mg/kg BMA) were orally administered to them, respectively. Serial blood samples (500 μL) were taken from the orbital sinus venous plexus at 0, 5, 15, 30, 45, 60, 90, 120, 180, 240, 420 and 600 min after gavage dosing. Each collected blood sample was centrifuged at 8000 rpm for 8 min. The plasma fractions were transferred into a disposable tube and then frozen at −80 °C until analysis.

3.7. Biosample Preparation

To detect BMA in the rat plasma, 80 μL of plasma was mixed with 320 μL methanol containing 100 nM testosterone. The mixture was centrifuged at 13,000 rpm for 30 min. A total of 300 μL of the supernatant was transferred to a disposable tube and then evaporated to dryness under a stream of nitrogen at room temperature. The residue was reconstituted with 100 μL of methanol-water (v:v = 1:1) and then injected into the UPLC-MS/MS system for analysis.

3.8. Method Validation

The specificity of the method was evaluated using blank plasma samples. Blank rat plasma was spiked with the working solutions of BMA (6 ng/mL) and testosterone (100 nM). The linearity of BMA in rat plasma was evaluated with the calibration curve based on the UPLC-MS/MS analysis of blank plasma spiked with different concentrations of BMA. This curve was obtained by plotting the peak area ratios *versus* the different concentrations of BMA. The LLOQ was defined based on the minimum concentration with a signal-to-noise ratio of ≥10. The precision and accuracy were evaluated by analyzing quality control (QC) samples with different concentrations. Intra-day and inter-day precision and accuracy were evaluated for 3 consecutive days with five replicates at each concentration per day.

The recoveries were determined by comparing the peak areas from the extracted samples with those from post-extracted blank plasma spiked with the analytes at the same concentration. The matrix effect was measured by comparing the peak areas of analytes added into the post-extracted blank with analytes dissolved in the matrix component-free reconstitution solvent. The stability of BMA in rat plasma was evaluated at room temperature for 12 h, after storage at −80 °C for 15 days, after three freeze-thaw cycles or in an auto-sampler for 12 h, respectively.

3.9. Data Analysis

The PK parameters were determined using the standard non-compartmental method and calculated using WinNonlin 5.2 (Pharsight, Mountain View, CA, USA). The C_{max} and the corresponding T_{max} were directly obtained from the raw data. The AUC to the last measurable concentration ($AUC_{(0-t)}$) was calculated using the linear trapezoidal method. The relative bioavailability of BMA after oral administration of Wutou decoction was calculated as $AUC_{(decoction)}/AUC_{(BMA)} \times 100\%$. An independent-sample *t*-test was performed twice to evaluate the differences of pharmacokinetic parameters between the two groups.

4. Conclusions

To the best of our knowledge, this study is the first to compare the pharmacokinetics of BMA after oral administration of pure BMA and Wutou decoction. The absorption of BMA in Wutou decoction was significantly reduced due to faster elimination, in comparison with that of pure BMA administration, which might contribute to toxicity reduction. Better understanding of the interactions between BMA and the coexisting ingredients in the Wutou decoction is still needed for clinical application of Wutou decoction.

Acknowledgments

We would like to thank the Sichuan Institute for Food and Drug Control for the help of the identification of the plant materials. This work was supported in part by the Key International Joint Research Project of National Natural Science Foundation of China (Grant No. 81120108025), the Science and Information Technology of Guangzhou (Grant No. 2011Y1-00017-6) and the Fund of the Natural Science Foundation of Guangdong Province (Grant No. S2013040012007).

Author Contributions

Conceived of and designed the experiments: Zhong-Qiu Liu, Pei-Min Dai, Ling Ye. Performed the experiments: Pei-Min Dai, Shan Zeng, Ying Wang, Zhi-Jie Zheng, Qiang Li, Lin-Lin Lu. Analyzed the data: Pei-Min Dai, Shan Zeng, Ying Wang, Zhi-Jie Zheng. Wrote the paper: Zhong-Qiu Liu, Pei-Min Dai, Ling Ye, Ying Wang.

References

1. Fraenkel, L.; Cunningham, M. High disease activity may not be sufficient to escalate care. *Arthrit. Care Res.* **2014**, *66*, 197–203.

2. Wisniacki, N.; Amaravadi, L.; Galluppi, G.R.; Zheng, T.S.; Zhang, R.; Kong, J.; Burkly, L.C. Safety, tolerability, pharmacokinetics, and pharmacodynamics of anti-TWEAK monoclonal antibody in patients with rheumatoid arthritis. *Clin. Ther.* **2013**, *35*, 1137–1149.
3. Van Landeghem, A.A.; de Letter, E.A.; Lambert, W.E.; van Peteghem, C.H.; Piette, M.H. Aconitine involvement in an unusual homicide case. *Int. J. Legal Med.* **2007**, *121*, 214–219.
4. Singh, S.; Fadnis, P.P.; Sharma, B.K. Aconite poisoning. *J. Assoc. Phys. India* **1986**, *34*, 825–826.
5. Bisset, N.G. Arrow poisons in China. Part II. Aconitum—Botany, chemistry, and pharmacology. *J. Ethnopharmacol.* **1981**, *4*, 247–336.
6. Ye, L.; Yang, X.; Yang, Z.; Gao, S.; Yin, T.; Liu, W.; Wang, F.; Hu, M.; Liu, Z. The role of efflux transporters on the transport of highly toxic aconitine, mesaconitine, hypaconitine, and their hydrolysates, as determined in cultured Caco-2 and transfected MDCKII cells. *Toxicol. Lett.* **2013**, *216*, 86–99.
7. Singhuber, J.; Zhu, M.; Prinz, S.; Kopp, B. Aconitum in traditional Chinese medicine: A valuable drug or an unpredictable risk? *J. Ethnopharmacol.* **2009**, *126*, 18–30.
8. Elliott, S.P. A case of fatal poisoning with the aconite plant: Quantitative analysis in biological fluid. *Sci. Justice* **2002**, *42*, 111–115.
9. Pullela, R.; Young, L.; Gallagher, B.; Avis, S.P.; Randell, E.W. A case of fatal aconitine poisoning by Monkshood ingestion. *J. Forensic Sci.* **2008**, *53*, 491–494.
10. Jiang, Z.H.; Xie, Y.; Zhou, H.; Wang, J.R.; Liu, Z.Q.; Wong, Y.F.; Cai, X.; Xu, H.X.; Liu, L. Quantification of aconitum alkaloids in aconite roots by a modified RP-HPLC method. *Phytochem. Anal.* **2005**, *16*, 415–421.
11. Suzuki, Y.; Hayakawa, Y.; Oyama, T.; Isono, T.; Ohmiya, Y.; Ikeda, Y.; Asami, A.; Noguchi, M. Analgesic effect of benzoylmesaconine. *Nihon Yakurigaku Zasshi* **1993**, *102*, 399–404.
12. Suzuki, Y.; Oyama, T.; Ishige, A.; Isono, T.; Asami, A.; Ikeda, Y.; Noguchi, M.; Omiya, Y. Antinociceptive mechanism of the aconitine alkaloids mesaconitine and benzoylmesaconine. *Planta Med.* **1994**, *60*, 391–394.
13. Kobayashi, M.; Mori, K.; Kobayashi, H.; Pollard, R.B.; Suzuki, F. The regulation of burn-associated infections with herpes simplex virus type 1 or Candida albicans by a non-toxic aconitine-hydrolysate, benzoylmesaconine. Part 1: Antiviral and anti-fungal activities in thermally injured mice. *Immunol. Cell Biol.* **1998**, *76*, 202–208.
14. Kobayashi, M.; Takahashi, H.; Herndon, D.N.; Pollard, R.B.; Suzuki, F. Therapeutic effects of IL-12 combined with benzoylmesaconine, a non-toxic aconitine-hydrolysate, against herpes simplex virus type 1 infection in mice following thermal injury. *Burns* **2003**, *29*, 37–42.
15. Ye, L.; Yang, X.S.; Lu, L.L.; Chen, W.Y.; Zeng, S.; Yan, T.M.; Dong, L.N.; Peng, X.J.; Shi, J.; Liu, Z.Q. Monoester-diterpene aconitum alkaloid metabolism in human liver microsomes: Predominant role of CYP3A4 and CYP3A5. *Evid.-Based Complement. Altern. Med.* **2013**, *2013*, doi:10.1155/2013/941093.
16. Ye, L.; Gao, S.; Feng, Q.; Liu, W.; Yang, Z.; Hu, M.; Liu, Z. Development and validation of a highly sensitive UPLC-MS/MS method for simultaneous determination of aconitine, mesaconitine, hypaconitine, and five of their metabolites in rat blood and its application to a pharmacokinetics study of aconitine, mesaconitine, and hypaconitine. *Xenobiotica* **2012**, *42*, 518–525.

17. Zhang, F.; Tang, M.H.; Chen, L.J.; Li, R.; Wang, X.H.; Duan, J.G.; Zhao, X.; Wei, Y.Q. Simultaneous quantitation of aconitine, mesaconitine, hypaconitine, benzoylaconine, benzoylmesaconine and benzoylhypaconine in human plasma by liquid chromatography-tandem mass spectrometry and pharmacokinetics evaluation of "SHEN-FU" injectable powder. *J. Chromatogr. B* **2008**, *873*, 173–179.
18. Liu, X.; Li, H.; Song, X.; Qin, K.; Guo, H.; Wu, L.; Cai, H.; Cai, B. Comparative pharmacokinetics studies of benzoylhypaconine, benzoylmesaconine, benzoylaconine and hypaconitine in rats by LC-MS method after administration of Radix Aconiti Lateralis Praeparata extract and Dahuang Fuzi Decoction. *Biomed. Chromatogr.* **2014**, *28*, 966–973.
19. Peng, W.W.; Li, W.; Li, J.S.; Cui, X.B.; Zhang, Y.X.; Yang, G.M.; Wen, H.M.; Cai, B.C. The effects of Rhizoma Zingiberis on pharmacokinetics of six Aconitum alkaloids in herb couple of Radix Aconiti Lateralis-Rhizoma Zingiberis. *J. Ethnopharmacol.* **2013**, *148*, 579–586.
20. Usui, K.; Hayashizaki, Y.; Hashiyada, M.; Nakano, A.; Funayama, M. Simultaneous determination of 11 aconitum alkaloids in human serum and urine using liquid chromatography-tandem mass spectrometry. *Legal Med.* **2012**, *14*, 126–133.
21. Tang, L.; Gong, Y.; Lv, C.; Ye, L.; Liu, L.; Liu, Z. Pharmacokinetics of aconitine as the targeted marker of Fuzi (Aconitum carmichaeli) following single and multiple oral administrations of Fuzi extracts in rat by UPLC/MS/MS. *J. Ethnopharmacol.* **2012**, *141*, 736–741.
22. Song, J.Z.; Han, Q.B.; Qiao, C.F.; But, P.P.; Xu, H.X. Development and validation of a rapid capillary zone electrophoresis method for the determination of aconite alkaloids in aconite roots. *Phytochem. Anal.* **2010**, *21*, 137–143.
23. Izzo, A.A.; Ernst, E. Interactions between herbal medicines and prescribed drugs: An updated systematic review. *Drugs* **2009**, *69*, 1777–1798.
24. Saxena, A.; Tripathi, K.P.; Roy, S.; Khan, F.; Sharma, A. Pharmacovigilance: Effects of herbal components on human drugs interactions involving cytochrome P450. *Bioinformation* **2008**, *3*, 198–204.
25. Peter, K.; Schinnerl, J.; Felsinger, S.; Brecker, L.; Bauer, R.; Breiteneder, H.; Xu, R.; Ma, Y. A novel concept for detoxification: Complexation between aconitine and liquiritin in a Chinese herbal formula ('Sini Tang'). *J. Ethnopharmacol.* **2013**, *149*, 562–569.
26. Chen, L.; Yang, J.; Davey, A.K.; Chen, Y.X.; Wang, J.P.; Liu, X.Q. Effects of diammonium glycyrrhizinate on the pharmacokinetics of aconitine in rats and the potential mechanism. *Xenobiotica* **2009**, *39*, 955–963.

4

Phosphosite Mapping of *HIP-55* Protein in Mammalian Cells

Ning Liu [2], Ningning Sun [2], Xiang Gao [2] and Zijian Li [1,]*

[1] Institute of Vascular Medicine, Peking University Third Hospital,
Key Laboratory of Cardiovascular Molecular Biology and Regulatory Peptides, Ministry of Health,
Key Laboratory of Molecular Cardiovascular Sciences, Ministry of Education and Beijing Key Laboratory of Cardiovascular Receptors Research, Beijing 100191, China

[2] Central Laboratory, Jilin University Second Hospital, Changchun 130041, China;
E-Mails: liu_ning@jlu.edu.cn (N.L.); sunnn13@mails.jlu.edu.cn (N.S.); gaoxiang13@mails.jlu.edu.cn (X.G.)

* Author to whom correspondence should be addressed; E-Mail: lizijian@bjmu.edu.cn

---

Abstract: In the present study, hematopoietic progenitor kinase 1 (HPK1)-interacting protein of 55 kDa (*HIP-55*) protein was over-expressed in HEK293 cells, which was genetically attached with 6x His tag. The protein was purified by nickel-charged resin and was then subjected to tryptic digestion. The phosphorylated peptides within the *HIP-55* protein were enriched by TiO_2 affinity chromatography, followed by mass spectrometry analysis. Fourteen phosphorylation sites along the primary structure of *HIP-55* protein were identified, most of which had not been previously reported. Our results indicate that bio-mass spectrometry coupled with manual interpretation can be used to successfully identify the phosphorylation modification in *HIP-55* protein in HEK293 cells.

Keywords: mass spectrometry; *HIP-55*; phosphorylation

---

1. Introduction

Protein phosphorylation is a fundamental type of post-translational modification, which plays a significant role in a wide range of cellular processes. Reversible phosphorylation results in a conformational change in the structure of many enzymes, receptors and adaptor proteins, triggering cellular signaling transduction, and modulating protein function, stability, interaction and localization. [1,2] Phosphorylation usually occurs on serine, threonine and tyrosine in eukaryotic proteins. [3] Phosphorylation on serine is the most common, followed by threonine and tyrosine. Determining protein phosphorylation sites is often the first step in the elucidation of a biological mechanism. Within a protein, phosphorylation can occur on several amino acids. The different phosphorylation sites mediate different biological processes. Therefore, the identification of *in vivo* phosphorylated sites of proteins is extremely important for understanding biological function and processes.

HIP-55, also called *SH3P7*, *mAbp1* and *DBNL*, is a multi-domain adaptor protein, with an actin-binding domain at its *N*-terminus and an SH3 domain at its *C*-terminus [4]. *HIP-55* acts as adaptor protein in many cellular processes such as cell signaling transduction and receptor endocytosis [5,6]. *HIP-55* also plays important roles in T-cell proliferation, immune responses and the development of cerebellar architecture [7–10]. Some important functions of the *HIP-55* protein are reported as mediated and regulated by its phosphorylation. *HIP-55* was identified as a tyrosine kinase substrate using anti-phosphotyrosine antibodies [11]. *HIP-55* is phosphorylated by Syk, Lyn, and Blk and further links antigen receptor signaling to components of the cytoskeleton [12]. *HIP-55* is also identified as a novel MELK substrate and is important for stem-cell characteristics and invasiveness [13]. Furthermore, src-mediated phosphorylation of *HIP-55* regulates podosome rosette formation in transformed fibroblasts [14]. Clearly, the *HIP-55* protein is involved in many signal transduction processes; thus, investigation of its phosphorylation modifications is important to better understand the role of *HIP-55* protein in the precise regulation of signal transduction.

Traditionally, three approaches are used to determine phosphorylation sites: the bioinformatics approach, the biochemical approach, and the genetics approach. In the last decade, advancements in mass spectrometry have redefined conventional biochemical approaches for the identification of phosphorylation sites. Mass spectrometry (MS) has become currently the most powerful technique for analysis of phosphorylation sites [15,16]. In the present study, mass spectrometry combined with phosphopeptide enrichment techniques were employed to identify the phosphorylation sites of *HIP-55* protein in mammalian cell. *HIP-55* protein carrying a tag of six histidines was over-expressed in HEK293 cells, which was enriched by Ni-NTA resin. Reverse capillary high efficiency liquid chromatography (HPLC) coupled with mass spectrometry was employed to profile phosphorylation modifications in the purified *HIP-55* protein. Several novel sites of phosphorylation in *HIP-55* protein were identified.

2. Results and Discussion

2.1. Expression and Purification of His-Tagged HIP-55 Protein

HIP-55, as an adaptor protein, was found to be expressed in various mammalian cells. The distribution of *HIP-55* was examined with an immunofluorescence assay. The result showed that *HIP-55* was observed throughout the cytosol and appeared enriched in the perinuclear area (Figure 1A). pDEST-His-*HIP-55* plasmid with full length *HIP-55* gene attached to a six-histidine tag were transfected into HEK293 cells which were maintained in DMEM containing 10% fetal bovine serum for 36 h at 37 °C. After the medium was removed, cells were washed with PBS three times.

Then the cells were harvested and lyzed. *HIP-55* protein was purified with Ni-NTA resin and further precipitated with chilled acetone to remove high amounts of salts such as guanidine HCl and imidazole. The protein pellet was collected and dissolved in SDS-containing buffer. The purified *HIP-55* protein was checked by western blot analysis as indicated in Figure 1B.

Figure 1. Expression and purification of *HIP-55* protein. (**A**) Cells were fixed and labeled with antibodies to *HIP-55*. Subcellular location of *HIP-55* was shown by laser scanning confocal microscopy; (**B**) After His-pull-down, the overexpression of *HIP-55* in the HEK293 cell line was detected with Western blot assay by *HIP-55* antibody. Actin was used as a loading control.

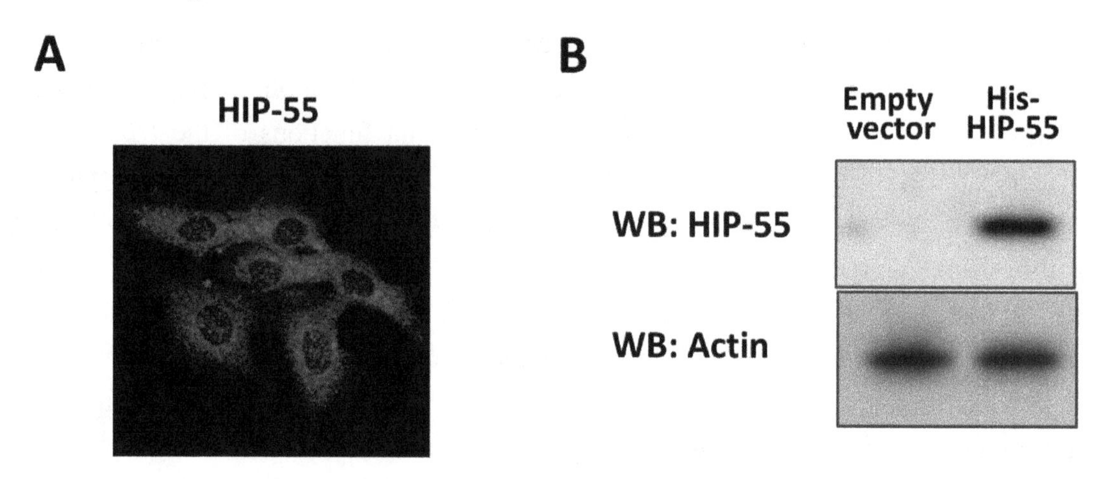

2.2. Identification of HIP-55 Protein by MS

The eluted proteins from Ni-NTA resin were desalted by acetone precipitation and then subjected to tryptic digestion. Prior to the analysis of phosphorylation, the peptide mixture was directly subjected to LC-MS/MS analysis, and the obtained data were inputted into a protein database for searching. Results showed a series of tryptic peptides from *HIP-55* protein with a coverage rate of 45.1%, in which only one phosphopeptide (LRS*PFLQK) was identified when Ser, Thr and Tyr phosphorylations were considered during the database searching. Figure 2 shows the distribution of peptides in the identification of *HIP-55* protein.

Figure 2. Primary structure of *HIP-55* protein. The tryptic peptides identified by LC-MS/MS analysis are highlighted.

MAANLSRNGP ALQEAYVRVV TEKSPTDWAL FTYEGNSNDI RVAGTGEGGL EEMVEELNSG
KVMYAFCRVK DPNSGLPKFV LINWTGEGVN DVRKGACASH VSTMASFLKG AHVTINARAE
EDVEPECIME KVAKASGANY SFHKESGRFQ DVGPQAPVGS VYQKTNAVSE IKRVGKDSFW
AKAEKEEENR RLEEKRRAEE AQRQLEQERR ERELREAARR EQRYQEQGGE ASPQRTWEQQ
QEVVSRNRNE QESAVHPREI FKQKERAMST TSISSPQPGK LRSPFLQKQL TQPETHFGRE
PAAAISRPRA DLPAEEPAPS TPPCLVQAEE EAVYEEPPEQ ETFYEQPPLV QQQGAGSEHI
DHHIQGQGLS GQGLCARALY DYQAADDTEI SFDPENLITG IEVIDEGWWR GYGPDGHFGM
FPANYVELIE

2.3. MS Analysis of Phosphorylation Sites of HIP-55 Protein

By enrichment of phosphopeptides from tryptic peptides mixture with TiO_2, a total of fourteen phosphopeptides from *HIP-55* protein were identified by LC-MS/MS analysis and database searching, including the one previously detected without TiO_2 enrichment (Table 1).

A few spectra were given below as examples for the identification of some of these phosphorylation sites in *HIP-55* protein.

Table 1. Identification of phosphopeptides within *HIP-55* protein over-expressed in HEK293 cells.

Phosphopeptides	Experimental *m/z* (mono)	Theoretical *m/z* (mono)	Charge	XCorr	Position in protein
GACAS *HVSTMASFLK	795.353	795.346	2	1.98	95–109
GACAS *HVSTM *ASFLK	803.351	803.344	2	1.94	95–109
GAHVT *INAR	509.744	509.745	2	2.37	110–118
ASGANY *SFHKES *GR	835.828	835.822	2	2.21	135–148
TNAVS *EIK	471.224	471.221	2	3.06	165–172
VGKDS *FWAK	559.253	559.258	2	2.57	174–182
REQRY *QEQGGEAS *PQR	1039.924	1039.926	2	1.94	220–235
AMS *TTSISSPQPGK	736.333	736.329	2	3.63	267–280
AMS *TT *SISSPQPGK	776.318	776.312	2	2.56	267–280
AM *S *TTSISSPQPGK	744.324	744.326	2	3.23	267–280
LRS *PFLQK	534.785	534.784	2	2.41	281–288
QLT *QPETHFGR	697.321	697.317	2	2.01	289–299
QLT *QPET *HFGR	737.304	737.300	2	1.98	289–299
EPAAAIS *RPR	574.289	574.285	2	2.49	300–309

NOTE: T *, S * and Y * refer to the phosphorylated form of threonine (Thr, T), serine (Ser, S) and tyrosine (Tyr, Y), respectively; M * refers to the oxidized methionine (Met, M).

Figure 3A showed the MS/MS spectrum of a doubly-charged molecular ion peak with *m/z* at 736.333. Database searching identified a phosphopeptide as AMpSTTSISSPQPGK (267–280) from *HIP-55* protein. A series of fragment ions such as fragment ions of *b* and *y* series can be clearly identified. The detection of *y* series ions at *m/z* 1102.5695, 1269.5607, 1171.5980 and 1302.6601 indicated that phosphorylation occurred exclusively at Serine 269 in this peptide. Detection of *b* series ions also confirmed this identification. Figure 3B shows the spectrum of a doubly-charged molecular ion peak with *m/z* at 776.318, which is similar to Figure 3A. The corresponding peptide is identified as the same peptide sequence as in Figure 3A, but with an additional phosphorylation site at Threonine 271. As shown in Figure 3B, *y* series ions at *m/z* 900.4679, 1081.4956, 1182.5466 and 1349.5401 indicated that phosphorylation occurred exclusively at Serine 269 in this peptide. Detection of b series ions also confirmed this identification. As shown in both mass spectra, the base peak at *m/z* 301.19 was observed, which resulted from fragmentation at Proline 278. It was noticed that most of the b series ions were subject to neutral loss of 98 (a phosphate molecule).

Figure 3. (**A**) MS/MS spectrum of a doubly-charged peak at *m/z* 736.333. The corresponding peptide is identified as AMpSTTSISSPQPGK (267–280), of which Serine 269 is phosphorylated; (**B**) MS/MS spectrum of a doubly-charged peak at *m/z* 776.318. The corresponding peptide is identified as AMpSTpTSISSPQPGK (267–280), of which both Serine 269 and Threonine 271 are phosphorylated.

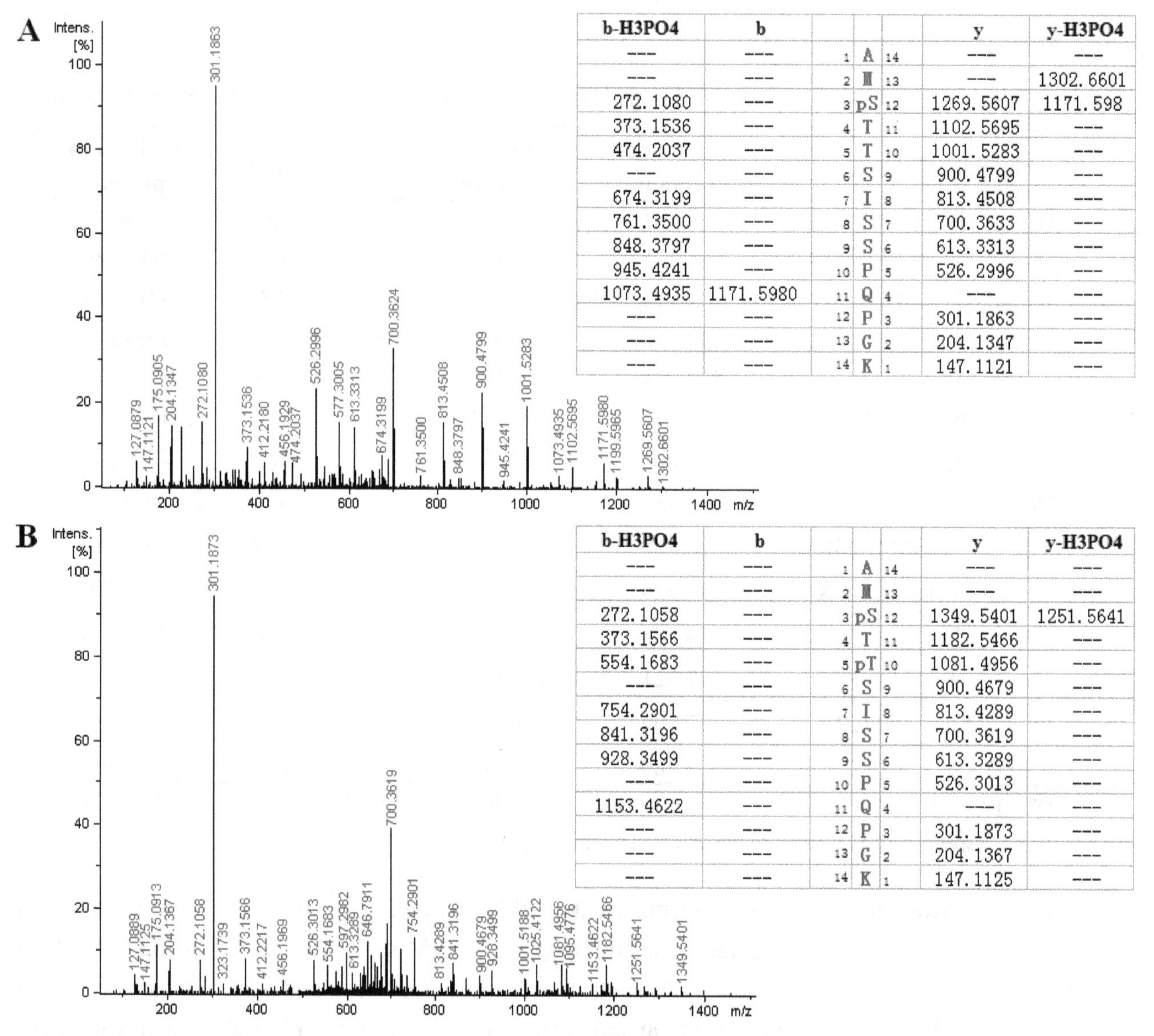

b-H3PO4	b				y	y-H3PO4
---	---	1	A	14	---	---
---	---	2	M	13	---	1302.6601
272.1080	---	3	pS	12	1269.5607	1171.598
373.1536	---	4	T	11	1102.5695	---
474.2037	---	5	T	10	1001.5283	---
---	---	6	S	9	900.4799	---
674.3199	---	7	I	8	813.4508	---
761.3500	---	8	S	7	700.3633	---
848.3797	---	9	S	6	613.3313	---
945.4241	---	10	P	5	526.2996	---
1073.4935	1171.5980	11	Q	4	---	---
---	---	12	P	3	301.1863	---
---	---	13	G	2	204.1347	---
---	---	14	K	1	147.1121	---

b-H3PO4	b				y	y-H3PO4
---	---	1	A	14	---	---
---	---	2	M	13	---	---
272.1058	---	3	pS	12	1349.5401	1251.5641
373.1566	---	4	T	11	1182.5466	---
554.1683	---	5	pT	10	1081.4956	---
---	---	6	S	9	900.4679	---
754.2901	---	7	I	8	813.4289	---
841.3196	---	8	S	7	700.3619	---
928.3499	---	9	S	6	613.3289	---
---	---	10	P	5	526.3013	---
1153.4622	---	11	Q	4	---	---
---	---	12	P	3	301.1873	---
---	---	13	G	2	204.1367	---
---	---	14	K	1	147.1125	---

As another example, Figure 4A showed the MS/MS spectrum of a doubly-charged molecular ion peak with *m/z* at 697.321, from which a phosphopeptide as QLpTQPETHFGR(289–299) from *HIP-55* protein was identified by database searching. The assignment of *y* series ions at *m/z* 971.4701, 1054.5074, 1152.4874 and 1167.5844 clearly indicated phosphorylation at Threonine 291, which was confirmed by detection of b series ions at *m/z* 242.1500, 325.1871, 423.1541 and 453.2197. The base peak at *m/z* 843.4137 was from the fragmentation at Proline 293. Figure 4B shows a similar spectrum of a doubly-charged molecular ion peak with *m/z* at 737.304, which was identified as the same peptide sequence (289–299) with two phosphorylation sites at Threonine 291 and 295. The detection of *y* series ions at *m/z* 697.3052, 826.3933, as well as the base peak ion at *m/z* 923.4073, confirmed the identification.

Figure 4. (**A**) MS/MS spectrum of a doubly-charged peak at *m/z* 697.321. The corresponding peptide is identified as QLpTQPETHFGR (289–299), in which Threonine 291 is phosphorylated; (**B**) MS/MS spectrum of a doubly-charged peak at *m/z* 737.304. The corresponding peptide is identified as QLpTQPEpTHFGR (289–299), in which both Threonine 291 and Threonine 295 are phosphorylated.

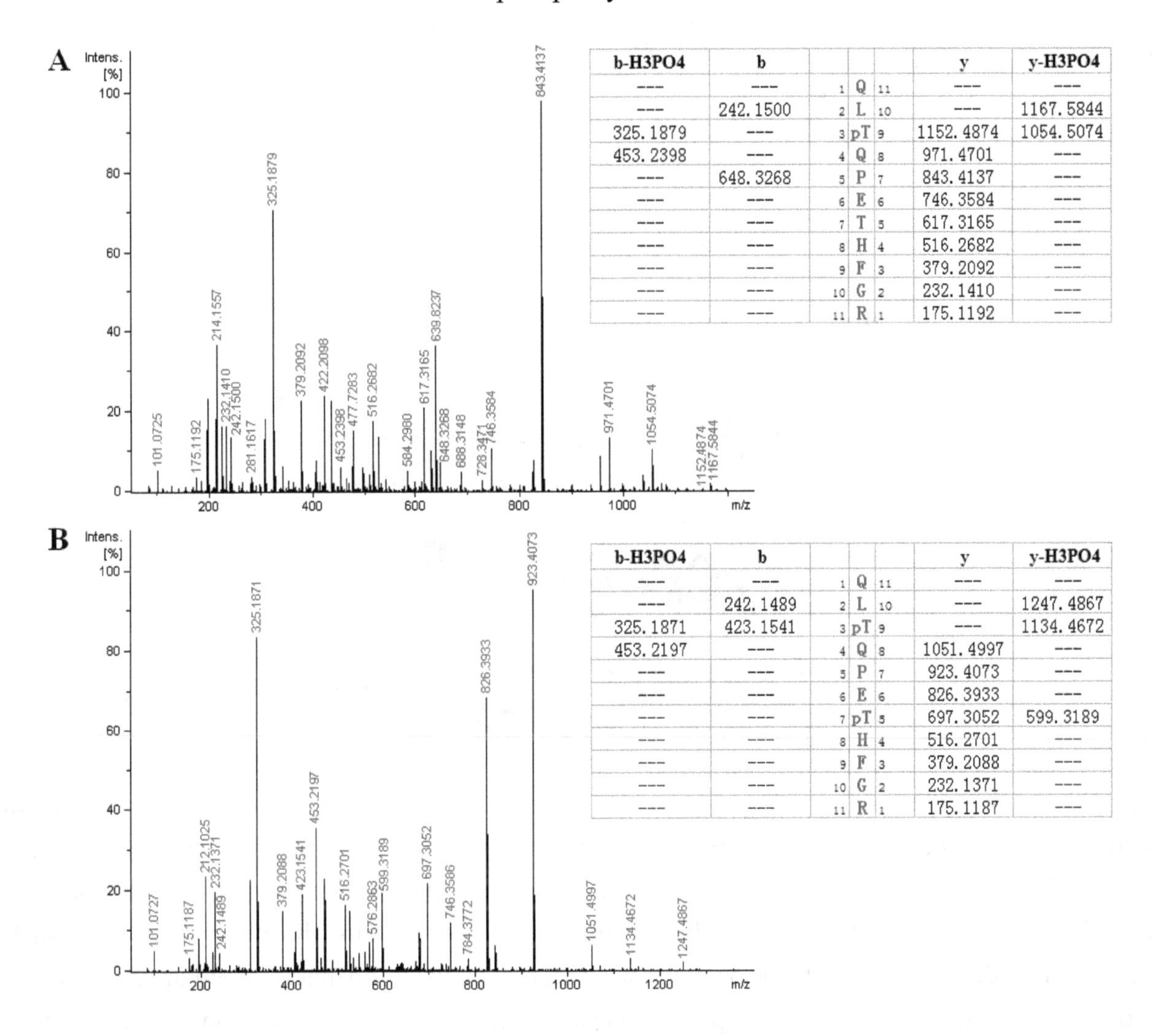

A

b-H3PO4	b				y	y-H3PO4
---	---	1	Q	11	---	---
---	242.1500	2	L	10	---	1167.5844
325.1879	---	3	pT	9	1152.4874	1054.5074
453.2398	---	4	Q	8	971.4701	---
---	648.3268	5	P	7	843.4137	---
---	---	6	E	6	746.3584	---
---	---	7	T	5	617.3165	---
---	---	8	H	4	516.2682	---
---	---	9	F	3	379.2092	---
---	---	10	G	2	232.1410	---
---	---	11	R	1	175.1192	---

B

b-H3PO4	b				y	y-H3PO4
---	---	1	Q	11	---	---
---	242.1489	2	L	10	---	1247.4867
325.1871	423.1541	3	pT	9	---	1134.4672
453.2197	---	4	Q	8	1051.4997	---
---	---	5	P	7	923.4073	---
---	---	6	E	6	826.3933	---
---	---	7	pT	5	697.3052	599.3189
---	---	8	H	4	516.2701	---
---	---	9	F	3	379.2088	---
---	---	10	G	2	232.1371	---
---	---	11	R	1	175.1187	---

The phosphosite mapping of *HIP-55* protein were graphically displayed as Figure 5A. Furthermore, motif analyses of phosphosites were created and displayed using the Weblogo server (http://weblogo.berkeley.edu/). A modification site consists of the modified residue at the 0 position, plus the seven flanking amino acids *N*-terminal (positions −7 to −1) and *C*-terminal (positions +1 to +7) to the modification site. For the analysis of sequence features adjacent to the identified phosphosites of *HIP-55* protein, the 14 amino acids surrounding each phosphosite were extracted and aligned. The motif logo shows the difference in the frequency of amino acids surrounding the different phosphosites (serine, threonine, and tyrosine), suggesting the kinase recognition motif of these three kinds of phosphorylation sites are different in mammalian cells (Figure 5B). In addition, our results showed that serine and threonine residues undergo phosphorylation more often than tyrosine residues. The phospho-amino acid content ratio (pSer:pThr:pTyr) was 9:4:2 within *HIP-55* protein, which was consistent with previous reports of that in the whole-cell [17,18].

Figure 5. Graphical display and motif analysis of Phosphosites of *HIP-55* Protein. (**A**) Graphical display of phosphosite mapping of *HIP-55* Protein; (**B**) Frequency distribution of amino acid residues surrounding phosphorylation sites at positions −7 to +7.

A

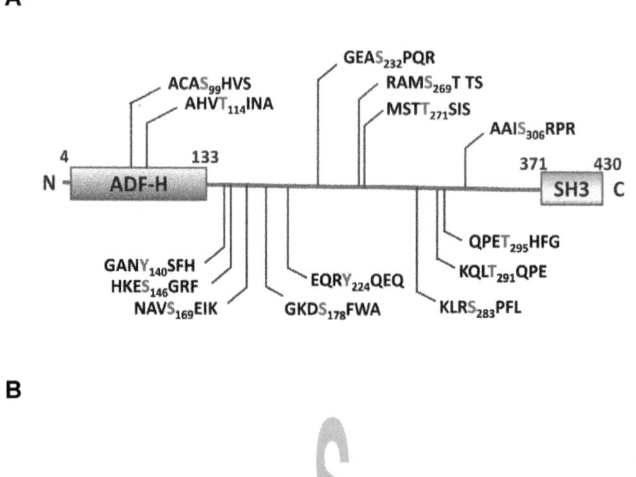

B

Protein phosphorylation plays a significant role in a wide range of cellular processes. It is estimated that approximately one-third of all proteins in eukaryotic cells are phosphorylated at any given time [17]. Therefore, the mapping of phosphorylation sites of special interest proteins is the subject of a large body of research. Because of the prominent role of protein phosphorylation in signaling transduction, traditional studies focus on enzymes, receptors and ion channels which are switched "on" or "off" by phosphorylation and dephosphorylation in signaling transduction process [19–22]. However, one of the major goals of studying signal transduction is to determine the mechanisms that control cross-talk between signaling cascades and to determine how specificity in signaling is achieved. An emerging class of proteins that are major contributors to these processes are adaptor proteins [23]. By linking specific binding proteins together, adaptor proteins control cellular signals appropriately. Most adaptor proteins binding to partners function as signaling regulators in a phosphorylation-dependent manner. Therefore, identification of phosphorylation sites of adaptor proteins is of critical importance in the field of cellular signal transduction. *HIP-55*, as a signaling adaptor protein, has been shown as a common effector of antigen receptor-signaling pathways and regulates T-cell activation by bridging TCRs [7,9]. In our previous studies, we found Ser269/Thr291-phosphosites of *HIP-55* mediated the interaction with 14-3-3τ (data not shown), which was confirmed by a recent research of 14-3-3-phosphoproteome [24]. Further, we also found pro-oncogenic function of *HIP-55* through Ser269/Thr291-phospho-sensor motifs (data not shown).

3. Experimental Section

3.1. Chemicals and Materials

Sequencing-grade TPCK-modified trypsin was obtained from Promega (Madison, WI, USA). ACN, formic acid, TCEP, Supel-Tips Ti Pipette Tips were purchased from Sigma-Aldrich (St. Louis, MO, USA). EDTA-free protease inhibitor cocktail tablets were from Roche (Basel, Switzerland). Phosphatase inhibitor cocktail and Ni-NTA resin were from Pierce Biotechnology (Rockford, IL, USA). Anti-*HIP-55* antibody was purchased from BD Bioscience (San Diego, CA, USA). All other chemicals, as well as the Bradford protein assay kit, were from Bio-Rad (Hercules, CA, USA); the *HIP-55* plasmid was kindly provided by Prof. Haian Fu from Emory University. Ultra-pure water was prepared by a MilliQ water purification system (Millpore, Bedford, MA, USA).

3.2. Expression and Purification of HIP-55 Protein

The pDEST-His-*HIP-55* plasmid with full length human *HIP-55* gene was transfected transiently into HEK293 cells using Fugene HD (Roche) following the manufacturer's protocol. Cells were lysed 36 h post-transfection in His pull-down lysis buffer (1% Nonidet P-40, 137 mM NaCl, 1 mM $MgCl_2$, 40 mM Tris-Cl, 60 mM imidazole, 5 mM $Na_4P_2O_7$, 5 mM NaF, 2 mM Na_3VO_4, 1 mM phenylmethylsulfonyl fluoride, 10 mg/L aprotinin, 10 mg/L leupeptin). The lysate was cleared by centrifugation and a small aliquot of the supernatant was removed for Western Blot analysis. Then the supernatant was incubated with Ni-NTA resin at 4 °C for 2 h, which was pre-equilibrated with a buffer containing 100 mmol/L PBS, 6 mol/L guanidine HCl and 10 mmol/L imidazole; pH 7.4. After washes by the same buffer, the proteins bound to the Ni-NTA resin were eluted with a buffer containing 100 mmol/L PBS, 6 mol/L guanidine HCl and 250 mmol/L imidazole; pH 7.4.

3.3. Tryptic Digestion

The eluted proteins were precipitated by adding four volumes of −40 °C acetone and kept at −20 °C for at least 2 h. Then the protein pellet was redissolved with a buffer containing 0.1% SDS and 50 mmol/L ammonium bicarbonate, pH 8.5. The protein sample was reduced at 60 °C for 30 min by adding TCEP to a final concentration of 5 mmol/L, followed by alkylation with iodoacetamide at a final concentration of 4 mmol/L in the dark at room temperature for 60 min. Then the protein samples were cooled down to room temperature and mixed with freshly prepared sequencing-grade TPCK-modified trypsin buffer (15 μg/mL in 25 mmol/L ammonium bicarbonate, pH 8.5) at an enzyme/protein ratio of 1:100. Digestion was performed at 37 °C for at least 15 h and stopped by adding 10% formic acid to a final concentration of 1%. The tryptic digests were cleared by centrifugation and the supernatant was purified over C18 tips. For enrichment of phosphopeptides, the supernatant that had not been processed by C18 tips was loaded onto TiO_2 tips that were fully equilibrated with 3% formic acid. After washed with pure water three times, the bond peptides were eluted with 0.5% piperidine and then lyophilized.

3.4. Mass Spectrometry

The tryptic peptide samples were dissolved in buffer A (0.1% formic acid, 1% ACN), which were separated on a C18 reverse capillary column (150 mm × 0.17 mm) with buffer B (0.1% formic acid, 99% ACN) over a 60 min gradient of 2%–40%. The eluted peptides were then delivered into an LTQ Orbitrap mass spectrometer (Thermo Fisher Scientific, Waltham, MA, USA) equipped with a nanospray source. The mass spectrometer was operated in a data dependent mode, in which five MS/MS scans on the most abundant ions detected in the MS scan (400–1500 *m/z*) were acquired. The dynamic exclusion time was set as 1 min. Singly-charged ions were excluded for MS/MS.

3.5. Peptide and Protein Identification

The collected spectra were analyzed by Thermo Proteome Discoverer 1.1 software (Thermo Fisher Scientific, Waltham, MA, USA), in which the built-in SEQUEST as used to search the data against IPI protein database (International Protein Index, ipi.HUMAN.v3.84, downloaded at ftp://ftp.ebi.ac.uk/pub/databases/IPI) with the following parameters: trypsin was selected with two maximum missed cleavage allowed; precursor mass tolerance was 15 ppm; fragment mass tolerance was 0.8 Da; carbamidomethylation of cysteine was set as static modification; phosphorylations of serine, threonine and tyrosine, as well as oxidation of methionine, were set as dynamic modifications. The results from SEQUEST searches were filtered with Peptide Confidence as High, in which XCorr threshold was set as 1.9, 2.3 for doubly, triply-charged ion peaks, respectively.

3.6. Motif Analysis of Phosphosites

Sequence features adjacent to the identified phosphosites (serine, threonine, and tyrosine) were created and displayed using the Weblogo server (http://weblogo.berkeley.edu/).

3.7. SDS-PAGE and Immunoblotting

Cell lysates (25 μg) were subjected to electrophoresis in 12% Tris-glycine-SDS polyacrylamide gel using a Mini-Cell system (Bio-Rad, Hercules, CA, USA). Gels were electroblotted to 0.2 μm pore size polyvinylidene fluoride (PVDF) membranes (Millipore, Bedford, MA, USA). Then the membranes were blocked with 1% nonfat dry milk in a buffer containing 25 mM Tris, pH 7.5, 150 mM NaCl, 0.05% Tween 20 for 1 h at room temperature. Membranes were then incubated with the primary antibody against *HIP-55* (BD Bioscience, San Diego, CA, USA) for 2 h at room temperature. After washing for 10 min in TBST solution, membranes were incubated with properly diluted secondary antibody conjugated with horseradish peroxidase (Sigma-Aldrich, St. Louis, MO, USA) for 1 h at room temperature. Western signals were developed with ECL chemiluminescent reagents (GE, Boston, MA, USA).

4. Conclusions

In the present study, HEK293 cells over-expressing human *HIP-55* protein were used to establish a profile of phosphorylation sites within *HIP-55*. Ni-NTA resin enrichment and gel electrophoresis were utilized to collect the target protein and TiO_2 affinity chromatography and biological mass spectrometry were employed to investigate phosphorylation. Investigation of the mass spectra revealed 14 phosphorylation sites of *HIP-55* protein, several of which were identified for the first time. Our findings provide evidence for future investigations of the phosphorylation status of *HIP-55* protein of different subtypes and facilitate further studies on their biological functions in health and disease.

Acknowledgments

This work was supported by Major State Basic Research Development Program of China (973 Program) (2011CB503900), Natural Science Foundation of China (81070078, 21175055, 81270157), Beijing Municipal Natural Science Foundation (7102158), Jilin Province Science and Technology Department (20110739), Jilin University Bethune Project B (2012210).

References

1. Fischer, E.H. Cellular regulation by protein phosphorylation. *Biochem. Biophys. Res. Commun.* **2013**, *430*, 865–867.
2. Cohen, P. The regulation of protein function by multisite phosphorylation—A 25 year update. *Trends Biochem. Sci.* **2000**, *25*, 596–601.
3. Via, A.; Diella, F.; Gibson, T.J.; Helmer-Citterich, M. From sequence to structural analysis in protein phosphorylation motifs. *Front. Biosci. (Landmark Ed.)* **2011**, *16*, 1261–1275.
4. Ensenat, D.; Yao, Z.; Wang, X.S.; Kori, R.; Zhou, G.; Lee, S.C.; Tan, T.H. A novel src homology 3 domain-containing adaptor protein, HIP-55, that interacts with hematopoietic progenitor kinase 1. *J. Biol. Chem.* **1999**, *274*, 33945–33950.
5. Kessels, M.M.; Engqvist-Goldstein, A.E.; Drubin, D.G.; Qualmann, B. Mammalian Abp1, a signal-responsive F-actin-binding protein, links the actin cytoskeleton to endocytosis via the GTPase dynamin. *J. Cell Biol.* **2001**, *153*, 351–366.
6. Fenster, S.D.; Kessels, M.M.; Qualmann, B.; Chung, W.J.; Nash, J.; Gundelfinger, E.D.; Garner, C.C. Interactions between Piccolo and the actin/dynamin-binding protein Abp1 link vesicle endocytosis to presynaptic active zones. *J. Biol. Chem.* **2003**, *278*, 20268–20277.
7. Han, J.; Kori, R.; Shui, J.W.; Chen, U.R.; Yao, Z.; Tan, T.H. The SH3 domain-containing adaptor HIP-55 mediates c-Jun *N*-terminal kinase activation in T cell receptor signaling. *J. Biol. Chem.* **2003**, *278*, 52195–52202.
8. Han, J.; Shui, J.W.; Zhang, X.; Zheng, B.; Han, S.; Tan, T.H. HIP-55 is important for T-cell proliferation, cytokine production, and immune responses. *Mol. Cell. Biol.* **2005**, *25*, 6869–6878.
9. Onabajo, O.O.; Seeley, M.K.; Kale, A.; Qualmann, B.; Kessels, M.; Han, J.; Tan, T.H.; Song, W. Actin-binding protein 1 regulates B cell receptor-mediated antigen processing and presentation in response to B cell receptor activation. *J. Immunol.* **2008**, *180*, 6685–6695.
10. Haag, N.; Schwintzer, L.; Ahuja, R.; Koch, N.; Grimm, J.; Heuer, H.; Qualmann, B.; Kessels, M.M. The actin nucleator Cobl is crucial for Purkinje cell development and works in close conjunction with the F-actin binding protein Abp1. *J. Neurosci.* **2012**, *32*, 17842–17856.
11. Lock, P.; Abram, C.L.; Gibson, T.; Courtneidge, S.A. A new method for isolating tyrosine kinase substrates used to identify fish, an SH3 and PX domain-containing protein, and Src substrate. *EMBO J.* **1998**, *17*, 4346–4357.

12. Larbolette, O.; Wollscheid, B.; Schweikert, J.; Nielsen, P.J.; Wienands, J. SH3P7 is a cytoskeleton adapter protein and is coupled to signal transduction from lymphocyte antigen receptors. *Mol. Cell. Biol.* **1999**, *19*, 1539–1546.
13. Chung, S.; Suzuki, H.; Miyamoto, T.; Takamatsu, N.; Tatsuguchi, A.; Ueda, K.; Kijima, K.; Nakamura, Y.; Matsuo, Y. Development of an orally-administrative MELK-targeting inhibitor that suppresses the growth of various types of human cancer. *Oncotarget* **2012**, *3*, 1629–1640.
14. Boateng, L.R.; Cortesio, C.L.; Huttenlocher, A. Src-mediated phosphorylation of mammalian Abp1 (DBNL) regulates podosome rosette formation in transformed fibroblasts. *J. Cell Sci.* **2012**, *125*, 1329–1341.
15. St-Denis, N.; Gingras, A.C. Mass spectrometric tools for systematic analysis of protein phosphorylation. *Prog. Mol. Biol. Transl. Sci.* **2012**, *106*, 3–32.
16. Dephoure, N.; Gould, K.L.; Gygi, S.P.; Kellogg, D.R. Mapping and analysis of phosphorylation sites: A quick guide for cell biologists. *Mol. Biol. Cell* **2013**, *24*, 535–542.
17. Mann, M.; Ong, S.E.; Grønborg, M.; Steen, H.; Jensen, O.N.; Pandey, A. Analysis of protein phosphorylation using mass spectrometry: deciphering the phosphoproteome. *Trends Biotechnol.* **2002**, *20*, 261–268.
18. Matthias, M.; Ole, N.J. Proteomic analysis of post-translational modifications. *Nat. Biotechnol.* **2003**, *21*, 255–261.
19. Maaty, W.S.; Lord, C.I.; Gripentrog, J.M.; Riesselman, M.; Keren-Aviram, G.; Liu, T.; Dratz, E.A.; Bothner, B.; Jesaitis, A.J. Identification of *C*-terminal phosphorylation sites of *N*-formyl peptide receptor-1 (FPR1) in human blood neutrophils. *J. Biol. Chem.* **2013**, *288*, 27042–27058.
20. Rudashevskaya, E.L.; Ye, J.; Jensen, O.N.; Fuglsang, A.T.; Palmgren, M.G. Phosphosite mapping of P-type plasma membrane H^+-ATPase in homologous and heterologous environments. *J. Biol. Chem.* **2012**, *287*, 4904–4913.
21. Baek, J.H.; Cerda, O.; Trimmer, J.S. Mass spectrometry-based phosphoproteomics reveals multisite phosphorylation on mammalian brain voltage-gated sodium and potassium channels. *Semin. Cell Dev. Biol.* **2011**, *22*, 153–159.
22. Roberts, J.A.; Bottrill, A.R.; Mistry, S.; Evans, R.J. Mass spectrometry analysis of human P2X1 receptors; insight into phosphorylation, modelling and conformational changes. *J. Neurochem.* **2012**, *123*, 725–735.
23. Flynn, D.C. Adaptor proteins. *Oncogene* **2001**, *20*, 6270–6272.
24. Johnson, C.; Tinti, M.; Wood, N.T.; Campbell, D.G.; Toth, R.; Dubois, F.; Geraghty, K.M.; Wong, B.H.; Brown, L.J.; Tyler, J.; *et al.* Visualization and biochemical analyses of the emerging mammalian 14-3-3-phosphoproteome. *Mol. Cell. Proteomics* **2011**, *10*, doi:10.1074/mcp.M110.005751.

5

Radiosensitization of Human Leukemic HL-60 Cells by ATR Kinase Inhibitor (VE-821): Phosphoproteomic Analysis

Barbora Šalovská [1], Ivo Fabrik [2], Kamila Ďurišová [3], Marek Link [2], Jiřina Vávrová [3], Martina Řezáčová [1] and Aleš Tichý [3,*]

[1] Institute of Medical Biochemistry, Faculty of Medicine in Hradec Králové, Charles University in Prague, Hradec Kralove 500 00, Czech Republic; E-Mails: salob5aa@lfhk.cuni.cz (B.S.); RezacovaM@lfhk.cuni.cz (M.R.)

[2] Institute of Molecular Pathology, Faculty of Health Sciences in Hradec Králové, University of Defense in Brno, Hradec Kralove 500 01, Czech Republic; E-Mails: fabrik@pmfhk.cz (I.F.); link@pmfhk.cz (M.L.)

[3] Department of Radiobiology, Faculty of Health Sciences in Hradec Králové, University of Defense in Brno, Hradec Kralove 500 01, Czech Republic; E-Mails: durisovak@pmfhk.cz (K.D.); vavrova@pmfhk.cz (J.V.)

* Author to whom correspondence should be addressed; E-Mail: tichy@pmfhk.cz

---

Abstract: DNA damaging agents such as ionizing radiation or chemotherapy are frequently used in oncology. DNA damage response (DDR)—triggered by radiation-induced double strand breaks—is orchestrated mainly by three Phosphatidylinositol 3-kinase-related kinases (PIKKs): Ataxia teleangiectasia mutated (ATM), DNA-dependent protein kinase (DNA-PK) and ATM and Rad3-related kinase (ATR). Their activation promotes cell-cycle arrest and facilitates DNA damage repair, resulting in radioresistance. Recently developed specific ATR inhibitor, VE-821 (3-amino-6-(4-(methylsulfonyl)phenyl)-*N*-phenylpyrazine-2-carboxamide), has been reported to have a significant radio- and chemo-sensitizing effect delimited to cancer cells (largely p53-deficient) without affecting normal cells. In this study, we employed SILAC-based quantitative phosphoproteomics to describe the mechanism of the radiosensitizing effect of VE-821 in human promyelocytic leukemic cells HL-60 (p53-negative). Hydrophilic interaction liquid chromatography (HILIC)-prefractionation with TiO_2-enrichment and nano-liquid chromatography—tandem mass spectrometry (LC-MS/MS) analysis revealed 9834 phosphorylation sites. Proteins with differentially up-/down-regulated phosphorylation were mostly localized in the nucleus and were involved in cellular processes such as DDR, all phases of the cell cycle, and cell division. Moreover, sequence motif analysis revealed significant changes in the activities of kinases involved in these processes. Taken together, our data indicates that ATR kinase has multiple roles in response to DNA damage throughout the cell cycle and that its inhibitor VE-821 is a potent radiosensitizing agent for p53-negative HL-60 cells.

Keywords: small-molecule kinase inhibitors; VE-821; ATR kinase; DNA damage response; radio-sensitization; quantitative phosphoproteomics; titanium dioxide chromatography; SILAC; leukemia; HL-60 cells

---

1. Introduction

One of the treatment modalities in oncology is radiotherapy. It often employs chemical agents increasing sensitivity towards ionizing radiation (IR). IR induces the most deleterious lesions of DNA, double strand breaks (DSB), and their repair is regulated by ataxia telangiectasia-mutated kinase (ATM), DNA-dependent protein kinase (DNA-PK), and ATM and Rad3-related kinase (ATR). While activation of ATM and DNA-PK is triggered by DNA double stranded breaks, ATR kinase responds to a broad spectrum of agents inducing single stranded DNA [1]. ATR acts primarily in S- and G2-phases and responds to replication and genotoxic stress, however, a recent report has shown that ATR is activated also in irradiated G1 phase cells [2].

In 2011, a specific inhibitor of ATR, VE-821 (3-amino-6-(4-(methylsulfonyl)phenyl)-*N*-phenylpyrazine-2-carboxamide), was developed [3]. Its excellent selectivity in regards to sensitization of cancer cells towards various types of DNA damaging agents leaving healthy cells unaffected has been recently reported [4–7]. VE-821 selectivity is based on the concept that (i) more than 50% of cancer cells have lost their G1-phase checkpoint for example due to p53 mutation/deletion and thus rely on S- and G2-checkpoints; which are known to be regulated by ATR and (ii) cancer cells with activated oncogenes generate replication stress at much higher levels than normal cells, thus activating ATR [8,9]. This hypothesis is based on well-established "classical" methodology in molecular biology comprising western-blotting, confocal microscopy, flow-cytometry and so forth. However, this approach is often focused on one or two particular mechanisms not reflecting the complexity of cellular signaling pathways. Therefore it fails to give a comprehensive view of the mechanisms of radiosensitization by ATR inhibition in the context of cancer cells.

In the DNA damage response as well as in the other molecular processes (such as transcriptional and translational regulation, proliferation, differentiation, apoptosis, cell survival and many others) phosphorylation frequently initiates and propagates signal transduction pathways [10]. Therefore we decided to apply a phosphoproteomic approach in order to study the effect of specific inhibition of ATR by the small molecule inhibitor VE-821.

Our goal was to describe the changes in the phosphoproteome in radiosensitized tumor cells lacking functional protein p53. Although mass spectrometry of phosphopeptides obtained from tryptic protein digests has become a powerful tool for characterization of phosphoproteins involved in cellular processes, there is an inevitable part of the protocol: phosphopeptide enrichment. It compensates the low abundance, insufficient ionization, and suppression effects of non-phosphorylated peptides [11]. Hence, we optimized the metal oxide affinity chromatography (MOAC) enrichment using titanium dioxide and employed it in our experiments [12].

We have previously compared the effects of inhibitors of ATM (KU55933) and ATR kinases (VE-821) on the radiosensitization of human promyelocytic leukemia cells (HL-60), lacking functional protein p53. The inhibition of ATR by its specific inhibitor VE-821 resulted in a more pronounced radiosensitizing effect in HL-60 cells compared to the inhibition of ATM. In contrast to KU55933, the VE-821 treatment prevented HL-60 cells from undergoing G2 cell cycle arrest [13].

In this paper we characterize the radiosensitizing effect of VE-821 from a new perspective and we report the mechanisms and signaling pathways involved in the processes triggered by IR in p53-negative leukemic cells.

2. Results

Our goal was to describe the changes in the phosphoproteome in radiosensitized tumor cells, since in the DNA damage response (as well as in other molecular processes) phosphorylation frequently initiates and propagates signal transduction pathways. We aimed to characterize the effect of specific inhibition of ATR kinase by VE-821 and to report the mechanisms and signaling pathways involved in the processes triggered by IR in p53-negative human promyelocytic leukemic cells HL-60 that are ATR-dependent. To reach our goal, we employed SILAC-based quantitative phosphoproteomics together with metal oxide affinity chromatography using TiO_2 microparticles to specifically enrich for phosphorylated peptides. To increase the number of identified and quantified phosphorylation sites, peptides were prefractionated by hydrophilic interaction liquid chromatography (HILIC) chromatography prior to the enrichment. Additionally, to make sure that the changes in the phosphoproteome are not based on changes of protein abundance caused by alterations in gene expression or protein degradation that could occur because of the relatively long incubation time (1 h after irradiation, and 1.5 h after VE-821 treatment), we also analyzed a non-phosphorylated complement of phosphorylated peptides obtained from flow-through fractions of the phosphopeptide enrichment.

2.1. Overall Phosphoproteomic Analysis Reveals Thousands of Phosphorylation Sites

In summary, we identified 9834 phosphorylation sites from 3210 protein groups, among them 4809 were quantified in all three experimental replicates. The phosphosite ratios were shown to correlate well between the replicates (Pearson correlation coefficient was between 0.8 and 0.87). Moreover, the correlation between the normalized H/L ratios of phosphopeptides and the normalized protein H/L ratios was also calculated (the calculation was based on 1941 phosphorylation sites from 557 proteins), and the low correlation coefficient value ($R = 0.0162$) indicated that the quantified changes in the phosphoproteome were presumably not caused by changes in the protein abundance. The overview of all identified phosphorylation sites is given in the supplementary (Table S1).

To determine whether a phosphorylation site exhibits an appropriate alteration in abundance in all of the replicates to be considered as differentially regulated after the treatment, we employed the Global rank test (GRT) with the non-parametric estimate of the false discovery rate (FDR; see material and methods). At the 1% FDR level, 336 phosphorylation sites (260 proteins) were evaluated as differentially up-regulated, whereas 202 phosphorylation sites (168 proteins) were differentially down-regulated (some of the proteins contained both down- and up-regulated phosphorylation sites). Since a kinase inhibitor was applied to modulate the response to IR induced DNA damage, the number of down-regulated sites would be expected to be larger than the number of up-regulated sites. However, the number of up-regulated sites increased and possibly reflected relatively long time period that passed from the initial treatment, and allowed the cells to eventually trigger a compensatory phosphorylation response in order to overcome the inhibitory effect of VE-821.

2.2. Combination of Ionizing Radiation and VE-821 Treatment Abrogated Phosphorylation of Checkpoint Kinase-1 on Ser345

In order to confirm the inhibitory effect of VE-821 on ATR in our model, we assessed phosphorylation of checkpoint kinase-1 (CHK1) on Ser345. Since it reports active and functional ATR, this particular direct downstream target of ATR was found to be phosphorylated upon irradiation. As shown in Figure 1, CHK1 phosphorylation was abrogated after VE-821 pre-incubation.

Figure 1. In order to confirm the inhibitory effect of VE-821 on ATM and Rad3-related kinase (ATR) we assessed phosphorylation of CHK1 on Ser345 (a direct downstream target of ATR) and we found this particular phosphorylation to be inhibited. Beta-actin was used as a gel loading control. The representative blots from at least three independent experiments are shown.

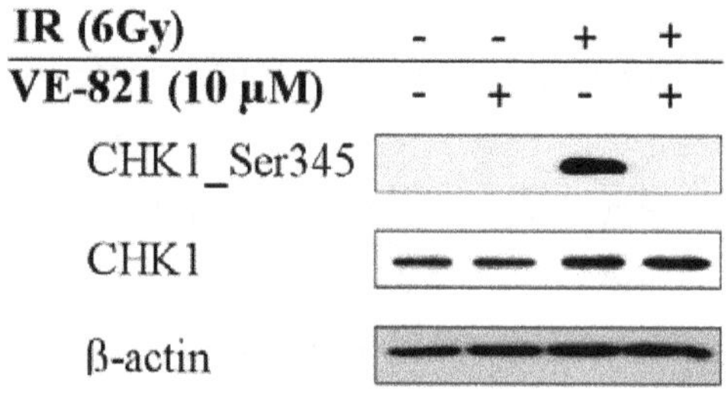

IR (6Gy) - - + +
VE-821 (10 μM) - + - +
CHK1_Ser345
CHK1
ß-actin

2.3. Most of the Regulated Phosphoproteins Were Localized in the Nucleus and Related to Mitosis, Cell Cycle Regulation, DNA Damage Response and Gene Expression

To functionally and spatially interpret the list of 396 phosphorylated proteins resulting from the GRT, Gene Ontology (GO) annotation was performed using the ConsensusPathDB, over-representation analysis "web tool" [14,15] and the summary of the results is given in Figure 2. The analysis of cellular component GO terms (level 4) revealed that most of the proteins were localized in the nucleus, many of them included in chromatin, localized at the centromeric region of a chromosome, replication fork or as a part of histone acetyltransferase or deacetylase complexes. Nevertheless, there was also a considerable fraction of proteins annotated to be a part of cytosol. The most over-represented molecular functions (level 4) of regulated phosphoproteins were binding of both nucleic acids (RNA binding and DNA binding), transcriptional co-activator and co-repressor activities, as well as DNA ligase or topoisomerase II activities. The list of overrepresented biological process terms (level 4) contained processes related to mitosis (nuclear envelope disassembly, sister chromatid cohesion, spindle localization or cytokinesis), cell cycle regulation (cell cycle progression, cell cycle phase transition, cell cycle checkpoint, regulation of transcription involved in G1/S transition of the mitotic cell cycle, cell cycle DNA replication), cellular response to DNA damage stimulus (base- and nucleotide-excision repair, double-strand break repair) or gene expression (mRNA transport, histone acetylation).

Figure 2. Selected results of Gene Ontology terms over-representation analysis (FDR < 0.01): (**A**) Overrepresented Gene Ontology (GO) Biological process level 4 terms; (**B**) Overrepresented GO Cellular component level 4 terms; (**C**) Overrepresented GO Molecular function level 4 terms. *x*-axis contains the number of proteins involved in a particular pathway that were found differentially phosphorylated in our study.

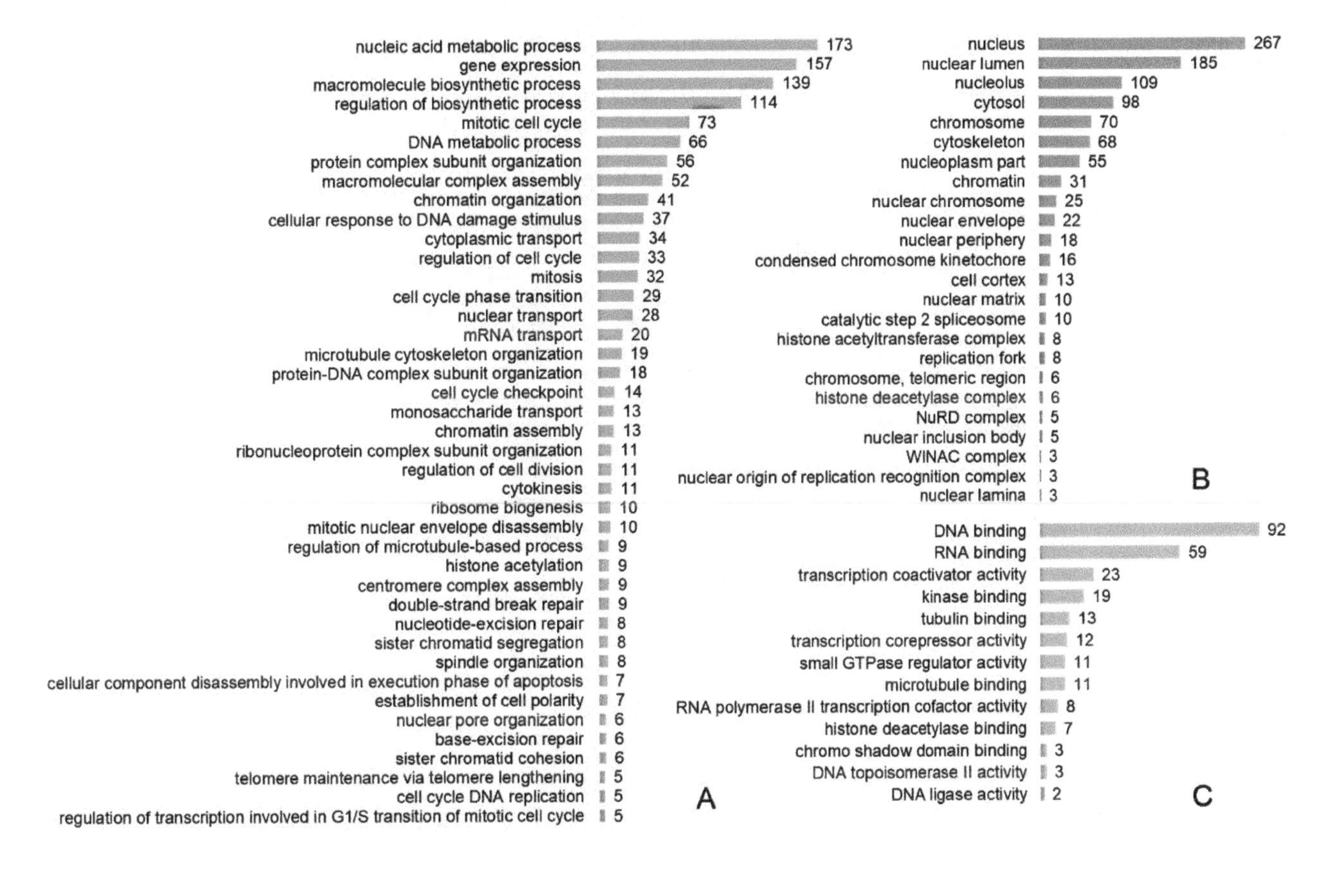

2.4. The Pathways Affected by Co-Treatment Were Involved in Cell Cycle Progression, Stimulus-Based Changes in Gene Expression, DDR and Apoptosis

To assess which signaling pathways contained differentially phosphorylated proteins and thus could be considered as triggered or modulated by ATR inhibition, we performed a pathway over-representation analysis using the same web-tool as in the case of GO enrichment, however, we addressed the signaling pathways stored in the Reactome pathway database [16] and Pathway Interaction Database (PID; [17]). The overrepresented pathways are shown in Figure 3. It is apparent that most of the pathways were related to different phases of the cell cycle and transitions between them, stimulus-based changes in gene expression, DNA damage repair and DNA damage-induced programmed cell death.

Figure 3. Selected results of Reactome pathways (blue bars) and Pathway Interaction Database (PID) pathways (orange bars) over-representation analysis (FDR < 0.01): *x*-axis shows the number of proteins involved in particular pathways that were differentially phosphorylated in our study.

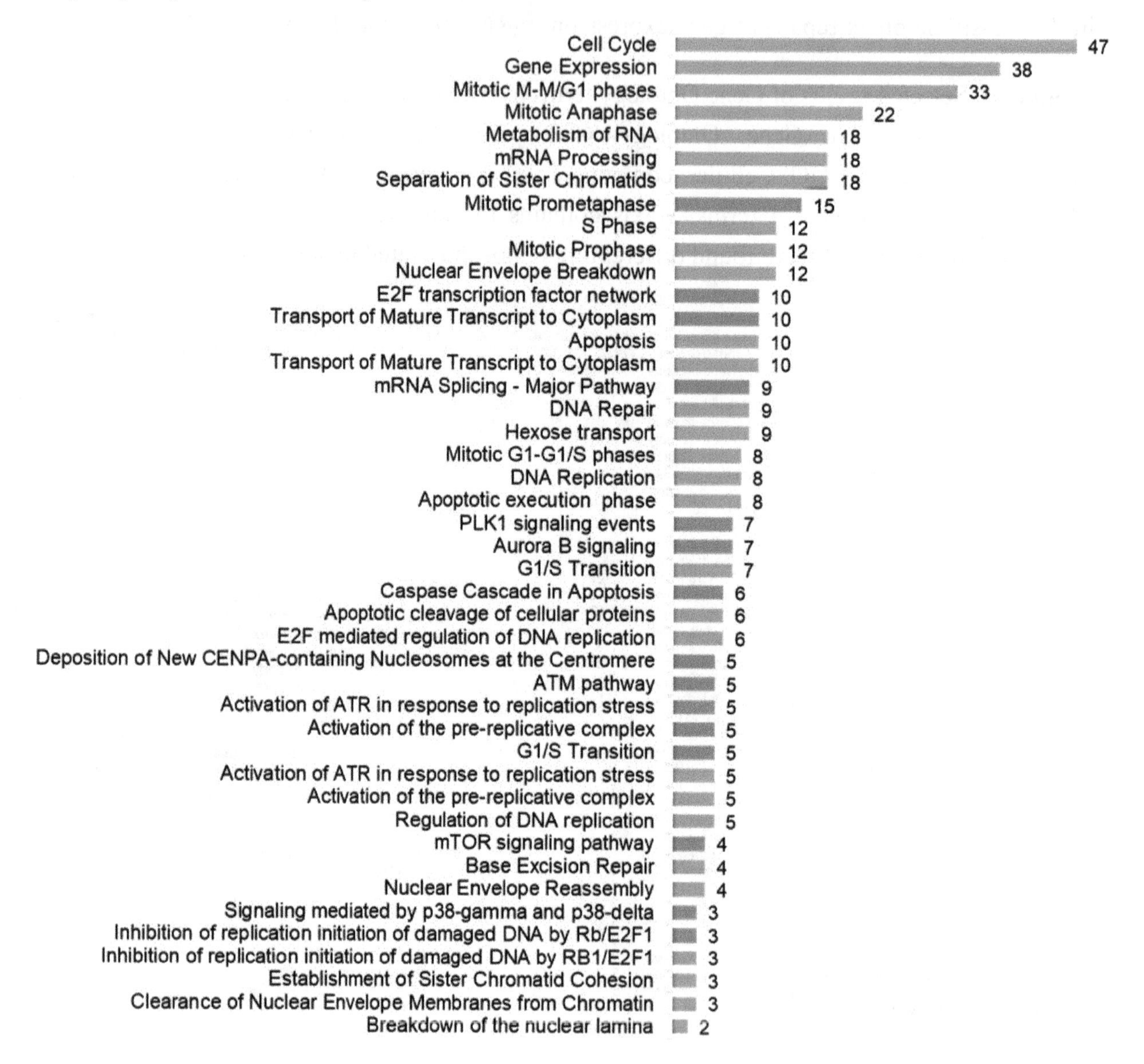

2.5. Significantly Down-Regulated Phosphorylation Sites Comprised SP/TP Motifs, while SQ/TQ Were amongst Up-Regulated Ones

To analyze and visualize sequence motifs that were enriched among significantly regulated class I phosphorylation sites (see material and methods), we utilized the iceLogo tool [18] and the motif-x algorithm [19,20]. Amino acid sequences surrounding either up- or down-regulated phosphorylation sites were tested against a background reference set composed of sequences surrounding all phosphorylation sites revealed in our study that reached the minimal localization probability 0.75. Results of these analyses are shown in Figure 4.

Figure 4A,C depicts that among up-regulated sites the proline-directed and basic motifs were overrepresented whereas acidic motifs were significantly underrepresented (typical e.g., for casein kinases). SP/TP motifs followed by basic amino acids are known to be typical consensus motifs of cyclin-dependent kinases (CDKs), mitogen-activated protein kinases (MAPKs) or glycogen synthase kinase 3 (GSK-3). On the other hand, SP/TP motifs were underrepresented amongst significantly down-regulated phosphorylation sites (Figure 4B,D) while SQ/TQ motifs (typical for DNA repair enzymes), TL, SR, and GT motifs were significantly overrepresented.

Figure 4. Sequence motifs analyses and visualization performed using iceLogo tool (**A**,**B**) and motif-x algorithm (**C**,**D**). The amino acid sequences of significantly differentially up- or down-regulated phosphorylation sites (**A**,**C** or **B**,**D** respectively) were analyzed against a statistical background comprising all class I sites revealed in our study. In **A** and **B**, amino acids that were more frequently observed in the proximity of a regulated phosphorylation site are indicated over the middle line, whereas the amino acids with lower frequency are indicated below the line; phosphorylated amino acid is located at position 8; **C** and **D** depict sequence motifs extracted by motif-x algorithm at a significance level of $p < 0.00003$ (which approximately corresponds to a q-value of 0.01 after Bonferroni correction for multiple hypothesis testing).

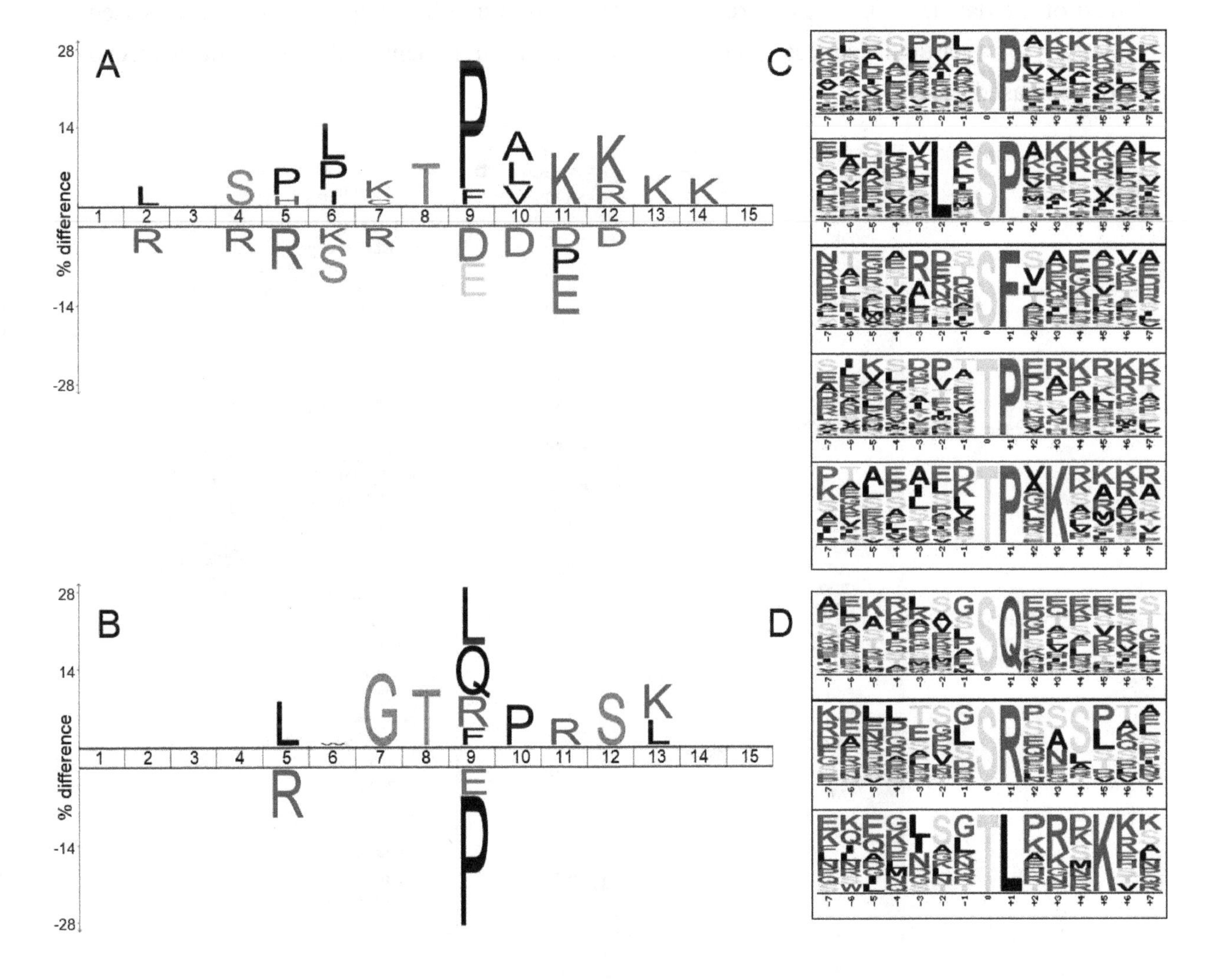

2.6. Kinase Activity Analyses Predicts Various Kinase Substrates to Be Regulated

There are several tools available to predict kinase-specific phosphorylation sites. Since each suffers its own limitations, the employment of different prediction algorithms seems to be a profitable approach to describe the alterations of intracellular signaling. Based on this assumption, at first we applied a very common approach relying on simple assigning of consensus motifs downloaded from HPRD (Human proteome reference database) to phosphorylated sequences. Consecutively, we employed two kinase predictors that apply consensus motif scoring together with the network context of kinases and their potential substrate, *i.e.*, NetworKIN 3 [21] and iGPS [22], differing in their motif scoring algorithms. The results are depicted in Figure 5. Only those kinases that reached the significance level of $p < 0.05$ are reported. The position score indicates the degree of regulation.

Figure 5. Analysis of kinases activity. Kinase-substrate relations predicted by iGPS (**A**) NetworKIN 3; and (**B**) linear motifs analysis derived from motifs stored in Human proteome reference database (HPRD) database; (**C**) The color shade corresponds to the value of the "position score" that indicates if the ratios (normalized *H*/*L* ratios) of sites phosphorylated by a particular kinase tend to be larger (0 to 1) or lower (−1 to 0) than the rest of the data ($p < 0.05$), *i.e.*, green color of a region in a heatmap indicates an increased activity of a particular kinase after IR and VE-821 co-treatment, red color indicates trend of decreased activity.

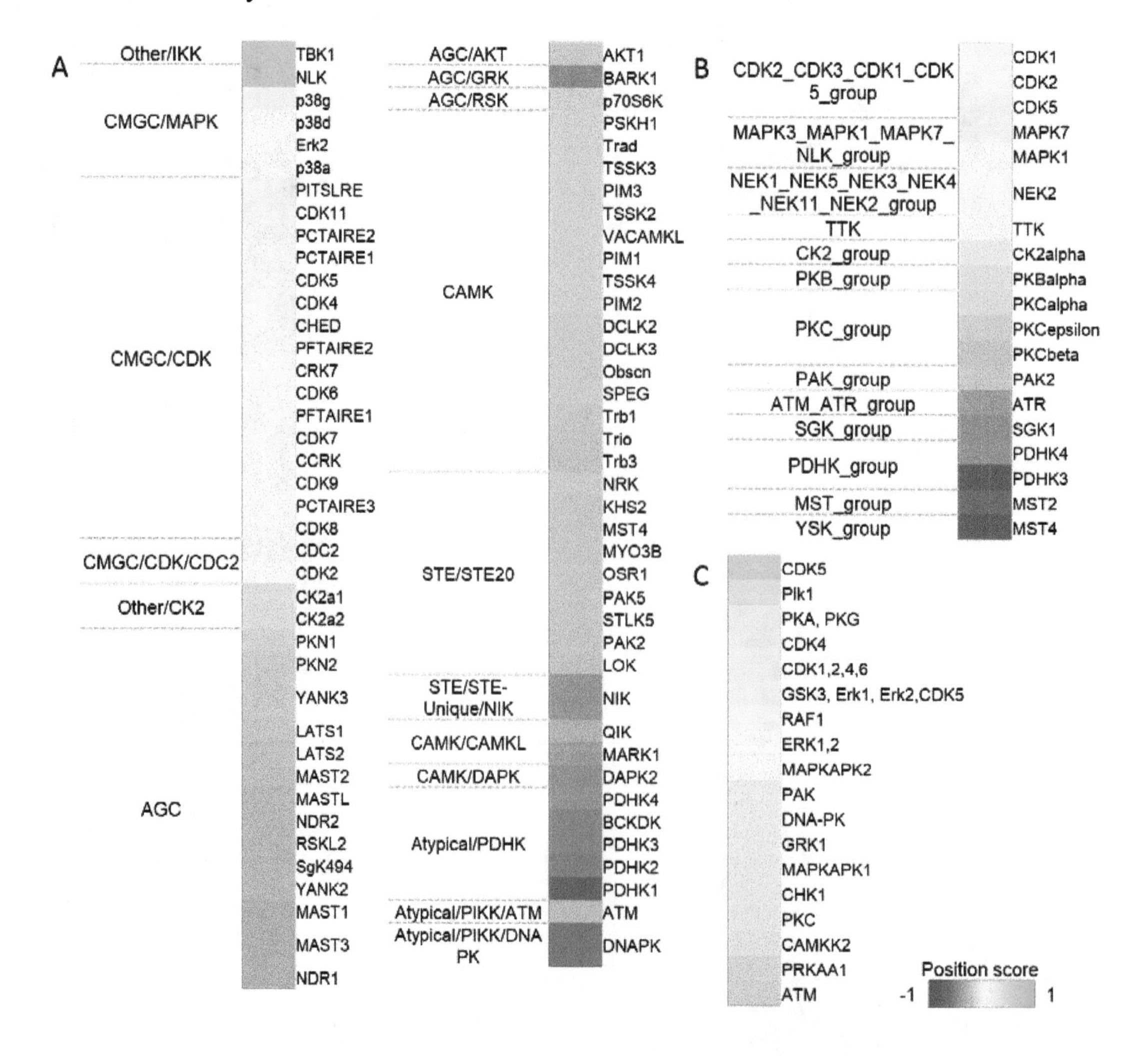

Our analyses confirmed the decrease in activity of the group of enzymes sharing the SQ/TQ motif (*i.e.*, ATM, ATR and DNA-PK). The remaining down-regulated kinases composed a quite heterogenic group of protein kinases including Casein kinase 2 (CK2), Protein kinase B (AKT) a subgroup of Protein kinases C (PKCs) or various isoforms of Pyruvate dehydrogenase kinase (PDK).

On the contrary, the increase in kinases with SP/TP specificity (*i.e.*, CDKSs, MAPKs, and GSK-3 *etc.*) was revealed. Additional groups of kinases that were indicated to be up-regulated mostly comprised kinases that were involved in mitosis, e.g., Polo-like kinase 1 (PLK1), Never in mitosis A-related kinase 2 (NEK2) or Dual specificity protein kinase TTK (TTK).

3. Discussion

DNA damage response (DDR) is an essential mechanism to maintain the genome integrity of cells. Moreover, its functionality implicates the efficiency of radio- or chemotherapeutic agents administered during cancer therapy and thus underlies cell resistance to genotoxic stress in normal as well as in cancer cells. Therefore, the therapeutic inhibition of DNA repair enzymes represents a promising strategy how to increase sensitivity towards genotoxic insults in cancer cells. One of the repair enzymes is the ATR kinase that is considered to primarily act in the S and G2 phases of the cell cycle and suppress the "physiological" replication stress as well as genotoxic stress induced by agents such as hypoxia, UV- and ionizing radiation, and hydroxyurea (HU) [8].

A recent discovery of the first selective inhibitors of ATR kinase facilitated investigation of the cellular functions of ATR and launched the development of new potential anti-cancer drugs targeting this kinase to sensitize cancer cells to chemo- or radiotherapy [3]. Reaper *et al.* confirmed that ATR inhibition using VE-821 that belongs to the group of 3-amino-6-pyrazines may offer great promise in cancer treatment. In concordance with previous results suggesting that disruption in ATM-p53 pathway should enhance the sensitivity of cells to the disruption of the ATR pathway [23], VE-821 was shown to selectively induce cytotoxicity in cancer cells without affecting normal cells. Treatment with 20 μM VE-821 alone induced irreversible growth arrest and apoptosis in a large fraction of different cancer cell lines that were p53-deficient (human colorectal carcinoma cell lines HT23, HT29 or human malignant melanoma cell line HT144) or p53-proficient but with a defective ATM-mediated pathway (human colorectal carcinoma cells HCT116). On the other hand, growth arrest induced by VE-821 in normal human lung fibroblasts (HFL-1) was fully reversible with minimal death [4].

In our previous study, we have confirmed the radiosensitizing effects of VE-821 in relatively radioresistant p53 negative cells of promyelocytic leukemia (HL-60). The inhibition of ATR (10 μM VE-821) in combination with IR (3 Gy) resulted in pronounced decreases in clonogenic survival and significant changes in the proportions of cells in S and G2 phases of the cell cycle as compared to irradiated control cells [13]. In our recent work, we employed the phosphoproteomic approach to describe the changes in intracellular signaling caused by radiosensitization in more detail. Our goal was to characterize signaling pathways triggered by irradiation and modulated by inhibition of an essential DNA repair enzyme ATR kinase. We aimed to reveal which kinases showed alterations in their activities in the presence of the ATR inhibitor VE-821 and thus were probably functionally linked to ATR-mediated response upon genotoxic stress. Moreover, deeper insight into processes triggered in cancer cells sensitive to ATR inhibition may in the future lead to identification of potential markers of sensitivity to VE-821 treatment.

Both "motif-network" kinase analyses confirmed an expected decrease in the activity of the ATR kinase that was inhibited in order to radiosensitize the HL-60 cells. However, a decrease in ATM kinase activity was also evaluated to be statistically significant. Since it was reported, that the activity of the ATM kinase is not affected by VE-821 in 10 μM concentration ([4,13]; assessed by monitoring of CHK2 Thr68 phosphorylation which is a widely accepted specific marker of ATM activity), we account it to the decrease in phosphorylation of the consensus SQ/TQ motif which is shared between ATR, ATM and also DNA-PK. Additionally, ATM and ATR kinases share overlapping substrate specificity and hence the addition of protein-protein interaction scoring into the analysis could not improve the scoring algorithm to further distinguish between them. Interestingly, the activity of CHK1, which is a direct downstream target of ATR, was revealed to be significantly decreased in the HPRD motifs analysis (*p*-value lower than 0.05), however the activity shift was evaluated to be lower than expected after ATR/CHK1 pathway disruption (position score of only −0.07, where position score of 0 indicates non-regulated activity and −1 strongly downshifted activity). Moreover, Ser296 autophosphorylation of CHK1 and importantly Ser216 phosphorylation of CDC25C (which is an accepted marker of functional CHK1 dependent signaling) were not significantly altered in our work. On the other hand, we detected down-regulation of phosphorylation at Ser345 by immunoblotting, which serves to localize CHK1 to the nucleus following checkpoint activation [24]. Another important CHK1 phosphorylation at Ser317 was not detected in our phosphoproteomic analysis. Altogether, these findings underline the important concepts that have been reported recently: (i) the abrogation of different phosphorylation sites at CHK1—and thus the ATR/CHK1 pathway-by VE-821—is cell line—and treatment-dependent [7]; (ii) the radiosensitizing mechanism of VE-821 cannot be attributed merely to the abrogation of the ATR/CHK1 pathway [2,7].

All analyses further confirmed an increase in the activity of CDKs (SP/TP consensus motifs) that are inhibited by DNA repair enzymes in response to genotoxic stress in order to induce the cell cycle arrest to allow the repair of the damaged DNA lesions. It has been reported previously, that p53-negative HL-60 cells irradiated by lower doses of IR (up to 10 Gy) cannot undergo the p53-dependent G1 cell cycle arrest and therefore transiently accumulate in S phase, arrest at G2/M cell cycle checkpoint and eventually die by a so-called postmitotic apoptosis [24]. In the presence of VE-821, the CDKs inhibitory effect of ATR kinase (that is normally activated by replication and genotoxic stress in S and the G2 phase of the cell cycle) diminished, which was proven by the cell cycle analysis; the proportion of cells in G2 phase significantly decreased when compared to irradiated cells indicating that this cell population proceeded into mitosis [13]. Together, these anticipated trends in ATR and CDKs activities revealed by bioinformatic processing of our data, proved their biological reliability and relevance in addition to calculated statistical significance.

IR induces various types of DNA damage in irradiated cells including base damage, DNA crosslinks, single strand breaks (SSBs) and importantly, the most deleterious double strand breaks (DSBs). While ATM is activated exclusively in response to DSBs, ATR responds to DSBs as well as to SSBs [25], however the response of ATR to DSBs that are not part of the replication process is ATM-dependent [26]. GO biological process (GOBP) over-representation analysis revealed a significantly overrepresented term "cellular response to DNA damage stimulus" that included 37 proteins with significantly changed phosphorylation measured in our experiments. Moreover, GOBP terms such as "base-excision repair", "nucleotide-excision repair" or "double-strand break repair" were also significantly enriched. Overrepresented pathways comprised for example "Activation of ATR in response to replication stress" from Reactome database or "ATM pathway" from PID.

Naturally, the most desirable effect of ATR inhibition in cancer treatment is increased cell death. Our data indicated, that phosphorylation changes in signaling pathways involved in apoptosis were apparent 1 h after irradiation, which can be illustrated by results of pathways overrepresentation analysis (e.g., "Apoptosis", "Apoptotic cleavage of cellular proteins"). Again, increased apoptosis induction in HL-60 cells after combined treatment with IR (3 Gy) and VE-821 has been reported previously [13].

The disruption of the G2/M cell cycle checkpoint by VE-821 can be monitored by the cell cycle analysis conducted by flow-cytometry 24 h after irradiation as a significant decrease of G2 phase cells in comparison to a control group of cells irradiated without VE-821 and was published in our previous work [13]. In our recent experiment, we observed multiple over-represented GOBP level 4 terms related to mitotic processes (e.g., "mitosis", "centromere complex assembly", "cytokinesis", "sister chromatid cohesion", or "spindle localization") together with GO Cellular Component (GOCC) terms revealing the localization of some of the phosphorylated proteins at centromeric regions of condensed chromosomes ("condensed chromosome kinetochore") which occur during mitosis. Some of the pathways related to cell division were also significantly over-represented, comprising for example "Mitotic M-M/G1 phases" or "Separation of sister chromatids" from the Reactome database, and "Aurora B signaling" or "PLK1 signaling events" from PID. Moreover, multiple protein kinases with known functionality in mitotic entry, spindle checkpoints, regulation of chromosome segregation and cytokinesis, such as PLK1 or NEK2, were more activated in the presence of VE-821 and both of them are known to be inhibited in an ATM-dependent manner in response to IR [27–30]. Notably, the regulation of PLK1 by ATR kinase after UV-but not gamma-irradiation has been very recently described [31].

In addition to these expected markers of G2/M checkpoint disruption, we further focused on the mechanisms involved in the G1/S transition and their possible modulation by ATR kinase. The direct involvement of ATR in regulation of transcription at G1/S transition after replication stress (*i.e.*, genotoxic stress in S-phase; caused by HU-treatment) has been reported recently [32]. Our results indicate the role of ATR in these processes after IR-induced DNA damage since a substantial portion of regulated phosphoproteins revealed in our study are involved in over-represented Reactome pathways such as: "Mitotic G1-G1/S phases", "G1/S Transition", "E2F mediated regulation of DNA replication", and "Inhibition of replication initiation of damaged DNA by RB1/E2F1".

The transcriptional regulators at G1/S comprise multiple proteins with different functions and thus are often grouped into four categories: activators (E2F1, E2F2, and E2F3), repressors (E2F4, E2F5, E2F6, E2F7, and E2F8), inhibitors (RB) and co-repressors (RBL1 and RBL2). Besides transcriptional regulation (e.g., E2F1 induces transcription of E2F6, E2F7, and E2F8 which in turn function as a negative feedback loop), phosphorylation mediated by CDK-Cyclin complexes, DNA repair kinases (ATM, ATR), and checkpoint kinases (CHK1 and CHK2) also have essential roles in regulation of G1/S [32]. Regarding the activators, cells exposed to IR in G1 phase were shown to accumulate E2F1 in an ATM-dependent manner (E2F1 stabilizing phosphorylation), which led to induction of S-phase and subsequently to apoptosis [33]. On the other hand, E2F6 repressor inhibition by ATR-CHK1-phosphorylation in response to replication stress (HU-treatment) resulted in pronounced E2F-dependent G1/S transcription, which in this particular scenario led to cell survival [32]. In summary, these two phosphorylation-mediated contradictory roles of the same transcriptional mechanism highlight the importance of phosphorylation in tight regulation of this process. Since we identified some significantly regulated phosphorylation sites of these proteins after ATR inhibition, we suggest that ATR kinase may contribute considerably to regulation of G1/S after gamma-irradiation at least in regards to HL-60 cells. Importantly, while the role of phosphorylation in modulation of pocket proteins (RB, RBL1, RBL2) activity is a well-described mechanism, the impact of posttranslational modifications on "atypical repressors" E2F7 and E2F8 functionality have not been described yet. We found two significantly increased phosphorylation sites in E2F8 and E2F4 repressors after ATR inhibition, however, further experiments are necessary to validate these phosphorylations as biologically relevant, which is our next goal.

Moreover, proteins involved in the pre-replicative complex (pre-RC) formation were significantly modified by phosphorylation after VE-821 treatment (e.g., "Activation of the pre-replicative complex" in over-represented Reactome pathways). Pre-RC formation and licensing, occurs in the G1 phase and the RC is activated at the G1/S transition by phosphorylation of its components, mainly ORCs (Origin recognition complex subunits) and MCMs (minichromosome maintenance complex subunits) complexes, CDC6, and CDT1, to initiate DNA-replication and thus progress into S phase [34]. We observed multiple phosphorylation sites with increased phosphorylation in ORCs and MCM proteins; most of them contained the SP/TP consensus motif typical for CDKs. Previously, ATR has been described to suppress origin firing in response to replication stress [9], which was confirmed here in the context of IR-induced DNA damage.

In conclusion, our analyses conducted in asynchronous p53 negative cells allowed us to observe a comprehensive impact of ATR kinase inhibition on cellular response to IR-induced genotoxic stress in all phases of the cell cycle. While the described work was conducted on a single cell line and the data need further validation, we found that ATR kinase inhibition modulated the mechanisms at the G1/S transition, impacted the intra-S-checkpoint, disrupted the G2/M checkpoint and altered the activity of kinases involved in mitosis.

4. Experimental Section

4.1. Cell Culture and Culture Conditions

HL-60 cells were obtained from the European Collection of Animal Cells Cultures (Porton Down, Salisbury, UK). Cells were cultured in Iscove's modified Dulbecco's media (IMDM) for SILAC containing 20% dialyzed fetal calf serum, 2 mM glutamine, 100 UI/mL penicillin, and 0.1 mg/mL streptomycin (all purchased from Sigma-Aldrich, St. Louis, MO, USA) at 37 °C, under controlled 5% CO_2 and humidified atmosphere. Media were further supplemented with either unlabeled L-lysine (100 mg/L, K0) and L-arginine (84 mg/L, R0) or equimolar amounts of L-$^{13}C_6$-lysine (K6) and L-$^{13}C_6$-arginine (R6; isotopicaly labelled amino acids were purchased from Invitrogen, Carlsbad, CA, USA). L-proline (300 mg/L) was added to avoid metabolic conversion of arginine to proline [35]. For complete incorporation of labelled amino acids, cells were cultured for at least 6 doublings [36].

Cell counts were assessed by a hemocytometer and the cell membrane integrity was determined by the Trypan Blue exclusion technique (Sigma-Aldrich, St. Louis, MO, USA).

4.2. Gamma-Ray Irradiation

The selective inhibitor of ATR kinase, 3-amino-6-(4-(methylsulfonyl)phenyl)-*N*-phenylpyrazine-2-carboxamide (VE-821, APIs Chemical Co., Ltd. , Shanghai, China) was dissolved in DMSO in 10 mM aliquots and stored at −80 °C. Thirty min before irradiation, the inhibitor was added to the "heavy" cells (K6/R6) at concentration of 10 μM, the "light" cells (K0/R0) were treated with DMSO, whose final concentration in culture was lower than 0.1% to avoid the DMSO-induced differentiation of HL-60 cells. Both groups were irradiated by the dose of 6 Gy using a ^{60}Co gamma-ray source (VF, Cerna Hora, Czech Republic) with a dose rate 0.5 Gy/min. After irradiation, the flasks were placed in an incubator. Three experimental replicates were analyzed.

4.3. Cell Lysis and Protein Digest

One hour after irradiation, the cells were washed with cold PBS and lysed as was published [37]: the cells were resuspended in ice-cold lysis buffer (50 mM ammonium bicarbonate, 1% sodium deoxycholate, phosphatase inhibitor cocktail 2). The lysate was immediately placed into boiling water bath and after 5 min incubation the samples were cooled to room temperature. To decrease viscosity, bensonase nuclease (2.5 U/μL) and $MgCl_2$ (1.5 mM) were added to samples. The lysate was then clarified by centrifugation and the protein concentration was measured by bicinchoninic acid assay. Sample volumes corresponding to 1.75 mg of "light" proteins and 1.75 mg of "heavy" proteins were pooled together, reduced with dithiothreitol, alkylated with iodoacetamide and digested O/N with trypsin at an enzyme-to-substrate ratio of 1:60 (sequence grade modified trypsin, Promega Corporation, Madison, WI, USA). Sodium deoxycholate was then extracted by ethyl acetate [38] and peptides were desalted via 500 mg Supelco C18 SPE cartridges (Supelco Analyticals, Bellefonte, PA, USA). All other chemicals for cell lysis and protein digestion were purchased from Sigma-Aldrich.

4.4. Electrophoresis and Western Blotting

One hour after irradiation by 6 Gy, the HL-60 cells were washed with PBS and lysed. Whole cell extracts were prepared by lysis in 500 μL of lysis buffer (137 mM NaCl; 10% glycerol; 1% *n*-octyl-β-glucopyranoside; 50 mM NaF; 20 mM Tris, pH = 8; 1 mM Na_3VO_4; 1 tablet of protein inhibitors Complete™ Mini, Roche, Manheim, Germany). The lysates containing equal amount of protein (30 μg) were loaded onto a 12% SDS polyacrylamide gel. After electrophoresis, proteins were transferred to a polyvinylidene difluoride membrane (BioRad, Hercules, CA, USA), and hybridized with an appropriate antibody: anti-CHK1 and anti-CHK1 phosphorylated at serine 345 (1:500) from Cell Signaling (Danvers, MA, USA); anti-β-actin (1:2000) from Sigma. After washing, the blots were incubated with secondary peroxidase-conjugated antibody (Dako, Glostrup, Denmark) and the signal was developed with ECL detection kit (BM Chemiluminescence-POD, Roche) by exposure to a film.

4.5. Hydrophilic Interaction Liquid Chromatography Fractionation and Enrichment of Phosphopeptides

Dried peptide samples were fractionated by HILIC chromatography according to a protocol that has been published previously [39] using the 4.6 × 25 cm TSKgel® (Tosoh Biosciences, Stuttgart, Germany) Amide-80 HR 5 μm particle column with the TSKgel® Amide-80 HR 5 μm 4.6 × 1 cm guard column operated with Waters Separations Module 2695 at 0.5 mL/min. Briefly, 3.5 mg of evaporated samples were reconstituted in 80% B (98% acetonitrile (ACN)/0.1% trifluoroacetic acid (TFA), mobile phase A consisted of 2% acetonitrile with 0.1% TFA) and loaded onto the HILIC column. Peptides were then separated by a gradient of A over B from 80% to 60% B in 40 min and from 60% to 0% B in 5 min. Across the gradient, 22 fractions were collected (2 × 2 and 20 × 1 mL) from each replicate. Each fraction was then enriched for phosphopeptides using titanium dioxide chromatography [40]. At first, each fraction was supplemented with TFA and glutamic acid to reach final concentrations of 2% TFA and 100 mM glutamic acid. Titanium dioxide particles (Titansphere® 5 μm particles, GL Sciences, Torrance, CA, USA) were suspended in the loading solution (65% acetonitrile, 2% TFA, 100 mM glutamic acid) and a particular volume of titanium dioxide suspension depending on an expected amount of peptides and phosphopeptides in a particular fraction was added to each sample microtube. Microparticles with bound phosphopeptides were then washed with 200 μL of loading solution, 200 μL of washing solution 1 (65% acetonitrile with 0.5% TFA), 200 μL of washing solution 2 (65% acetonitrile with 0.1% TFA) and 100 μL of washing solution 2. Phosphopetides were then eluted by 150 μL of elution solution (20% acetonitrile/NH_4OH, pH 11.5) in two sequential elutions. Late fractions were subjected to the second enrichment. Eluates from the first and second enrichment were pooled together, acidified with 100% formic acid and placed in a SpeedVac until all crystals of ammonium formate were evaporated.

4.6. Mass Spectrometric Analysis

On-line LC-MS analyses were performed on Thermo Scientific Dionex Ultimate™ 3000 RSLCnano system (Thermo Scientific, Bremen, Germany) coupled through Nanospray Flex ion source with Q Exactive mass spectrometer (Thermo Scientific, Bremen, Germany). TiO_2-enriched HILIC fractions were dissolved in 18 μL of 2% ACN/0.05% TFA and 2 μL were injected into RSLCnano system. Peptides were loaded on capillary trap column (C18 PepMap100, 3 μm, 100 A, 0.075 × 20 mm; Dionex) by 2% ACN/0.05% TFA mobile phase at flow rate 5 μL/min for 5 min and then eluted and separated on capillary column (C18 PepMap RSLC, 2 μm, 100 A, 0.075 × 150 mm; Dionex). Elution was carried out by step linear gradient of mobile phase B (80% ACN/0.1% FA) over mobile phase A (0.1% FA); from 4% to 36% B in 19 min and from 36% to 55% B in 6 min at flow rate 300 nL/min. Temperature of the column was 40 °C and eluent was monitored at 215 nm during the separation. Spraying voltage was 1.7 kV and heated capillary temperature was 220 °C. The mass spectrometer was operated in the positive ion mode performing survey MS (range 300 to 1800 *m/z*) and data-dependent MS/MS scans performed on the six most intense precursors with dynamic exclusion window of 40 s. MS scans were acquired with 70,000 resolution at 200 *m/z* from 1×10^6 accumulated charges (maximum fill time was 100 ms). The lock mass at *m/z* 445.12003 ($[(C_2H_6SiO)_6 + H]^+$) was used for internal calibration of mass spectra. Intensity threshold for triggering MS/MS was set at 1×10^5 for ions with $z \geq 2$ with a 3 Da isolation window. Precursor ions were accumulated with AGC of 1×10^5 (maximum fill time was 100 ms) and normalized collisional energy for HCD fragmentation was 27 units. MS/MS spectra were acquired with 17,500 resolution (at 200 *m*/*z*).

4.7. Data Processing

The raw files were processed with MaxQuant software version 1.3.0.5 [41]. Peak lists were searched against the human UniProt database (release February 2014) using Andromeda search engine [42]. Minimum peptide length was set to six amino acids and two missed cleavages were allowed. Carbamidomethylation of cysteines was set as fixed modification while oxidation of methionine, protein N-terminal acetylation, deamidation of glutamine and arginine and phosphorylation of serine, threonine, and tyrosine residues were used as variable modifications. Additionally, appropriate SILAC labels were selected (Arg6, Lys6). A mass tolerance of 6 and 20 ppm was allowed for MS and MS^2 peaks, respectively. Only proteins, peptides and phosphorylation sites with FDR lower than 0.01 were accepted. For protein quantification only unmodified peptides, peptides oxidized at methionine residues, acetylated at *N*-terminus, or deamidated were accepted, both razor and unique peptides were used for the calculation of protein *H*/*L* ratios.

4.8. Bioinformatic Analysis

Contaminants and reversed hits were removed before further data processing and data were further manually inspected. The GRT was used to find differentially regulated phosphorylation sites. Phosphorylation sites quantified in all three replicates were included in GRT and the FDR was estimated non-parametrically as described by Zhou *et al.* [43]. The significance level of GRT was set to FDR < 0.01.

GO and signaling pathways over representation analyses were performed using ConsensusPathDB, over-representation analysis web tool [14,15]. Only those proteins containing differentially regulated phosphorylation sites evaluated in GRT were included in the analyses. Significance was estimated using hypergeometric testing with Benjamini-Hochberg FDR correction (FDR < 0.01). All three GO classes (molecular function, cellular component and biological process) were statistically tested as well as the enrichment of proteins involved in signaling pathways stored in the Reactome pathway database [16] and in the PID [17]. The background reference set for the statistical analysis comprised all ConsensusPathDB entities: (i) annotated with a UniProt identifier that were included in at least one signaling pathway or (ii) that were annotated with a GO category which comprised at least one protein from the “regulated” dataset.

To analyze and visualize sequence motifs surrounding phosphorylation sites identified and quantified in our study, we employed the iceLogo tool and motif-x algorithm. In both motif analyses, the amino acid sequences (±7 residues) surrounding either significantly up- or down-regulated phosphorylation sites (determined using GRT) were tested against a background reference set composed of sequences surrounding all phosphorylation sites revealed in our study that reached the minimal localization probability 0.75 (*i.e.*, "class I phosphosites" [44]). In the iceLogo analysis the significance level was set to $p < 0.01$. Using motif-x, the significantly enriched linear motifs were extracted. Search parameters were set to at least 10 occurrences of a particular motif and the significance level of $p < 0.00003$ (which approximately corresponds to a q-value of 0.01 after Bonferroni correction for multiple hypothesis testing).

The changes in kinase activities were evaluated using three different tools for the predictions of phosphorylation site-specific kinases together with the evaluation of significance based on the Wilcoxon-Mann-Whitney test ("1D annotation enrichment" tool available in Perseus software, version 1.4.0.20 [45]). Input data were filtered to contain only those class I phosphorylation sites that were quantified in at least 2 of 3 replicates. In the first analysis, the consensus sequence motifs downloaded from the HPRD database were simply matched to sequences motifs of phosphorylation sites identified in our study (using "Add linear motifs" tool integrated in Perseus program). To further specify kinase predictions by combining motif scoring together with contextual motif (*i.e.*, protein-protein interaction scoring downloaded from the STRING database [46]) we employed two freely available predictors NetworKIN 3 [21] and iGPS [22]. Kinase-substrate relations predicted by NetworKIN 3 algorithm were further filtered according to the "NetworKIN score" (>2). For iGPS predictions, the "high" significance threshold was chosen (*i.e.*, FPR of 2% for S/T kinases and FPR of 4% for Y kinases). The significance threshold for the Wilcoxon-Mann-Whitney test was set to $p < 0.05$.

5. Conclusions

Defects in the DSB-induced DDR such as ATM and/or p53 deletion/mutation are common in human tumors, occur in up to 70% of cancer cells [47,48] and have been proposed to enable the proliferation of cancer cells with DNA lesions [49]. Our previous study suggested that in a p53-negative environment, the most effective inhibition involves ATR rather than ATM and/or DNA-PK [13]. This work provides for the first time a complex insight into the mechanism of radiosensitization of the p53-deficient model cell line (HL-60, human promyelocytic leukemia) by the highly specific ATR inhibitor, VE-821. We show here a powerful phosphoproteomic approach for investigation of the role of ATR-mediated signaling in the context of radiosensitization of cancer cells. We found that significantly regulated phosphorylation sites were involved in cellular processes related to mitosis, cell cycle regulation, and DNA repair. Furthermore, understanding these processes by applying the presented strategy might facilitate characterization of biomarkers for patients susceptible to selective ATR inhibition. Although the experiments presented here were performed only in one cell line, this study provides valuable information regarding complex ATR-mediated signaling changes, and should be further validated in other cell lines as well as in primary tumor samples from patients, which is our next goal. Our data underline the concept of radiosensitization in tumor cells as a promising new way to substantially increase efficacy of current cancer therapy.

Acknowledgments

This work was supported by the Grant Agency of the Charles University in Prague (GAUK 1220313) and by the Grant Agency of Czech Republic (project No. P206/12/338) and we thank Lenka Mervartova for her excellent technical support.

Author Contributions

Aleš Tichý, Barbora Šalovská, and Martina Řezáčová conceived and designed the experiments; Barbora Šalovská performed cell cultivation, SILAC labeling, cell treatment, lysis, protein digest and phopsphopeptide enrichment; Kamila Ďurišová and Jiřina Vávrová performed Western-blotting experiments; Barbora Šalovská, Ivo Fabrik and Marek Link performed HILIC prefractionation and MS analysis; Barbora Šalovská performed data processing and bioinformatical analysis; Barbora Šalovská and Aleš Tichý wrote and revised the manuscript.

References

1. Tichý, A.; Vávrová, J.; Pejchal, J.; Rezácová, M. Ataxia-telangiectasia mutated kinase (ATM) as a central regulator of radiation-induced DNA damage response. *Acta Med. Hradec Králove* **2010**, *53*, 13–17.
2. Gamper, A.M.; Rofougaran, R.; Watkins, S.C.; Greenberger, J.S.; Beumer, J.H.; Bakkenist, C.J. ATR kinase activation in G1 phase facilitates the repair of ionizing radiation-induced DNA damage. *Nucleic Acids Res.* **2013**, *41*, 10334–10344.
3. Charrier, J.-D.; Durrant, S.J.; Golec, J.M.C.; Kay, D.P.; Knegtel, R.M.A.; MacCormick, S.; Mortimore, M.; O'Donnell, M.E.; Pinder, J.L.; Reaper, P.M.; *et al.* Discovery of potent and selective inhibitors of ataxia telangiectasia mutated and Rad3 related (ATR) protein kinase as potential anticancer agents. *J. Med. Chem.* **2011**, *54*, 2320–2330.
4. Reaper, P.M.; Griffiths, M.R.; Long, J.M.; Charrier, J.D.; Maccormick, S.; Charlton, P.A.; Golec, J.M.C.; Pollard, J.R. Selective killing of ATM- or p53-deficient cancer cells through inhibition of ATR. *Nat. Chem. Biol.* **2011**, *7*, 428–430.
5. Pires, I.M.; Olcina, M.M.; Anbalagan, S.; Pollard, J.R.; Reaper, P.M.; Charlton, P.A.; McKenna, W.G.; Hammond, E.M. Targeting radiation-resistant hypoxic tumour cells through ATR inhibition. *Br. J. Cancer* **2012**, *107*, 291–299.
6. Prevo, R.; Fokas, E.; Reaper, P.M.; Charlton, P.A.; Pollard, J.R.; McKenna, W.G.; Muschel, R.J.; Brunner, T.B. The novel ATR inhibitor VE-821 increases sensitivity of pancreatic cancer cells to radiation and chemotherapy. *Cancer Biol. Ther.* **2012**, *13*, 1072–1081.
7. Huntoon, C.J.; Flatten, K.S.; Wahner Hendrickson, A.E.; Huehls, A.M.; Sutor, S.L.; Kaufmann, S.H.; Karnitz, L.M. ATR inhibition broadly sensitizes ovarian cancer cells to chemotherapy independent of BRCA status. *Cancer Res.* **2013**, *73*, 3683–3691.
8. Fokas, E.; Prevo, R.; Hammond, E.M.; Brunner, T.B.; McKenna, W.G.; Muschel, R.J. Targeting ATR in DNA damage response and cancer therapeutics. *Cancer Treat. Rev.* **2014**, *40*, 109–117.
9. Toledo, L.I.; Altmeyer, M.; Rask, M.B.; Lukas, C.; Larsen, D.H.; Povlsen, L.K.; Bekker-Jensen, S.; Mailand, N.; Bartek, J.; Lukas, J. ATR prohibits replication catastrophe by preventing global exhaustion of RPA. *Cell* **2013**, *155*, 1088–1103.
10. Johnson, L.N. The regulation of protein phosphorylation. *Biochem. Soc. Trans.* **2009**, *37*, 627–641.
11. Tichy, A.; Salovska, B.; Rehulka, P.; Klimentova, J.; Vavrova, J.; Stulik, J.; Hernychova, L. Phosphoproteomics: Searching for a needle in a haystack. *J. Proteomics* **2011**, *74*, 2786–2797.

12. Salovska, B.; Tichy, A.; Fabrik, I.; Rezacova, M.; Vavrova, J. Comparison of resins for metal oxide affinity chromatography with mass spectrometry detection for the determination of phosphopeptides. *Anal. Lett.* **2013**, *46*, 1505–1524.
13. Vávrová, J.; Zárybnická, L.; Lukášová, E.; Řezáčová, M.; Novotná, E.; Sinkorová, Z.; Tichý, A.; Pejchal, J.; Durišová, K. Inhibition of ATR kinase with the selective inhibitor VE-821 results in radiosensitization of cells of promyelocytic leukaemia (HL-60). *Radiat. Environ. Biophys.* **2013**, *52*, 471–479.
14. Kamburov, A.; Pentchev, K.; Galicka, H.; Wierling, C.; Lehrach, H.; Herwig, R. ConsensusPathDB: Toward a more complete picture of cell biology. *Nucleic Acids Res.* **2011**, *39*, D712–D717.
15. Kamburov, A.; Wierling, C.; Lehrach, H.; Herwig, R. ConsensusPathDB—A database for integrating human functional interaction networks. *Nucleic Acids Res.* **2009**, *37*, D623–D628.
16. Croft, D. Building models using Reactome pathways as templates. *Methods Mol. Biol.* **2013**, *1021*, 273–283.
17. Schaefer, C.F.; Anthony, K.; Krupa, S.; Buchoff, J.; Day, M.; Hannay, T.; Buetow, K.H. PID: The Pathway Interaction Database. *Nucleic Acids Res.* **2009**, *37*, D674–D679.
18. Colaert, N.; Helsens, K.; Martens, L.; Vandekerckhove, J.; Gevaert, K. Improved visualization of protein consensus sequences by iceLogo. *Nat. Methods* **2009**, *6*, 786–787.
19. Chou, M.F.; Schwartz, D. Biological sequence motif discovery using motif-x. *Curr. Protoc. Bioinforma.* **2011**, *2011*, doi:10.1002/0471250953.bi1315s35.
20. Schwartz, D.; Gygi, S.P. An iterative statistical approach to the identification of protein phosphorylation motifs from large-scale data sets. *Nat. Biotechnol.* **2005**, *23*, 1391–1398.
21. Linding, R.; Jensen, L.J.; Ostheimer, G.J.; van Vugt, M.A.; Jørgensen, C.; Miron, I.M.; Diella, F.; Colwill, K.; Taylor, L.; Elder, K.; *et al.* Systematic discovery of *in vivo* phosphorylation networks. *Cell* **2007**, *129*, 1415–1426.
22. Song, C.; Ye, M.; Liu, Z.; Cheng, H.; Jiang, X.; Han, G.; Songyang, Z.; Tan, Y.; Wang, H.; Ren, J.; *et al.* Systematic analysis of protein phosphorylation networks from phosphoproteomic data. *Mol. Cell. Proteomics* **2012**, *11*, 1070–1083.
23. Nghiem, P.; Park, P.K.; Kim, Y.; Vaziri, C.; Schreiber, S.L. ATR inhibition selectively sensitizes G1 checkpoint-deficient cells to lethal premature chromatin condensation. *Proc. Natl. Acad. Sci. USA* **2001**, *98*, 9092–9097.
24. Mareková, M.; Vávrová, J.; Vokurková, D. Monitoring of premitotic and postmitotic apoptosis in gamma-irradiated HL-60 cells by the mitochondrial membrane protein-specific monoclonal antibody APO2.7. *Gen. Physiol. Biophys.* **2003**, *22*, 191–200.
25. Toledo, L.I.; Murga, M.; Fernandez-Capetillo, O. Targeting ATR and Chk1 kinases for cancer treatment: A new model for new (and old) drugs. *Mol. Oncol.* **2011**, *5*, 368–373.
26. Myers, J.S.; Cortez, D. Rapid activation of ATR by ionizing radiation requires ATM and Mre11. *J. Biol. Chem.* **2006**, *281*, 9346–9350.
27. Smits, V.A.J.; Klompmaker, R.; Arnaud, L.; Rijksen, G.; Nigg, E.A.; Medema, R.H. Polo-like kinase-1 is a target of the DNA damage checkpoint. *Nat. Cell Biol.* **2000**, *2*, 672–676.

28. Tsvetkov, L.; Stern, D.F. Phosphorylation of Plk1 at S137 and T210 is inhibited in response to DNA damage. *Cell Cycle Georget. Tex.* **2005**, *4*, 166–171.
29. Fletcher, L.; Cerniglia, G.J.; Nigg, E.A.; Yend, T.J.; Muschel, R.J. Inhibition of centrosome separation after DNA damage: A role for Nek2. *Radiat. Res.* **2004**, *162*, 128–135.
30. Mi, J.; Guo, C.; Brautigan, D.L.; Larner, J.M. Protein phosphatase-1alpha regulates centrosome splitting through Nek2. *Cancer Res.* **2007**, *67*, 1082–1089.
31. Qin, B.; Gao, B.; Yu, J.; Yuan, J.; Lou, Z. Ataxia telangiectasia-mutated- and Rad3-related protein regulates the DNA damage-induced G2/M checkpoint through the Aurora A cofactor Bora protein. *J. Biol. Chem.* **2013**, *288*, 16139–16144.
32. Bertoli, C.; Klier, S.; McGowan, C.; Wittenberg, C.; de Bruin, R.A.M. Chk1 inhibits E2F6 repressor function in response to replication stress to maintain cell-cycle transcription. *Curr. Biol.* **2013**, *23*, 1629–1637.
33. Lin, W.C.; Lin, F.T.; Nevins, J.R. Selective induction of E2F1 in response to DNA damage, mediated by ATM-dependent phosphorylation. *Genes Dev.* **2001**, *15*, 1833–1844.
34. Lau, E.; Tsuji, T.; Guo, L.; Lu, S.H.; Jiang, W. The role of pre-replicative complex (pre-RC) components in oncogenesis. *FASEB J.* **2007**, *21*, 3786–3794.
35. Bendall, S.C.; Hughes, C.; Stewart, M.H.; Doble, B.; Bhatia, M.; Lajoie, G.A. Prevention of amino acid conversion in SILAC experiments with embryonic stem cells. *Mol. Cell. Proteomics.* **2008**, *7*, 1587–1597.
36. Ong, S.E.; Mann, M. A practical recipe for stable isotope labeling by amino acids in cell culture (SILAC). *Nat. Protoc.* **2006**, *1*, 2650–2660.
37. Rogers, L.D.; Fang, Y.; Foster, L.J. An integrated global strategy for cell lysis, fractionation, enrichment and mass spectrometric analysis of phosphorylated peptides. *Mol. Biosyst.* **2010**, *6*, 822–829.
38. Yeung, Y.G.; Stanley, E.R. Rapid detergent removal from peptide samples with ethyl acetate for Mass Spectrometry Analysis. *Curr. Protoc. Protein Sci.* **2010**, *2010*, doi:10.1002/0471140864.ps1612s59.
39. McNulty, D.E.; Annan, R.S. Hydrophilic interaction chromatography reduces the complexity of the phosphoproteome and improves global phosphopeptide isolation and detection. *Mol. Cell. Proteomics* **2008**, *7*, 971–980.
40. Larsen, M.R.; Thingholm, T.E.; Jensen, O.N.; Roepstorff, P.; Jorgensen, T.J.D. Highly selective enrichment of phosphorylated peptides from peptide mixtures using titanium dioxide microcolumns. *Mol. Cell. Proteomics* **2005**, *4*, 873–886.
41. Cox, J.; Mann, M. MaxQuant enables high peptide identification rates, individualized p.p.b.-range mass accuracies and proteome-wide protein quantification. *Nat. Biotechnol.* **2008**, *26*, 1367–1372.
42. Cox, J.; Neuhauser, N.; Michalski, A.; Scheltema, R.A.; Olsen, J.V.; Mann, M. Andromeda: A peptide search engine integrated into the MaxQuant environment. *J. Proteome Res.* **2011**, *10*, 1794–1805.
43. Zhou, Y.; Cras-Méneur, C.; Ohsugi, M.; Stormo, G.D.; Permutt, M.A. A global approach to identify differentially expressed genes in cDNA (two-color) microarray experiments. *Bioinformatics* **2007**, *23*, 2073–2079.

44. Olsen, J.V.; Blagoev, B.; Gnad, F.; Macek, B.; Kumar, C.; Mortensen, P.; Mann, M. Global, *in vivo*, and site-specific phosphorylation dynamics in signaling networks. *Cell* **2006**, *127*, 635–648.
45. Cox, J.; Mann, M. 1D and 2D annotation enrichment: A statistical method integrating quantitative proteomics with complementary high-throughput data. *BMC Bioinfom.* **2012**, *13*, doi:10.1186/1471-2105-13-S16-S12
46. Franceschini, A.; Szklarczyk, D.; Frankild, S.; Kuhn, M.; Simonovic, M.; Roth, A.; Lin, J.; Minguez, P.; Bork, P.; von Mering, C.; *et al.* STRING v9.1: Protein-protein interaction networks, with increased coverage and integration. *Nucleic Acids Res.* **2013**, *41*, D808–D815.
47. Ding, L.; Getz, G.; Wheeler, D.A.; Mardis, E.R.; McLellan, M.D.; Cibulskis, K.; Sougnez, C.; Greulich, H.; Muzny, D.M.; Morgan, M.B.; *et al.* Somatic mutations affect key pathways in lung adenocarcinoma. *Nature* **2008**, *455*, 1069–1075.
48. Greenman, C.; Stephens, P.; Smith, R.; Dalgliesh, G.L.; Hunter, C.; Bignell, G.; Davies, H.; Teague, J.; Butler, A.; Stevens, C.; *et al.* Patterns of somatic mutation in human cancer genomes. *Nature* **2007**, *446*, 153–158.
49. Halazonetis, T.D.; Gorgoulis, V.G.; Bartek, J. An oncogene-induced DNA damage model for cancer development. *Science* **2008**, *319*, 1352–1355.

6

Development of Laser Ionization Techniques for Evaluation of the Effect of Cancer Drugs Using Imaging Mass Spectrometry

Hiroki Kannen [1], Hisanao Hazama [1,*], Yasufumi Kaneda [2], Tatsuya Fujino [3] and Kunio Awazu [1,4,5]

[1] Graduate School of Engineering, Osaka University, 2-1 Yamadaoka, Suita, Osaka 565-0871, Japan; E-Mails: kannen-h@mb.see.eng.osaka-u.ac.jp (H.K.); awazu@see.eng.osaka-u.ac.jp (K.A.)

[2] Graduate School of Medicine, Osaka University, 2-2 Yamadaoka, Suita, Osaka 565-0871, Japan; E-Mail: kaneday@gts.med.osaka-u.ac.jp

[3] Graduate School of Science and Engineering, Tokyo Metropolitan University, 1-1 Minamiosawa Hachioji, Tokyo 192-0397, Japan; E-Mail: fujino@tmu.ac.jp

[4] Graduate School of Frontier Biosciences, Osaka University, 1-3 Yamadaoka, Suita, Osaka 565-0871, Japan

[5] The Center for Advanced Medical Engineering and Informatics, Osaka University, 2-2 Yamadaoka, Suita, Osaka 565-0871, Japan

* Author to whom correspondence should be addressed; E-Mail: hazama-h@see.eng.osaka-u.ac.jp

---

Abstract: Recently, combined therapy using chemotherapy and photodynamic therapy (PDT) has been proposed as a means of improving treatment outcomes. In order to evaluate the efficacy of combined therapy, it is necessary to determine the distribution of the anticancer drug and the photosensitizer. We investigated the use of imaging mass spectrometry (IMS) to simultaneously observe the distributions of an anticancer drug and photosensitizer administered to cancer cells. In particular, we sought to increase the sensitivity of detection of the anticancer drug docetaxel and the photosensitizer protoporphyrin IX (PpIX) by optimizing the ionization-assisting reagents. When we used a matrix consisting of equal weights of a zeolite (NaY5.6) and a conventional organic matrix (6-aza-2-thiothymine) in matrix-assisted laser desorption/ionization, the signal intensity of the sodium-adducted ion of docetaxel (administered at 100 μM) increased about 13-fold. Moreover, we detected docetaxel with the zeolite matrix using the droplet method, and detected PpIX by fluorescence and IMS with α-cyano-4-hydroxycinnamic acid (CHCA) using the spray method.

Keywords: imaging mass spectrometry; laser ionization; anticancer drug; drug resistance; zeolite matrix

1. Introduction

In chemotherapy for cancer, the curative effect of an anticancer drug decreases with time because the cancer cells acquire drug resistance [1,2]. For example, docetaxel is often used to treat prostate cancers, but the response rate of docetaxel is not high (44.2% when used with prednisolone [3]), and the drug-resistant cells that arise are refractory to treatment. Consequently, new methods for treating drug-resistant tumors are urgently needed. One proposal for increasing the efficacy of treatment involves combining chemotherapy with photodynamic therapy (PDT). In PDT, the patient is treated with a photosensitizer that selectively accumulates in the tumor tissue, and then the tumor is targeted with a laser or other light source in order to excite the photosensitizer. However, current techniques for PDT do not exert a sufficient curative effect [4,5]. When chemotherapy and PDT are combined, the anticancer drug–resistant cells are killed by PDT, resulting in treatment that is more effective than PDT alone [6,7].

In order to evaluate the efficacy of combined therapy, it is important to determine the distributions of the anticancer drug and photosensitizer. Autoradiography is the standard means for determining drug distributions, but conventional drug imaging by autoradiography cannot be used to simultaneously observe multiple drugs [8]. Such simultaneous observation can be achieved using imaging mass spectrometry (IMS) of drugs, their metabolites, and other molecules in the biological tissue. However, the major disadvantage of the IMS is that its detection sensitivity is inferior to that of autoradiography. Moreover, a large difference in the ionization efficiencies of the drugs of interest can result in ion suppression (*i.e.*, obstruction of the ionization of the drug with lower ionization efficiency).

To resolve this problem, we are investigating ways to use IMS to simultaneously observe the distributions of anticancer drug and photosensitizer administered to cancer cells. In IMS, matrix-assisted laser desorption/ionization (MALDI) is often used as the ionization method because of its sensitive solid sampling and soft ionization technique, which has been extensively applied to the analysis of macromolecules and low-molecular weight compounds [9–13]. The sensitivity of detection of a drug of interest by MALDI can be greatly enhanced by the use of ionization-assisting reagents. The purpose of this research was to increase the sensitivity of detection for the anticancer drug and photosensitizer administered to human prostate cancer cells, first by optimizing the ionization-assisting reagents used in the process, and then by investigating a new technique for simultaneous imaging of the anticancer drug and photosensitizer.

2. Results and Discussion

2.1. Ionization of Docetaxel Using Zeolite Matrix

With the goal of enhancing the sensitivity of detection for anticancer drugs and photosensitizers, we attempted to optimize the ionization-assisting reagents used in the detection process. To this end, we investigated four conventional matrices: α-cyano-4-hydroxycinnamic acid (CHCA), 2,5-dihydroxy benzoic acid (DHB), 6-aza-2-thiothymine (ATT), and 4-nitroaniline (4NA). Figure 1 shows the mass spectra obtained from a 1-μL spot of 100 μM docetaxel using each matrix. Protonated, sodium-adducted, and potassium-adducted ions of docetaxel were detected. The peak of the sodium-adducted ion of docetaxel was most intense when 4NA was used as the matrix, but the ionization efficiency was not sufficient to allow detection of docetaxel administered to human prostate cancer cells (PC-3). To enhance the detection sensitivity further, we investigated the use of an additive dissolved in the matrix solvent, e.g., sodium acetate or potassium acetate, because the peak intensity of the sodium-adducted ion of docetaxel was the highest among the three peaks we detected. Figure 2 shows the average peak intensities of the sodium-adducted and potassium-adducted ions of docetaxel when sodium acetate or potassium acetate were added to the matrix solvent. Although the peak intensity increased about 4-fold, the ionization efficiency was still not sufficient to allow detection of docetaxel administered to PC-3 cells.

Figure 1. Typical mass spectra obtained from a the spot of the docetaxel using (**a**) α-cyano-4-hydroxycinnamic acid (CHCA); (**b**) 2,5-dihydroxy benzoic acid (DHB); (**c**) 4-nitroaniline (4NA); and (**d**) 6-aza-2-thiothymine (ATT). M indicates docetaxel.

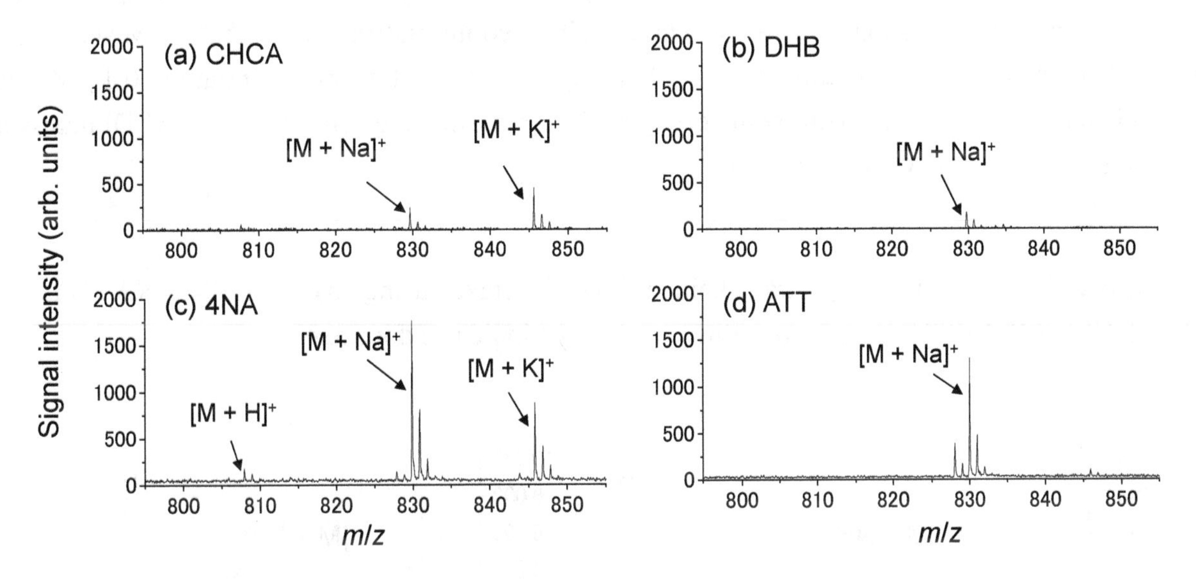

Figure 2. Relationship between the signal intensities of [M + Na]$^+$ and [M + K]$^+$. Black and gray bars mean that additive agent or no additive agent, respectively. M indicated docetaxel.

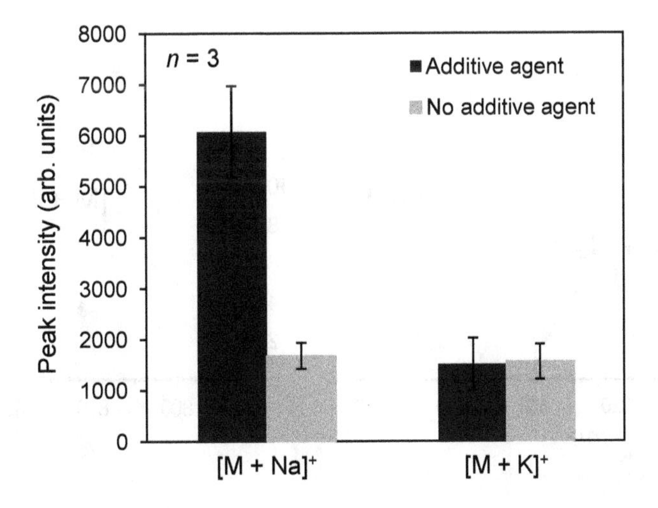

Next, in order to further enhance the detection sensitivity of docetaxel, we compared conventional matrix to a mixture of conventional matrix and zeolite matrix, a new ionization-assisting reagent that has been reported to increase the sensitivity of detection [14]. For this experiment, we selected NaY5.6 zeolite because it contains the sodium ion in its structure; as noted above, the peak intensity of sodium-adducted docetaxel was elevated when sodium acetate was added to the matrix solvent. Figure 3 shows typical mass spectra obtained from a spot of 100 μM docetaxel using the conventional matrix (ATT or 4NA) either alone or in combination with a zeolite matrix (NaY5.6). The peak intensity of the sodium-adducted ion $[M + Na]^+$ was increased. Figure 4 shows the relationship between the signal-to-noise (S/N) ratio of the docetaxel peak in the presence of zeolite matrix at each concentration and laser pulse energy. Zeolite matrix prepared at 5 mg/mL was the most suitable for ionization of docetaxel on a metal sample plate; at a zeolite matrix concentration of 1 or 3 mg/mL, docetaxel could not be detected. Figure 5 shows the average peak intensity of the sodium-adducted ion of docetaxel with 4NA or ATT, in the presence or absence of zeolite matrix. When zeolite matrix was used with ATT, the detection sensitivity of the sodium-adducted docetaxel peak increased 13-fold. By contrast, when zeolite matrix was used with 4NA, the detection sensitivity was not increased. Using the Voyager DE-PRO TOF instrument, the detection limit of docetaxel using zeolite matrix was 1 μM. However, another MALDI-TOF instrument, the Ultraflex III, could obtain a peak at a lower concentration, 0.1 μM. Thus, docetaxel and PpIX in the PC-3 cells could easily be detected using the conventional MALDI matrix and a modern time of flight (TOF) instrument.

Figure 3. Typical mass spectra obtained from docetaxel using (**a**) 4NA; (**b**) 4NA/NaY; (**c**) ATT; (**d**) ATT/NaY with no additive agents. M indicates docetaxel.

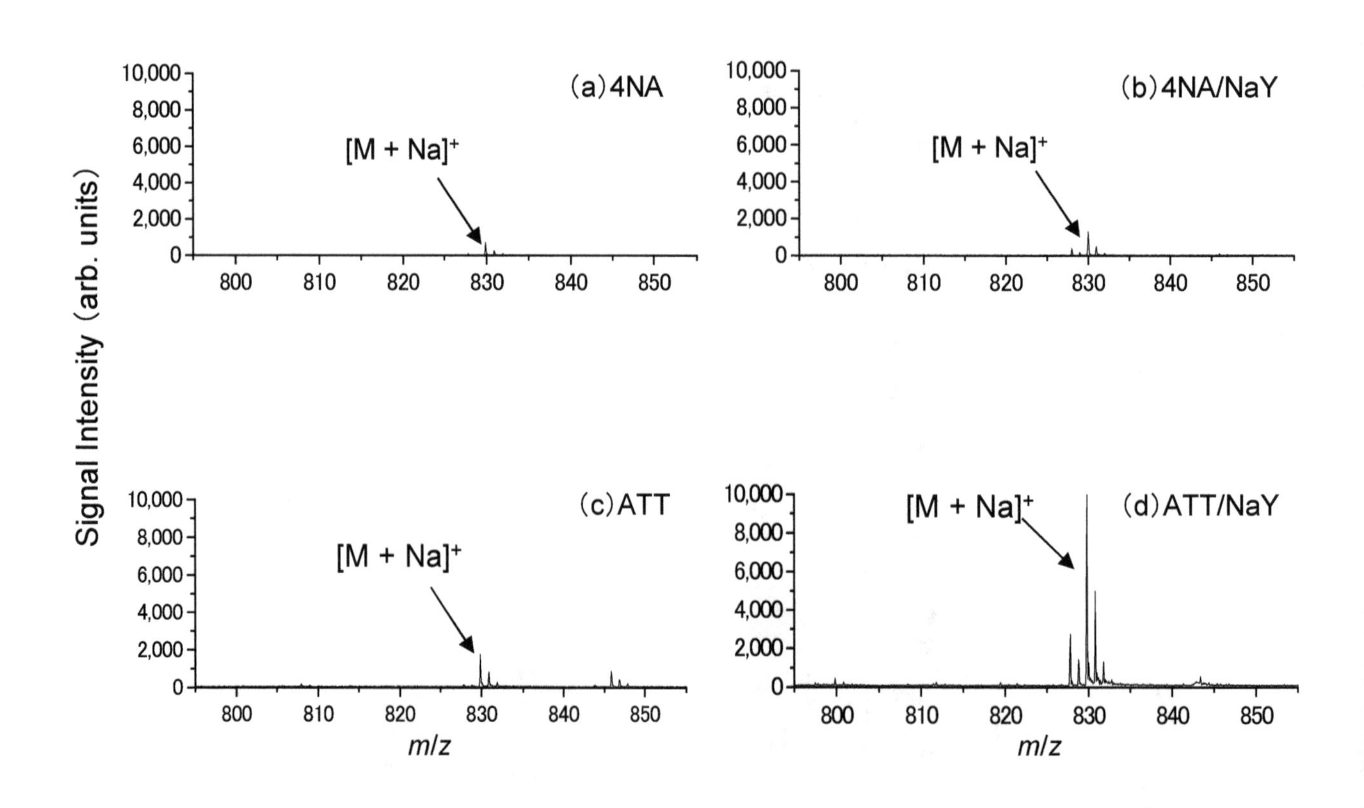

Figure 4. Relationship between S/N ratio of the docetaxel peak using zeolite matrix with no additive agents, at the indicated concentrations and laser pulse energies.

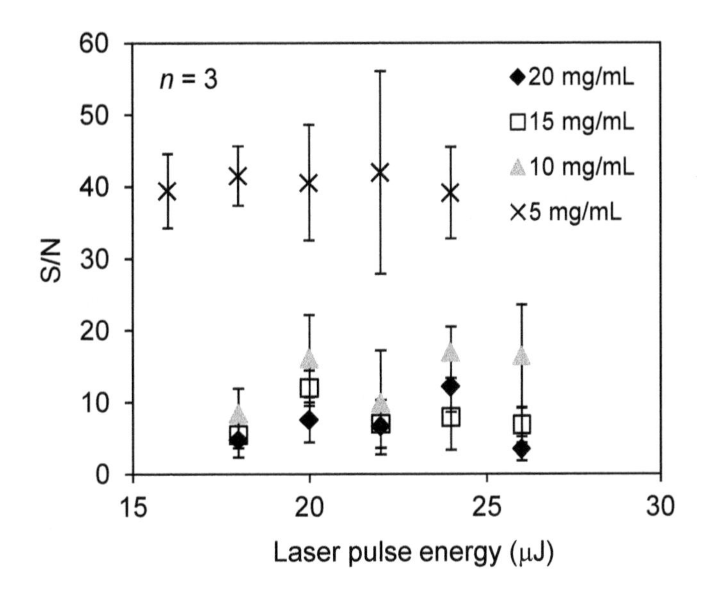

Figure 5. The average of peak intensity of the sodium-adducted ion of docetaxel using zeolite matrix with 4NA or ATT with no additive agents.

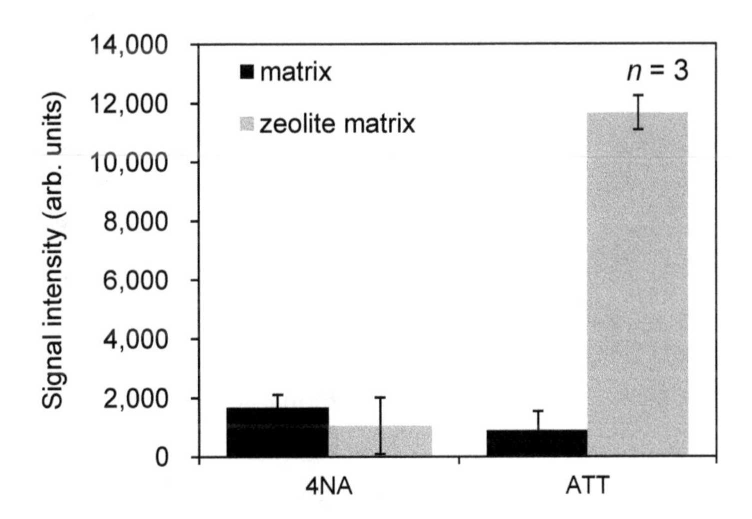

2.2. Detection of Docetaxel Administered to PC-3 Cells

As a first step toward acquiring images of docetaxel in prostate cancer tissue, we attempted to detect docetaxel administered to PC-3 cells using the zeolite matrix, ATT/NaY. Figure 6 shows typical mass spectra obtained from PC-3 cells administered 100 μM docetaxel using (a) zeolite matrix or (b) conventional matrix using the droplet method. Only in condition (a), the sodium-adducted ion of docetaxel $[M + Na]^+$ was detected from a spot of PC-3 cells. Figure 7 shows the S/N ratio of the $[M + Na]^+$ peak at concentrations of 0, 10, and 100 μM. $[M + Na]^+$ was detected from PC-3 cells administered 10 μM docetaxel. In this method, we observed the "sweet spot" effect. Therefore, each mass spectrum used for analysis was the average of 10 mass spectra.

Figure 6. Typical mass spectra obtained from the prostate cancer cells (PC-3) cells administered 100 μM docetaxel using (**a**) the zeolite matrix; ATT/NaY or (**b**) the conventional matrix, ATT.

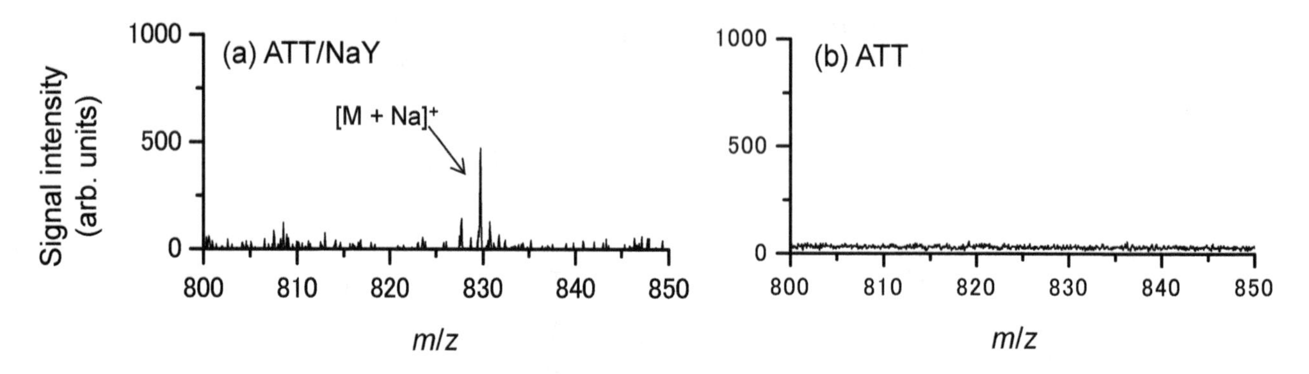

Figure 7. S/N ratio of the peak of $[M + Na]^+$ peak at 0, 10 and 100 μM docetaxel.

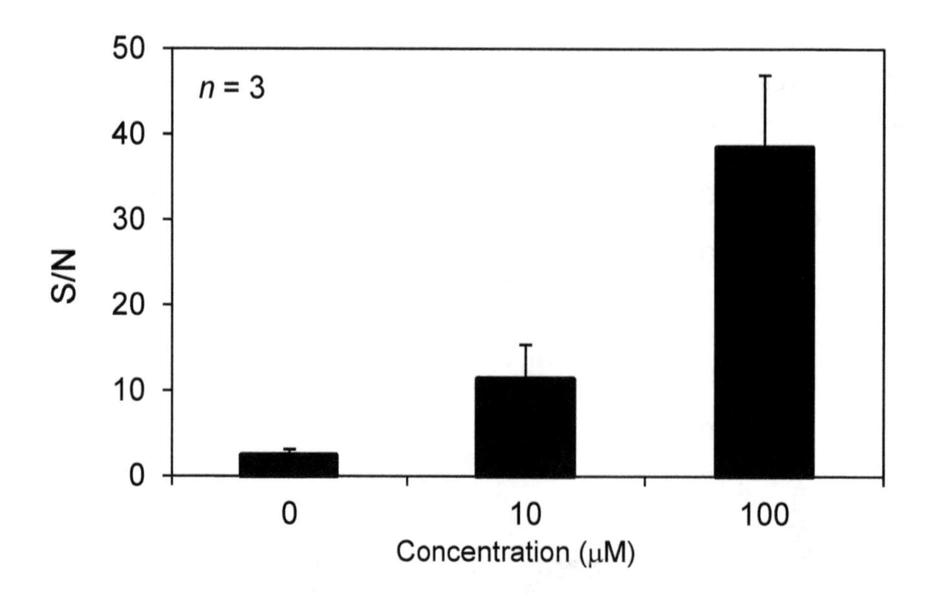

2.3. Imaging of Cancer Drugs Administered to PC-3 Cells

The use of the zeolite matrix enhanced the ionization efficiency of docetaxel sufficiently to allow detection of the drug in cancer cells. Next, we investigated ways to use IMS to simultaneously observe the distributions of anticancer drug and photosensitizer administered to cancer cells. First, we attempted IMS of docetaxel and PpIX administered to PC-3 with the zeolite matrix using the droplet method. Figure 8 shows the ion image of (a) *m/z* 829.5 (sodium-adducted docetaxel) and (b) *m/z* 563.1 (protonated PpIX) obtained from cultured PC-3 administered cells treated with 50 μM docetaxel and 10 μM PpIX, obtained with the zeolite matrix, ATT/NaY using the droplet method. In previous studies, the peak intensity of docetaxel was reduced when conventional matrices were used. In this study, however, docetaxel administered to cancer cells could be detected using zeolite matrix, whereas PpIX could not be detected in the same samples.

Figure 8. Ion image of (**a**) *m/z* 829.5 $[D + Na]^+$; and (**b**) *m/z* 563.3 $[P + H]^+$ obtained from PC-3 cells administered 50 μM docetaxel and 10 μM protoporphyrin IX (PpIX) using the zeolite matrix, ATT/NaY with the droplet method.

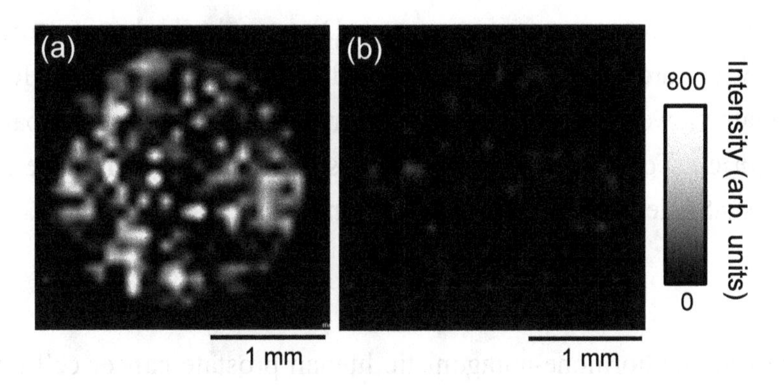

2.4. Imaging of PpIX Administered to PC-3 Cells

Although docetaxel administered to PC-3 cells could be detected with the zeolite matrix, PpIX could not be detected by IMS in the same samples. To address this issue, we investigated a new combined method. Initially, we had acquired optical and fluorescence images of the PC-3 cells, and also detected docetaxel by IMS. The new method was developed to allow acquisition of the ion image by IMS after acquisition of the fluorescence and optical images. Figure 9 shows (a) optical bright field image, (b) fluorescence image, and (c) ion image measured with CHCA using the spray method. All images were obtained from the same cultured PC-3 cells administered 10 μM PpIX. Under these conditions, PpIX could be detected, and the distribution of PpIX in the ion image approximately corresponded to that in the fluorescence image. In MALDI imaging, docetaxel and PpIX could not be detected in the same sample. However, using the new method, it was possible to obtain both fluorescence images and IMS from the same cells. Thus, using this method, docetaxel and PpIX could be detected in the same sample, by MALDI imaging and fluorescence imaging, respectively. We will seek to confirm this in a future study with cellular scale using a stigmatic imaging mass spectrometer, MULTUM-IMG2 [15,16].

Figure 9. (**a**) Optical bright field image; (**b**) fluorescence image; and (**c**) ion image measured with CHCA using the spray method, obtained from the same PC-3 cells administered 10 μM PpIX.

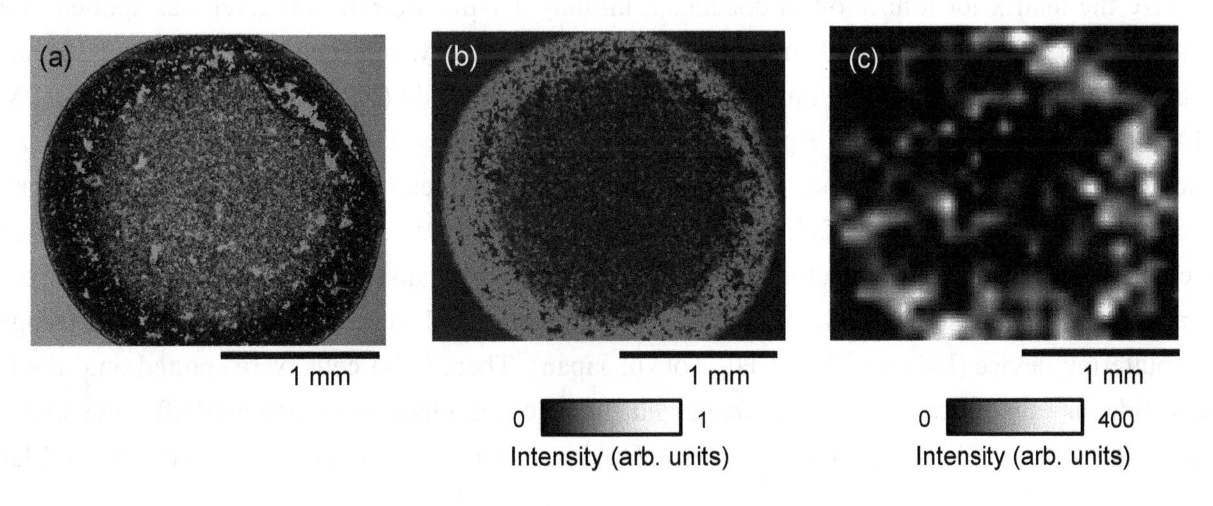

3. Experimental Section

3.1. Chemical and Reagents

Docetaxel is frequently used to treat prostate cancer. Docetaxel trihydrate was purchased from Wako (Osaka, Japan). PpIX was purchased from SIGMA-Aldrich (Tokyo, Japan). CHCA, DHB, 4NA, and ATT, used as the matrix, were purchased from Sigma-Aldrich (Tokyo, Japan). Acetonitrile and methanol (SIGMA-Aldrich, Tokyo, Japan) were used as the matrix solvent. The zeolite, NaY5.6, was supplied by the Catalysis Society of Japan (Tokyo, Japan).

3.2. Cell Culture

The cell line used was the hormone-antagonistic human prostate cancer cell line PC-3. Cells were cultured in Dulbecco's modified Eagle's medium (D-MEM, D6046, SIGMA-Aldrich, Tokyo, Japan) containing 10% fetal bovine serum (FBS, S1820, Biowest SAS, Nuaillé, France) and 100 units/mL each of penicillin and streptomycin (T4049, SIGMA-Aldrich, Tokyo, Japan). Cells were incubated at 37 °C in an atmosphere containing 5% CO_2.

3.3. Mass Spectrometer

All experiments were carried out in positive mode with reflectron-mode using Voyager DE-PRO TOF mass spectrometer (Applied Biosystems, Foster City, CA, USA) equipped with a 355-nm third-harmonic Nd:YAG laser (GAIA II 30-T, Rayture Systems Co., Ltd., Tokyo, Japan). Instrument parameters on the reflectron-mode of Voyager DE-PRO were as follows: accelerating voltage, +20 kV; voltage for the extraction grid, +13.6 kV; voltage for the guide wire, 0 V; extraction delay time, 100 ns.

3.4. Sample Preparation

For the conventional matrices, CHCA was dissolved in 1:1 (*v*/*v*) acetonitrile:water (*v*/*v*) at 10 mg/mL, and DHB, 4NA, and ATT were dissolved in 1:1 (*v*/*v*) methanol:water at 30, 5, and 10 mg/mL, respectively. Sodium acetate or potassium acetate was used as an additive agent and dissolved in the matrix solvent at 10 mM. For the zeolite matrices, 4NA or ATT was mixed with an equal weight of NaY5.6 zeolite with a mortar for over 10 min, and then suspended in 1:1 (*v*/*v*) methanol:water. To optimize the matrix for ionization of docetaxel, initially a 1-μL drop of docetaxel was spotted onto a metal sample plate, and then dried under vacuum; the each matrix was placed on dried spot of docetaxel. In this experiment, either indium tin oxide (ITO)-coated glass slide (# 237001, Bruker, Billerica, MA, USA) or a metal plate (4347686, Applied Biosystems, Foster City, CA, USA) was used as the sample plate. From experiments described in Sections 2.1–2.3, the metal sample plate was used. For the experiment described in Section 2.4, the ITO-coated glass slide was used to acquire the optical images. To compensate for the low conductivity of the glass slide, which caused a charge increase relative to the metal sample plate, the ITO glass slide was coated with gold at a thickness of 10 nm using an ion-sputtering device (E-1010, HITACHI, Tokyo, Japan). Then, PC-3 cells were spotted onto the ITO glass slide, and an optical bright field image and fluorescence image were acquired after the spot was dried under vacuum. Before matrix application, Gold was again coated at a thickness of 10 nm. Matrix was applied using an airbrush (PS-153: 3500 PRO SPRY Mk-2, Mr. hobby, GSI Creos Co., Tokyo, Japan). Because it was possible that the metal nozzle would be degraded by the acidity of the matrix, reducing the intensities of the signals of interest, the nozzle was replaced with a microchip (DIAMOND TIP 0.1–10 μL, F171100, Gilson, Middleton, WI, USA) and a glass tube (Calibrated Pipets, 2-000-001-90, Drummond Scientific Co., Broomall, PA, USA).

3.5. Drug Administrations into Cells

Docetaxel and PpIX were dissolved in dimethyl sulfoxide (DMSO, D4540, SIGMA-Aldrich, Tokyo, Japan). The solution of docetaxel or PpIX was diluted 100-fold in medium containing 10% fetal bovine serum or not, respectively. PC-3 cells were incubated for 3 h in the medium containing each drug. After each drug administration, the cells were washed once with phosphate-buffered saline (PBS, D8537, SIGMA-Aldrich, Tokyo, Japan), trypsinized, resuspended in PBS, and centrifuged. After removal of the supernatant, a suspension of PC-3 cells at 1.0×10^7 cells/mL was prepared by addition of distilled water.

3.6. Microscopy

The optical bright field and fluorescence images of PpIX administered to PC-3 cells were acquired using an inverted fluorescence microscope (LEITZ DMIRB, Leica Microscope and System GmbH, Wetzlar, Germany) equipped with a mercury lamp (02651000, IREM, Borgone Susa, Italy) as the excitation light source. Exposure time was set at 20 s.

4. Conclusions

In order to evaluate the efficacy of combined therapy, it is important to determine the distributions of the anticancer drug and photosensitizer. The purpose of this research was to enhance the sensitivities of detection of anticancer drug and photosensitizer administered to PC-3 human prostate cancer cells by improving the ionization assisting reagents used in the process. When we used a matrix consisting of equal weights of zeolite (NaY5.6) and the conventional organic matrix 6-aza-2-thiothymine (ATT) in matrix-assisted laser desorption/ionization, the signal intensity of the sodium-adducted ion of docetaxel (administered at 100 μM) increased about 13-fold. Moreover, docetaxel administered to PC-3 cells could be detected with zeolite matrix using the droplet method. Although the ion image of docetaxel and PpIX could not be acquired simultaneously, PpIX administered to PC-3 cells could be detected in a fluorescence image, after which we could successfully detect PpIX in the same samples by IMS with CHCA using the spray method. These results demonstrate that it is possible to detect docetaxel in IMS using a zeolite matrix, following the detection of PpIX by fluorescence imaging.

Acknowledgments

The authors thank Hiroshi Hatano, Hiroyuki Nakamura, Michisato Toyoda, and Jun Aoki for their technical support and advice. This work was supported by a Grant-in-Aid for Scientific Research on Innovative Areas from the Ministry of Education, Culture, Sports, Science and Technology of Japan (Grant Number 25109009).

Author Contributions

Hiroki Kannen and Hisanao Hazama carried out the analysis and interpretation of data, the collection and assembly of data and drafted the manuscript. Yasufumi Kaneda and Tatsuya Fujino designed the conception of this study from medical and chemical sides, respectively. Kunio Awazu supervised this work. All authors read and approved the final manuscript.

References

1. Kaufmann, S.; Vaux, D. Alterations in the apoptotic machinery and their potential role in anticancer drug resistance. *Oncogene* **2003**, *23*, 7414–7430.
2. Patterson, S.G.; WEI, S.; Chen, X.; Sallman, D.A.; Gilvary, D.L.; Zhong, B.; Pow-Sang, J.; Yeatman, T.; Djeu, J.Y. Novel role of Stat1 in development of docetaxel resistance in prostate tumor cells. *Oncogene* **2006**, *25*, 6112–6122.
3. The Interview Form of Taxotere® (Docetaxel for Injec-tion, Sanofi-Aventis) http://www.sec.gov/Archives/edgar/data/1121404/000119312511050947/d20f.htm (accessed on 16 June 2014)
4. Thompson, M.S.; Jonansson, T.; Andersson-Engels, S.; Svanberg, S.; Bendsoe, N.; Svanberg, K. Clinical system for interstitial photodynamic therapy with combined on-line dosimetry measurements. *Appl. Opt.* **2005**, *44*, 4023–4031.
5. Yamauchi, M.; Honda, N.; Hazama, H.; Tachikawa, S.; Nakamura, H.; Kaneda, Y.; Awazu, K. A novel photodynamic therapy for drug-resistant prostate cancer cells using porphyrus envelope as a novel photosensitizer. *Photodiagn. Photodyn. Ther.* **2014**, *11*, 48–54.
6. Zuluaga, M.F. Combination of photodynamic therapy with anti-cancer agents. *Curr. Med. Chem.* **2008**, *15*, 1655–1673.
7. Sinha, A.; Anand, S.; Ortel, B.; Chang, Y.; Mai, Z.; Hasan, T.; Maytin, E. Methotrexate used in combination with aminolaevulinic acid for photodynamic killing of prostate cancer cells. *Br. J. Cancer* **2006**, *95*, 485–495.
8. Hsieh, Y.; Casale, R.; Fukuda, E.; Chen, J.; Knemeyer, I.; Wingate, J.; Morrison, R.; Korfmacher, W. Matrix-assisted laser desorption/ionization imaging mass spectrometry for direct measurement of clozapine in rat brain tissue. *Rapid Commun. Mass Spectrom.* **2006**, *20*, 965–972.
9. Russell, D.H.; Edmondson, R.D. High-resolution mass spectrometry and accurate mass measurements with emphasis on the characterization of peptides and proteins by matrix-assisted laser desorption/ionization time-of-flight mass spectrometry. *J. Mass Spectrom.* **1997**, *32*, 263–276.
10. Gusev, A.I.; Wilkinson, W.R.; Proctor, A.; Hercules, D.M. Improvement of signal reproducibility and matrix/comatrix effects in MALDI analysis. *Anal. Chem.* **1995**, *67*, 1034–1041.
11. Billeci, T.M.; Stults, J.T. Tryptic mapping of recombinant proteins by matrix-assisted laser desorption/ionization mass spectrometry. *Anal. Chem.* **1993**, *65*, 1709–1716.
12. Cohen, S.L.; Chait, B.T. Influence of matrix solution conditions on the MALDI-MS analysis of peptides and proteins. *Anal. Chem.* **1996**, *68*, 31–37.

13. Duncan, M.W.; Matanovic, G.; Cerpa-Poljak, A. Quantitative analysis of low molecular weight compounds of biological interest by matrix-assisted laser desorption ionization. *Rapid Commun. Mass Spectrom.* **1993**, *7*, 1090–1094.
14. Komori, Y.; Shima, H.; Fujino, T.; Kondo, J.N.; Hashimoto, K.; Korenaga, T. Pronounced selectivity in matrix-assisted laser desorption-ionization mass spectrometry with 2,4,6-trihydroxyacetophenone on a zeolite surface: Intensity enhancement of protonated peptides and suppression of matrix-related ions. *J. Phys. Chem.* **2010**, *114*, 1593–1600.
15. Hazama, H.; Aoki, J.; Nagao, H.; Suzuki, R.; Tashima, T.; Fujii, K.; Masuda, K.; Awazu, K.; Toyoda, M.; Naito, Y. Construction of a novel stigmatic MALDI imaging mass spectrometer. *Appl. Surf. Sci.* **2008**, *255*, 1257–1263.
16. Aoki, J.; Ikeda, S.; Toyoda, M. Observation of accumulated metal cation distribution in fish by novel stigmatic imaging time-of-flight mass spectrometer. *J. Phys. Soc. Jpn.* **2014**, *83*, doi:10.7566/JPSJ.83.023001.

7

Phosphorylation Stoichiometries of Human Eukaryotic Initiation Factors

Armann Andaya, Nancy Villa, Weitao Jia, Christopher S. Fraser and Julie A. Leary *

Department of Molecular and Cellular Biology, University of California at Davis, Davis, CA 95616, USA; E-Mails: aaandaya@ucdavis.edu (A.A.); nvilla@ucdavis.edu (N.V.); wtjia@ucdavis.edu (W.J.); csfraser@ucdavis.edu (C.S.F.)

* Author to whom correspondence should be addressed; E-Mail: jaleary@ucdavis.edu

---

Abstract: Eukaryotic translation initiation factors are the principal molecular effectors regulating the process converting nucleic acid to functional protein. Commonly referred to as eIFs (eukaryotic initiation factors), this suite of proteins is comprised of at least 25 individual subunits that function in a coordinated, regulated, manner during mRNA translation. Multiple facets of eIF regulation have yet to be elucidated; however, many of the necessary protein factors are phosphorylated. Herein, we have isolated, identified and quantified phosphosites from eIF2, eIF3, and eIF4G generated from log phase grown HeLa cell lysates. Our investigation is the first study to globally quantify eIF phosphosites and illustrates differences in abundance of phosphorylation between the residues of each factor. Thus, identification of those phosphosites that exhibit either high or low levels of phosphorylation under log phase growing conditions may aid researchers to concentrate their investigative efforts to specific phosphosites that potentially harbor important regulatory mechanisms germane to mRNA translation.

Keywords: mass spectrometry; eukaryotic initiation factor; translation; phosphorylation quantification

---

1. Introduction

The process of converting nucleic acid to functional protein, a process involving the precise spatial and temporal arrangement of proteins, protein complexes, and nucleic acids, is known as translation [1–3]. Initiation, elongation, termination, and recycling constitute the four chronological phases of mRNA translation with initiation bearing the greatest regulation. The principal molecules that regulate initiation for eukaryotes are the eukaryotic initiation factors (eIFs), a family which is comprised of 12 members encompassing at least 25 individual proteins [1]. One of the most common means of regulating proteins involves phosphorylation, a well-established post-translational modification. Herein, we report our global investigation of phosphorylation quantification for three essential eukaryotic initiation factors, eIF2, eIF3, and eIF4G.

At the onset of eukaryotic translation initiation, the 40S ribosome binds several factors including eIF2 and eIF3 to form the 43S preinitiation complex (PIC). eIF2 (a heterotrimer consisting of the proteins eIF2α, eIF2β, and eIF2γ) binds as a ternary complex (TC) comprised of eIF2, the initiator methionyl tRNA (Met-tRNA$_i$), and GTP (guanosine triphosphate). The PIC can then bind to nascent messenger RNA (mRNA) through the direct interaction of eIF3 and eIF4G, a member of the eIF4F cap-binding complex. The largest of the eIFs, eIF3 is comprised of thirteen distinct subunits with a holoprotein mass of approximately 800 kDa. Numerous studies implicate the holoprotein eIF3 and its many subunits as an essential factor during translation initiation: It is the central assembly on which other factors, proteins, and nucleic acids bind [4–10]. The factor eIF4F is also an essential protein to translation initiation. It is comprised of three other eIFs: eIF4A, eIF4E, and the largest of the three, eIF4G with a mass of 176 kDa. eIF4G functions as a scaffolding protein onto which other factors, such as eIF3, eIF4A, and eIF4E functionally interact [11–13].

Although each subunit has a defined function during translation initiation, the regulatory mechanisms governing initiation have yet to be entirely solved. Protein phosphorylation, albeit a well-established post-translational modification, is still not completely understood, particularly in eIFs, as regards target specificity. However, the resulting effects once a protein is phosphorylated can be dramatic as that of ser-51 phosphorylation on eIF2α, which has been extensively studied and serves as a prime example of the effect phosphorylation has on the initiation process.

During initiation, eIF2 TC hydrolyzes GTP then releases Met-tRNA$_i$ following start codon recognition. eIF2B, a GTP exchange factor (GEF) for eIF2, exchanges GDP (guanosine diphosphate) for GTP, which then promotes binding of a new Met-tRNA$_i$ for further rounds of translation. Phosphorylation of eIF2α at ser-51 reduces the dissociation rate of eIF2 from eIF2B, effectively sequestering eIF2B and preventing eIF2 TC regeneration and thus globally repressing the rate of mRNA translation [14–18]. The largest protein of the three subunits, eIF2γ, functions as a scaffolding protein on which the remaining subunits bind. We recently identified eight novel sites of phosphorylation on eIF2γ and demonstrated the potential *in vitro* effects of eIF2γ phosphorylation via protein kinase C (PKC) [19].

In addition to the identification of novel phosphosites, determining those levels of phosphorylation on specific residues allows researchers to define potentially important phosphosites, thereby distinguishing those sites from potentially less biologically meaningful ones [20–26]. As with eIF2α, phosphorylation at ser-51 becomes more pronounced under conditions of cellular stress which demonstrates how the fluidity of the phosphorylation stoichiometry reveals intrinsic mechanisms implicated in regulating phosphorylation in response to cellular cues [14–18]. Establishment of phosphosite stoichiometry under specific biological conditions may help to focus ensuing investigations for deciphering biologically important phosphosites.

While identification of novel phosphosites is the essential first step in the eventual evaluation of their biological impact, follow-up studies, even on a few novel sites of phosphorylation, often presents a task too burdensome for subsequent in-depth investigations. Researchers identify phosphosites for further study through evaluation of their structural location. Phosphosites residing in structurally salient locations such as established binding sites, binding pockets, or areas of significant secondary, tertiary, and/or quaternary structure, are considered desirable candidates for further evaluation. However, without such informative data, further investigations into newly discovered phosphosites are often avoided.

In order to assess the amount of phosphorylation occurring at specific residues, we recently published a mass spectrometry method proficient at measuring phosphosite stoichiometry [27]. Our method relies on the measurement of dephosphorylation of phosphopeptides accomplished chemically via cerium oxide [28]. We have further optimized cerium oxide's capacity for dephosphorylating peptides and have subsequently developed a method to measure the amount of dephosphorylation via tandem mass tags (TMT) [25]. Use of the isobaric TMT reporter ions allows for assessment of phosphorylation stoichiometry. We have verified the use of this quantification protocol and have shown its efficacy in measuring the stoichiometry of previously established phosphosites. One of the subunits of eIF3, eIF3h, has one previously identified phosphorylation site, ser-183 [29], and we have now determined its level of phosphorylation to be 70% in log phase grown HeLa cells (data provided and discussed in the results section). This observation coincides with the critical *in vivo* effects of eIF3h's ser-183 phosphorylation during malignant transformation of NIH 3T3 cells [27,30].

This report is the first to highlight the phosphorylation levels of three heavily phosphorylated eukaryotic initiation factors. Additionally, our quantification analysis is specific to the analysis of cells grown at optimal conditions and thus underscores the importance of variability possible with these phosphosites under different environmental circumstances. Knowledge gained from this study provides a platform for future investigations not only for phosphorylated proteins, but also for the inherent variability of specific phosphosites within the protein itself. Hence, the purpose of this study is to add to the growing pool of knowledge of these factors and more importantly, to initiate an investigation primarily aimed at quantified phosphorylated residues and their implication on translation initiation. Future studies into the regulation of these factors may be based on the findings within this study.

2. Results and Discussion

2.1. Quantification of Phosphorylation for Eukaryotic Initiation Factor 2 (eIF2)

We isolated eIF2 from HeLa cell lysate in order to quantify its level of phosphorylation. The factor was purified from HeLa cells under optimal growth conditions (log phase growth). As eIF2 is a heterotrimer with a molecular mass of 126 kDa, we analyzed the two factors that have been previously reported as phosphorylated within HeLa cells, eIF2β and eIF2γ. While eIF2α has been heavily studied at its principally phosphorylated residue of ser-51, quantification of that residue does not lead to new information or developments, which is the principal aim of this study. Thus, we focused our efforts at quantifying the numerous sites on the remaining two subunits of eIF2 in order to gain further insight into their function and regulation.

Recently, we published the identification of eight sites of phosphorylation, seven of which were novel and reside on the largest subunit, eIF2γ, the core subunit of the heterotrimer [19]. All identified phosphosites were quantified. On eIF2γ, numerous functional domains exist dedicated to specific tasks during eukaryotic translation. Threonine 66, a phosphorylated residue, is located within the switch-1 region of the protein, and in this study, we observed its level of phosphorylation at 71% (Figure 1). Two sites adjacent to one another, ser-55 and thr-56, reside directly in the nucleotide binding pocket of eIF2γ and our quantitative phosphoanalysis revealed the levels of phosphorylation to be 85% for both ser-55 and thr-56 combined (Supplementary Information). Proximal to the *C*-terminal end of the protein, ser-412 and thr-413 exhibit phosphorylation at 7% while ser-418 and thr-435 have phosphorylation at 70% and 60%, respectively (Supplementary Information). Lastly, thr-109's phosphorylation level was observed at 30% and this residue sits adjacent to the zinc binding domain (Supplementary Information).

Figure 1. Quantification of phosphorylation on Thr-66 on eIF2γ. (**A**) Precursor ion mass scan of the $[M + 2H]^{2+}$ ion is shown; (**B**) MS/MS spectra of *m*/*z* ion 582.9 illustrating indicative b- and y-ions for peptide spanning residues 60–68 of eIF2γ; and (**C**) Zoomed in view of the *m*/*z* region containing the TMT (tandem mass tags) reporter ions. Calculation of the reporter ion ratio reveals a phosphorylation level of 70.5% for Thr-66.

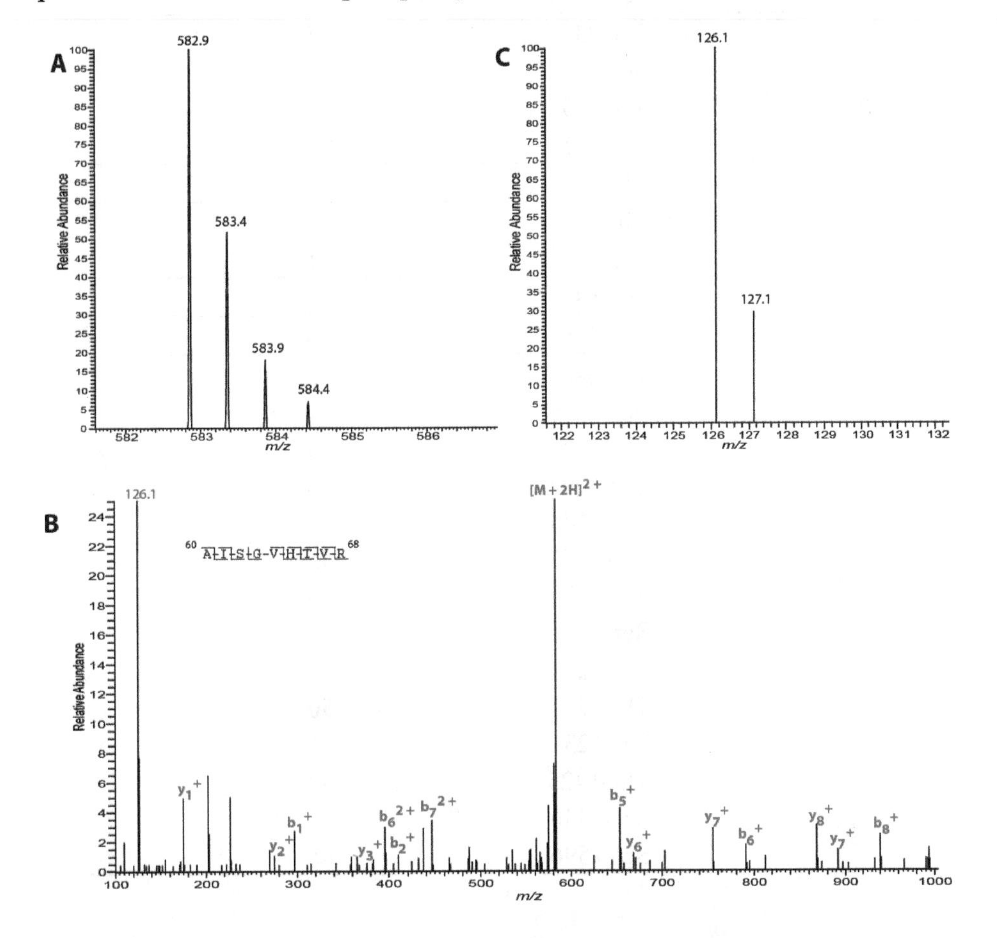

In contrast to eIF2γ, eIF2β has a lower abundance of phosphorylated residues within HeLa cells. Nevertheless, we quantified these sites and observed levels of 18% for thr-31 but near or less than 5% for the other reported eIF2β phosphosites (Table 1).

Table 1. Three of the factors, eIF2, eIF3, and eIF4G were analyzed as to their quantification of phosphorylation levels within log phase grown HeLa cells. Percentages of the phosphosites are shown.

Protein	Subunit	Residue Phosphorylated	% Phosphorylation
eIF2	β	Thr-31	18
		Ser-67	<5
		Ser-105	<5
		Thr-111	<5
		Ser-158	8
	γ	Ser-55	85
		Thr-56	85
		Thr-66	71
		Thr-109	30
		Ser-412	7
		Thr-413	7
		Ser-418	70
		Thr-435	66
eIF3	a	Ser-881	84
		Ser-1198	92
		Ser-1336	18
		Ser-1364	36
	b	Ser-83	85
		Ser-85	85
		Ser-119	70
		Ser-125	70
	c	Thr-524	95
	g	Thr-41	31
		Ser-42	31
	h	Ser-183	89
	j	Thr-109	88
eIF4G	-	Thr-647	65
		Ser-1028	5
		Ser-1077	15
		Ser-1092	40
		Ser-1144	45
		Ser-1147	45
		Ser-1185	70
		Ser-1187	70
		Ser-1209	50
		Thr-1211	50
		Ser-1231	75
		Thr-1425	62
		Ser-1430	62
		Ser-1596	<5

2.2. Quantification of Phosphorylation for Eukaryotic Initiation Factor 3 (eIF3)

Although the exact structure and placement of each subunit within the tridodecamer of eIF3 has yet to be fully solved, the number of phosphorylations derived from log phase HeLa cells is 29 [29]. We thus sought to quantify the levels of each of these 29 sites of phosphorylation. Given that each subunit is a distinct protein, the varying levels of phosphorylation from subunit to subunit or within each subunit should not be surprising.

The core subunits of eIF3 are widely believed to be eIF3a, eIF3b, and eIF3c, the three largest of the thirteen subunits. These three subunits house 21 of the 29 sites of phosphorylation for eIF3, 15 of which were accurately quantified. The largest subunit, eIF3a, has a varied landscape of phosphorylation ranging from 18% for ser-1336 to a high of 92% for ser-1198 (Figure 2). Intriguing are the two phosphorylated residues at the *C*-terminal end (or *C*-Terminus) of the protein, ser-1336 and ser-1364. While ser-1336 has a phosphorylation of 18%, ser-1364 has a level of 36% (Supplementary Information). The exact crystal structure of human eIF3a has yet to be solved, but from this study, we see that the *C*-terminal phosphorylated residues are lower in phosphorylation levels than that for ser-1198 at 92% and for that of ser-881 at 84%. This rather high level of phosphorylation for these residues may implicate functional significance for eIF3 quaternary structure given that the *C*-terminal phosphorylated residues possess a lower level of phosphorylation.

The second largest eIF3 subunit, eIF3b, has 7 sites of phosphorylation. Unfortunately, we did not attain full sequence coverage for all proteins analyzed which lead to a lack of quantification for some previously identified phosphosites. This was due in part to either the absence of a lysine or arginine within the vicinity of the phosphosite in question or an inadequacy of sufficient identifiable b- and y-ions. However, as with eIF3a's ser-1198 and ser-881, the levels of phosphorylation for eIF3b were relatively large ranging from 70% for both ser-119 and ser-125, and 85% for both ser-83 and ser-85 (Supplementary Information). Again, implications in eIF3 quaternary structure for these high levels of phosphorylation will be investigated further.

The third largest purported eIF3 core subunit is eIF3c. Again, due to the same limitations as previously mentioned, the *N*-terminal phosphorylated residues of ser-9, ser-11, ser-13, ser-15, ser-16, ser-18, and ser-39 could not be quantified. But as with subunits eIF3a and eIF3b, higher levels of phosphorylation were observed for thr-524 at 95% (Supplementary Information).

Regarding the subunit eIF3h, we have already published a report detailing its relatively high level of phosphorylation from log phase HeLa cells at 70% and within the current study, 89% [27]. This high level of phosphorylation is in direct correlation with biological data showing that mutation of this residue to alanine produced a decrease in growth rate of NIH-3T3 cells [30]. Residues not quantified for eIF3 due to previously mentioned experimental limitations were eIF3f (ser-258) and the *N*-terminal residues of eIF3j (ser-11, ser-13, and ser-20). A level of 31% phosphorylation was observed for eIF3g's two phosphorylated residues, thr-41 and ser-42, and a high level of phosphorylation comparable to that of eIF3h was observed for thr-109 on eIF3j (Table 1 and Supplementary Information).

Figure 2. Quantification of phosphorylation on eIF3. (**A**) MS/MS spectra of m/z ion 406.6, $z = 3$. Peptide spans residues 1329–1337 for eIF3a; (**B**) Zoomed in view of the m/z region containing the TMT reporter ions. Calculation of the reporter ion ratio reveals a phosphorylation level of 18% for Ser-1336; (**C**) MS/MS spectra of m/z ion 405.5, $z = 3$. Peptide spans residues 1194–1201 for eIF3a; and (**D**) Zoomed in view of the m/z region containing the TMT reporter ions. Calculation of the reporter ion ratio reveals a phosphorylation level of 92% for Ser-1198.

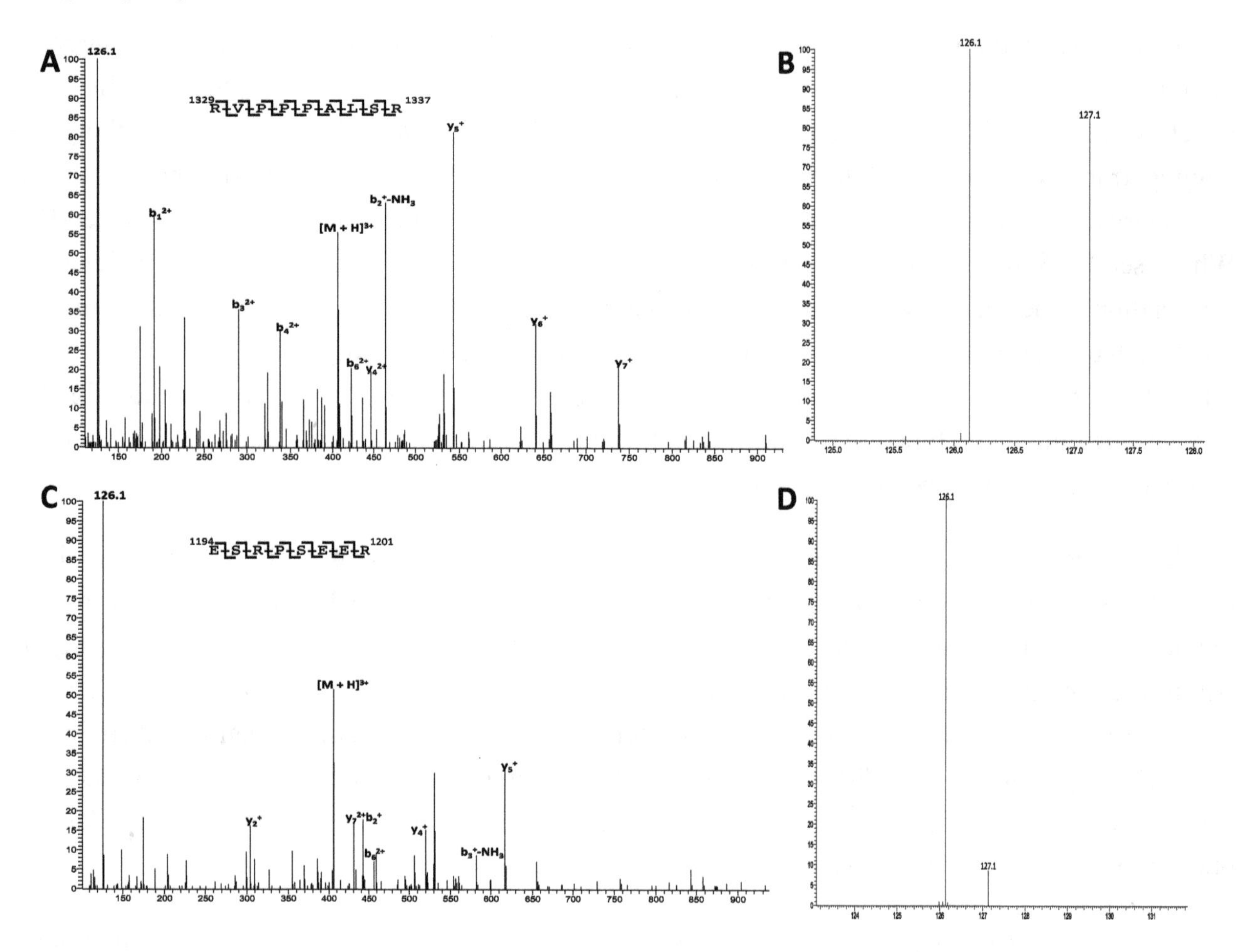

2.3. *Quantification of Phosphorylation for Eukaryotic Initiation Factor 4G (eIF4G)*

A large scaffolding protein, eIF4G commonly occurs with varying isoforms that are dependent on the location of the start codon. Nonetheless, the principal isoform of the protein is comprised of 1599 amino acids with a molecular mass of 176 kDa. Prior to translation initiation, eIF4G serves as a docking site to which other proteins bind prior to forming larger complexes necessary for the translation process. Known proteins that bind to eIF4G are poly A binding protein (PABP), the kinase Mnk1, and initiation factors eIF4E, eIF4A, and eIF3. The phosphorylation of eIF4G from HeLa cells has been well documented with evidence of up to 20 sites of phosphorylation corresponding to eIF4G derived from HeLa cells have been identified. However, as with all other eIFs, the quantification of this phosphorylation has yet to be reported.

Currently, considerable investigation has focused on the portion of eIF4G that has been mapped to bind to eIF3 [31]. This portion, corresponding to residues 1015 to 1105, contains three sites of phosphorylation identified from HeLa cells, ser-1028, ser-1077, and ser-1092. Quantification of each site of phosphorylation reveals 5%, 15%, and 40% for ser-1028, ser-1077, and ser-1092, respectively (Figure 3). The only other site of phosphorylation that maps to a known binding region on eIF4G is located at the *C*-terminus of the protein and is ser-1596 that corresponds to the Mnk1 binding region. Quantification of its phosphorylation reveals less than 5% levels of phosphorylation.

As for the remaining known sites of phosphorylation for eIF4G from HeLa cells, insufficiently scored peptides prevented our quantitative analysis of phosphosites residing in the *N*-terminal portion of the protein. However, high levels of phosphorylation were observed for thr-647 at 65%. Regions near but distinct from the eIF3 binding site had phosphorylation levels at or near 50%. The peptide mapped to ser-1144 and ser-1147 had a combined level of 45% and the peptide mapped to ser-1209 and ser-1211 was at 50%. The peptide mapping to ser-1185 and ser-1187 had a level of 70% and the peptide mapping to ser-1231 had a level of 75%. Peptides mapping in close proximity to the Mnk1 binding region, thr-1425 and ser-1430 were quantified at phosphorylation levels of 62%.

Figure 3. Quantification of phosphorylation on eIF4G. (**A**) MS/MS spectra of *m*/*z* ion 992.6, $z = 2$. Peptide spans residues 1072–1085; (**B**) Zoomed in view of the *m*/*z* region containing the TMT reporter ions. Calculation of the reporter ion ratio reveals a phosphorylation level of 15% for Ser-1077; (**C**) MS/MS spectra of *m*/*z* ion 520.8, $z = 2$. Peptide spans residues 1091–1095; and (**D**) Zoomed in view of the *m*/*z* region containing the TMT reporter ions. Calculation of the reporter ion ratio reveals a phosphorylation level of 40% for Ser-1092.

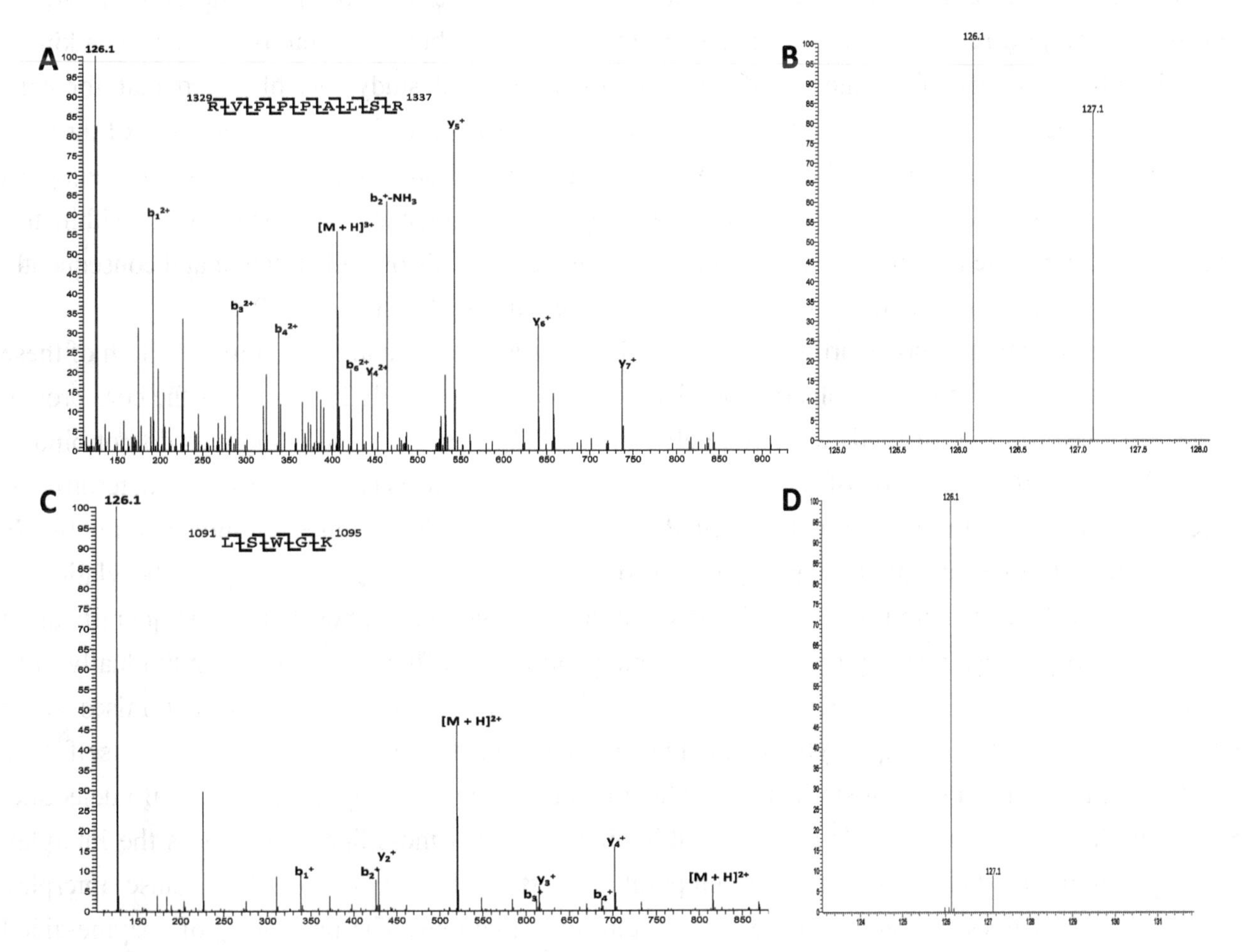

The entire process of translation initiation requires a concerted and choreographed effort from numerous proteins to effectively translate the nascent mRNA into functional protein. Although the process of translation may be separated into the four distinct phases of initiation, elongation, termination, and recycling, much of the regulation is centralized to the initiation phase. Translation initiation is the process by which the 40S ribosome, bound to Met-tRNA$_i$ and several initiation factors, binds to the 5' end of mRNA then scans in the 5' to 3' direction until the initiation codon is recognized. Recruitment of Met-tRNA$_i$ requires formation of a ternary complex also containing eIF2 and GTP, while mRNA recruitment to the ribosome complex requires binding of eIF4G and eIF3.

As the formation of the ternary complex is critical to translation, a better understanding of the major protein component, eIF2, would help define this process. A closer look into the function of eIF2 reveals that the heterotrimer consists of three subunits, eIF2α and eIF2β bound to eIF2γ but not each other. Classically, eIF2α's phosphorylation on ser-51 converts the eIF2 heterotrimer into an inhibitor of eIF2B, the heteropentamer responsible for exchanging GDP for GTP on eIF2. Since the phosphorylation of this one residue on eIF2α has a profound impact on the overall process of translation, we hypothesized that the other subunits of eIF2 may also contribute in some part to the overall process of translation.

Our phosphoanalysis of eIF2 centered on known phosphosites for both eIF2β and eIF2γ derived from HeLa cells. In our previous report for eIF2γ, one of the novel phosphosites identified was thr-66 [19]. Further investigations of this site were conducted based on previous structural studies placing it within eIF2γ's switch 1 region [32,33]. As eIF2 forms a ternary complex with met-tRNA$_i$ and GTP, the switch 1 region of the molecule undergoes a conformational change in order to allow for complexation [34,35]. Our previous report revealed not only phosphorylation of thr-66, but also established PKC as a kinase with the ability to phosphorylate thr-66 *in vitro*. In the current study, we observed that for cells propagated under log-phase growth conditions, thr-66 on eIF2γ is 71%. Given that a majority of thr-66 is phosphorylated for a cancer cell line under optimal conditions, future discernment of the physiological role of this phosphosite should be of paramount importance. Logically, phosphorylation within this region of the molecule may have a significant impact on the protein's overall structure and consequently its influence on the conformational change experienced upon GTP binding to eIF2γ.

Two other structurally noteworthy phosphorylation sites are ser-55 and thr-56 [36,37]. Both of these sites reside directly within the nucleotide binding pocket of eIF2γ. As eIF2γ represents the core protein of the eIF2 heterotrimer, it binds directly to eIF2α and eIF2β, yet the α and β subunits do not bind to each other [38,39]. Two-thirds of the ternary complex forms upon GTP binding to the heterotrimer, specifically, to eIF2γ. In its inactive form, GDP is bound along with a magnesium molecule in eIF2γ's binding pocket. Our quantitative phosphoanalysis revealed that the level of phosphorylation is approximately 85% for both ser-55 and thr-56 combined. As with most phosphorylation quantification protocols, phosphosites residing on the same tryptic peptide are difficult to quantify individually [40]. Nonetheless, a level of 85% phosphorylation within the nucleotide binding pocket raises some intriguing possibilities. Phosphorylation in this binding pocket may potentially present itself as a regulatory mechanism eliciting steric strain, which could prevent binding of an additional nucleotide. Such a high level of phosphorylation within this binding pocket most likely influences the interplay of magnesium and GTP binding. The temporal aspects of kinase and/or phosphatase interplay with GTP and magnesium binding necessitate future investigations. The possibility of enzyme-aided phosphorylation diminishes with such a small binding pocket thus raising questions of this event possibly occurring prior to protein folding. Clearly, future studies are needed to determine the effects of such high levels of phosphorylation within an important region of the protein.

In contrast to those phosphosites near the binding pocket and within the switch 1 region, phosphosites associated with the *C*-terminal portion of the protein, ser-412 and thr-413 have levels at 7% while ser-418 and thr-435 are at or near 70%. Thus, from the current study, we can see the dynamics of the phosphorylation landscape unfold. Residues 412 and 413 reside on the outer face of the molecule and despite the fact that they may be more readily available to kinases as compared to residues 55, 56, and 66, (which lie on the interior of the molecule), their levels of phosphorylation are significantly smaller. Future investigations will determine the significance of these phosphorylation stoichiometries, but for now, we can speculate that phosphorylation does not play a major role in the binding to other nucleotides and/or proteins. The same appears not to be true for either ser-418 or thr-435. Again, future studies will determine these residues' phosphorylation significance, but undoubtedly, different physiological circumstances and pressures exist on the phosphorylated residues of the *C*-terminal of eIF2γ. Such a variation of phosphorylation may imply a structural necessity adopted by the *C*-terminus during the binding processes. It would be interesting to observe phosphorylation differences not only under different growth conditions, but also when one of the four *C*-terminal sites is mutated and how this effect translates to the overall phosphorylation landscape across the *C*-terminal region.

Lastly, thr-109's phosphorylation level at 30% seems to neither weaken nor strengthen the importance of this phosphosite to the overall physiology of eIF2. However, as stated in our previous report, this residue does reside near a zinc binding domain. The electrostatics involved with the zinc cation and the negative charge inherent in a phosphate moiety could be responsible for an interaction that may determine the proper positioning of the zinc ion. Proteins containing zinc binding motifs have been well documented in their binding affinity for nucleotides. As eIF2 binds to met-tRNA$_i$ and is part of the ternary complex that shuttles this polynucleotide to the awaiting mRNA during protein translation initiation, the interplay of the phosphate on thr-109 and the zinc ion may have a meaningful impact on protein translation.

Considerable more research is required to decipher the significance of eIF2β phosphosite structural regions. Previous targeted investigations into eIF2β have established an interaction of the protein with the kinase CK2 and suggest numerous physiological pressures affecting the dynamics of eIF2β phosphorylation warranting further investigation [41]. Nonetheless, our report represents the first measurements of the levels of quantification of these phosphosites. Aside from ser-51 on eIF2α, the phosphosites of eIF2β and eIF2γ are not only dynamic, but may house important functions necessary for translation initiation [42].

Along with eIF2, eIF3 is a necessary initiation factor as it binds to many of the numerous eIFs including eIF4A, eIF1, eIF1A, eIF2, and eIF4G. A critical function of eIF3 is prevention of premature association to the 60S ribosome by binding to the 40S ribosome along with other initiation factors. Although its function and structure have been previously investigated, to date eIF3 has neither a definitive crystal structure nor have any of its phosphorylation sites been thoroughly investigated. Recent studies have shown the structure of the tridodecamer as a 5 lobed entity comprised of a core with five protrusions [43]. The exact structure of each of the lobes has yet to be fully determined, but mass spectrometry along with reconstitution of a recombinant tridodecamer has indicated the core subunits, eIF3a, eIF3b, and eIF3c, as being the three largest [10,44,45]. Interestingly, these same three subunits contain 21 of the 29 possible phosphorylation sites for eIF3. All three of the purported core subunits have high levels of phosphorylation. As the quaternary structure of eIF3 may rely heavily on the integrity of its core proteins, a high level of phosphorylation, especially from log phase grown cells, suggests an increasing role of electrostatic interactions that possibly strengthen the integrity of the entire complex. Levels of phosphorylation were not as elevated for the other phosphorylated subunits except for that of eIF3j. However, eIF3j has been known to disassociate from the other 12 subunits of eIF3 thus a high level of phosphorylation for thr-109 may be indicative of an increased affinity to the holoprotein during translation [7,8]. Future investigations will help determine the nature of each of these phosphosites.

The final protein whose phosphorylation was quantified in this study was eIF4G. A classic example of a scaffolding protein, much like the core proteins of eIF3, phosphorylation was present at its highest level for residues residing in known areas of protein binding. Three phosphorylated residues lie within the eIF3 binding region of eIF4G and all three exhibited high levels of phosphorylation. Other residues that also showed high levels of phosphorylation are residues between the eIF3 and eIF4A binding regions. Although future investigations will decipher the role of these specific phosphosites, phosphosite ser-1187 is known in HEK293 cells to function as a substrate for the kinase PKCα and modulates Mnk1 binding [30]. Again, phosphorylated residues implicated in binding appear to have the highest levels of phosphorylation.

3. Experimental Section

3.1. Purification of eIFs from HeLa Cell Lysate

All proteins for this study, eIF2, eIF3, and eIF4G were enriched from HeLa cell lysate prior to nano-LC-MS/MS according to previously published protocols [19,29]. Briefly, HeLa cell lysate (from approximately 30 L cells) was quickly thawed at 37 °C supplemented in a mixture including 10% glycerin, 1 mM EDTA, 1 mM EGTA, 50 mM NaF, 50 mM beta-glycerol phosphate, 10 mM benzamidine, 1 mM DTT, and 1× protease inhibitor mixture (Roche, Basel, Switzerland). After stirring for 10 min at 4 °C, the mixture was centrifuged at 20,000× *g* for 20 min at 4 °C. To the resulting supernatant, KCl was added to a final concentration of 450 mM followed by centrifugation in a Beckman Ti-45 rotor for 4 h at 4 °C at 45,000 rpm. The middle two-thirds of the supernatant was carefully removed and stirred at 4 °C while saturated ammonium sulfate was added to a final concentration of 40%. After stirring on ice for 1 h, the suspension was centrifuged at 20,000× *g* for 10 min at 4 °C and the pellet (referred to as the A cut) was frozen for future use. To the remaining supernatant, ammonium sulfate was added to a final concentration of 70%. After stirring for 1 h, the mixture was centrifuged at 20,000× *g* for 10 min at 4 °C. The pellet (the B cut) was resuspended in 50 mL of buffer A (20 mM HEPES, pH 7.5, 10% glycerol, 1 mM EDTA, 1 mM EGTA, 50 mM NaF, 50 mM β-glycerol phosphate, 10 mM benzamidine, 1 mM DTT) containing 50 mM KCl and dialyzed in two liters of the same buffer for 2.5 h at 4 °C. Following dialysis, the lysate was passed through a 0.2 μm syringe filter and loaded onto a MonoQ (10/10) column (GE Healthcare, Sunnyvale, CA, USA).

The column was eluted with a linear gradient of 100 to 500 mM KCl in buffer A at 2 mL/min with 3 mL fractions collected. The fractions that contained the protein of interest (either eIF2, eIF3, or eIF4G) were then pooled, dialyzed against buffer A containing 100 mM KCl for 2.5 h at 4 °C, and loaded onto a MonoS (10/10) column (GE Healthcare, Sunnyvale, CA, USA). The same gradient as that of the MonoQ column was applied and fractions from the MonoS column were analyzed using SDS-PAGE in a similar fashion. Fractions containing the protein of interest were pooled and loaded onto a hydroxyapatite column made in-house using commercial hydroxyapatite (Calbiochem, San Diego, CA, USA). Elution was performed using a linear gradient from 0% to 100% 0.5 M potassium phosphate buffer at pH 7.5. Fractions were again analyzed via SDS-PAGE in a similar manner to those eluting from either the MonoQ or MonoS columns. Fractions containing the now purified protein were pooled and concentrated using Amicon Ultra filtration devices with a MWCO of 10,000 Da to yield a final concentration between 1 and 2 mg/mL.

3.2. Tryptic Digestion of eIFs with Subsequent Tandem Mass Tag (TMT) Labeling

Each eIF protein was digested with trypsin following the same protocol. Approximately 1 μmol of either eIF2, eIF3, or eIF4G was first reduced at 56 °C for 45 min in 5.5 mM DTT final concentration followed by alkylation for one hour in the dark with iodoacetamide (IAA) added to a final concentration of 10 mM. Trypsin was added at a final enzyme:substrate mass ratio of 1:50 and digestion carried out overnight at 37 °C. The reaction was quenched by flash freezing in liquid nitrogen and the digest was lyophilized. Prior to TMT labeling, each lyophilized sample was dissolved into 100 μL of 2% acetonitrile supplemented with 0.1% trifluoroacetic acid.

All eIF peptides were prepared for TMT labeling following our published protocol [27]. Briefly, each eIF peptide mixture was divided equally; one population was treated with cerium oxide nanopowder (Sigma-Aldrich, St. Louis, MO, USA) to dephosphorylate the phosphopeptides, the other half was left untreated. Subsequent to dephosphorylation, the dephosphorylated samples were labeled with TMT-126. The remaining untreated half was labeled with TMT-127. Samples were then combined after labeling prior to nano-LC-MS/MS analysis.

3.3. Nano-LC-MS/MS of eIF Proteins

All samples underwent complete labeling and were combined as outlined in our protocol [19,27] prior to nano-LC-MS/MS analysis using an LTQ-Orbitrap XL (Thermo Fisher, San Jose, CA, USA) mass spectrometer equipped with an ADVANCE ion max source (Michrom Bioresources Inc., Auburn, CA, USA), a Surveyor MS pump (Thermo Fisher, San Jose, CA, USA), and a microautosampler (Thermo Fisher, San Jose, CA, USA). Samples were loaded onto the column for 30 min at 2% solvent B (0.1% (*v*/*v*) formic acid in acetonitrile) with 98% solvent A at a flow rate of 750 nL/min. Peptides were eluted off the column at 750 nL/min using the following gradient: 2%–10% solvent B for 5 min, 10%–35% solvent B for 65 min, 35%–70% solvent B for 5 min, 35%–70% solvent B for 5 min, 70%–90% solvent B for 5 min, 90% solvent B for 5 min, then reversed to 2% solvent B for 10 min. All ions with TMT labels were analyzed via higher energy collision dissociation (HCD), all others with collision-induced dissociation (CID). Settings for the LTQ-Orbitrap XL were set as follows: data-dependent scan of the top 5 most abundant ions, minimum signal threshold of 55,000, collision energy set to 35%, resolution set to 30,000, dynamic exclusion set to 60 s, repeat count set to 2, and repeat duration set to 30 s. All RAW data were deposited directly into SEQUEST Bioworks 3.3.1 (Thermo Fisher, San Jose, CA, USA) and manually validated. Searches were conducted against an in-house developed database consisting of 584 proteins containing the sequences of all known mammalian eukaryotic initiation factors as well as their reversed sequences. Common contaminants including human keratins, porcine trypsin, bovine serum albumin, bovine beta-casein, as well as their reversed sequences were also included. We performed searches with tryptic specificity and allowed for three missed cleavages at a tolerance of 20 ppm in MS mode and 0.2 Da in MS^2 mode. Possible structure modifications included for consideration were *N*-terminal and lysine TMT labeling, methionine oxidation, carbamidomethylation of cysteine, and serine, threonine, and tyrosine phosphorylation. Peptides with Xcorr scores greater than 2.5 were considered for further manual validation. All TMT ratios were measured individually and experiments were carried out in triplicate.

4. Conclusions

This study represents the first global attempt to elucidate the levels of phosphorylation for three integral protein complexes in the translation initiation pathway (Table 1). Prior studies have revealed the identification of new and novel phosphosites within proteins. In this current investigation, we have provided quantification of phosphosites that may aid in further research of these important sites of regulation. As phosphorylation is a highly dynamic modification itself involving enzymatic addition and removal, levels of phosphorylation may be indicative of a more subtle mechanism of regulation incapable of being observed by identification of novel phosphosites alone. We have concentrated on those sites within this investigation that exhibit high levels of phosphorylation; however, those that display lower levels should not be ignored as gain or absence of function at those residues may be less sensitive to the absolute level of their phosphorylation. Phosphorylation levels appear to be at their highest when the residues are within a region known to be involved in binding to other proteins, metal ions, or nucleic acids. Phosphomimetic studies will ultimately unravel the significance of these phosphosites, which may be critical to protein structure and function as well as to overall influence on the translation initiation process.

Acknowledgments

We thank J.W.B. Hershey for his insight and valuable suggestions during the study. This work was funded by the National Institutes of Health Program Project Grant GM073732.

Author Contributions

A.A.: designed the study, performed experiments, isolated and provided eIF2 and eIF3, interpreted data, and wrote the manuscript; N.V.: isolated and provided eIF4G, assisted in writing of the manuscript; W.J.: assisted in interpretation of mass spectra; C.S.F.: isolated and provided eIF4G, assisted in design of the study; J.A.L.: supervised the project, assisted in design of the study, and was involved in data interpretation and writing of the manuscript.

References

1. Hershey, J. W.; Sonenberg, N.; Mathews, M.B. Principles of translational control: An overview. *Cold Spring Harb. Perspect. Biol.* **2012**, *4*, doi:10.1101/cshperspect.a011528.
2. Hinnebusch, A.G.; Lorsch, J.R. The mechanism of eukaryotic translation initiation: New insights and challenges. *Cold Spring Harb. Perspect. Biol.* **2012**, *4*, doi:10.1101/cshperspect.a011544.
3. Fraser, C.S. The molecular basis of translational control. *Prog. Mol. Biol. Transl. Sci.* **2009**, *90*, 1–51.
4. Valasek, L.; Nielsen, K.H.; Zhang, F.; Fekete, C.A.; Hinnebusch, A.G. Interactions of eukaryotic translation initiation factor 3 (eIF3) subunit NIP1/c with eIF1 and eIF5 promote preinitiation complex assembly and regulate start codon selection. *Mol. Cell. Biol.* **2004**, *24*, 9437–9455.
5. Hinnebusch, A.G. eIF3: A versatile scaffold for translation initiation complexes. *Trends Biochem. Sci.* **2006**, *31*, 553–562.
6. Jivotovskaya, A.V.; Valasek, L.; Hinnebusch, A.G.; Nielsen, K.H. Eukaryotic translation initiation factor 3 (eIF3) and eIF2 can promote mRNA binding to 40S subunits independently of eIF4G in yeast. *Mol. Cell. Biol.* **2006**, *26*, 1355–1372.
7. Fraser, C.S.; Berry, K.E.; Hershey, J.W.; Doudna, J.A. eIF3j is located in the decoding center of the human 40S ribosomal subunit. *Mol. Cell* **2007**, *26*, 811–819.
8. Fraser, C.S.; Lee, J.Y.; Mayeur, G.L.; Bushell, M.; Doudna, J.A.; Hershey, J.W. The j-subunit of human translation initiation factor eIF3 is required for the stable binding of eIF3 and its subcomplexes to 40 S ribosomal subunits *in vitro*. *J. Biol. Chem.* **2004**, *279*, 8946–8956.
9. Chiu, W.L.; Wagner, S.; Herrmannova, A.; Burela, L.; Zhang, F.; Saini, A.K.; Valasek, L.; Hinnebusch, A.G. The *C*-terminal region of eukaryotic translation initiation factor 3a (eIF3a) promotes mRNA recruitment, scanning, and, together with eIF3j and the eIF3b RNA recognition motif, selection of AUG start codons. *Mol. Cell. Biol.* **2010**, *30*, 4415–4434.
10. Sun, C.; Todorovic, A.; Querol-Audi, J.; Bai, Y.; Villa, N.; Snyder, M.; Ashchyan, J.; Lewis, C.S.; Hartland, A.; Gradia, S.; *et al.* Functional reconstitution of human eukaryotic translation initiation factor 3 (eIF3). *Proc. Natl. Acad. Sci. USA* **2011**, *108*, 20473–20478.
11. Hilbert, M.; Kebbel, F.; Gubaev, A.; Klostermeier, D. eIF4G stimulates the activity of the DEAD box protein eIF4A by a conformational guidance mechanism. *Nucleic Acids Res.* **2011**, *39*, 2260–2270.
12. Nielsen, K.H.; Behrens, M.A.; He, Y.; Oliveira, C.L.; Jensen, L.S.; Hoffmann, S.V.; Pedersen, J.S.; Andersen, G.R. Synergistic activation of eIF4A by eIF4B and eIF4G. *Nucleic Acids Res.* **2011**, *39*, 2678–2689.
13. Ozes, A.R.; Feoktistova, K.; Avanzino, B.C.; Fraser, C.S. Duplex unwinding and ATPase activities of the DEAD-box helicase eIF4A are coupled by eIF4G and eIF4B. *J. Mol. Biol.* **2011**, *412*, 674–687.
14. Dever, T.E. Gene-specific regulation by general translation factors. *Cell* **2002**, *108*, 545–556.
15. Dever, T.E.; Dar, A.C.; Sicheri, F. 12 The eIF2α kinases. In *Translational Control in Biology and Medicine*, 3rd ed.; Mathews, M., Sonenberg, N., Hershey, J.W.B., Eds.; Cold Spring Harbor Laboratory Press: Cold Spring Harbor, NY, USA, 2007.
16. Kudlicki, W.; Wettenhall, R.E.; Kemp, B.E.; Szyszka, R.; Kramer, G.; Hardesty, B. Evidence for a second phosphorylation site on eIF-2 α from rabbit reticulocytes. *FEBS Lett.* **1987**, *215*, 16–20.
17. Proud, C.G. eIF2 and the control of cell physiology. *Semin. Cell Dev. Biol.* **2005**, *16*, 3–12.

18. Ron, D.; Harding, H.P. 13 eIF2α phosphorylation in cellular stress responses and disease. In *Translational Control in Biology and Medicine*; Mathews, M., Sonenberg, N., Hershey, J.W.B., Eds.; Cold Spring Harbor Laboratory Press: Cold Spring Harbor, NY, USA, 2007.
19. Andaya, A.; Jia, W.; Sokabe, M.; Fraser, C.S.; Hershey, J.W.; Leary, J.A. Phosphorylation of human eukaryotic initiation factor 2γ: Novel site identification and targeted PKC involvement. *J. Proteome Res.* **2011**, *10*, 4613–4623.
20. Goodlett, D.R.; Aebersold, R.; Watts, J.D. Quantitative *in vitro* kinase reaction as a guide for phosphoprotein analysis by mass spectrometry. *Rapid Commun. Mass Spectrom.* **2000**, *14*, 344–348.
21. Gygi, S.P.; Rist, B.; Gerber, S.A.; Turecek, F.; Gelb, M.H.; Aebersold, R. Quantitative analysis of complex protein mixtures using isotope-coded affinity tags. *Nat. Biotechnol.* **1999**, *17*, 994–999.
22. Mann, M. Functional and quantitative proteomics using SILAC. *Nat. Rev. Mol. Cell Biol.* **2006**, *7*, 952–958.
23. Steen, H.; Jebanathirajah, J.A.; Springer, M.; Kirschner, M.W. Stable isotope-free relative and absolute quantitation of protein phosphorylation stoichiometry by MS. *Proc. Natl. Acad. Sci. USA* **2005**, *102*, 3948–3953.
24. Stukenberg, P.T.; Lustig, K.D.; McGarry, T.J.; King, R.W.; Kuang, J.; Kirschner, M.W. Systematic identification of mitotic phosphoproteins. *Curr. Biol.* **1997**, *7*, 338–348.
25. Thompson, A.; Schafer, J.; Kuhn, K.; Kienle, S.; Schwarz, J.; Schmidt, G.; Neumann, T.; Johnstone, R.; Mohammed, A.K.; Hamon, C. Tandem mass tags: A novel quantification strategy for comparative analysis of complex protein mixtures by MS/MS. *Anal. Chem.* **2003**, *75*, 1895–1904.
26. Wright, C.A.; Howles, S.; Trudgian, D.C.; Kessler, B.M.; Reynard, J.M.; Noble, J.G.; Hamdy, F.C.; Turney, B.W. Label-free quantitative proteomics reveals differentially regulated proteins influencing urolithiasis. *Mol. Cell. Proteomics* **2011**, doi:10.1074/mcp.M110.005686.
27. Jia, W.; Andaya, A.; Leary, J.A. Novel mass spectrometric method for phosphorylation quantification using cerium oxide nanoparticles and tandem mass tags. *Anal. Chem.* **2012**, *84*, 2466–2473.
28. Tan, F.; Zhang, Y.; Wang, J.; Wei, J.; Cai, Y.; Qian, X. An efficient method for dephosphorylation of phosphopeptides by cerium oxide. *J. Mass Spectrom.* **2008**, *43*, 628–632.
29. Damoc, E.; Fraser, C.S.; Zhou, M.; Videler, H.; Mayeur, G.L.; Hershey, J.W.; Doudna, J.A.; Robinson, C.V.; Leary, J.A. Structural characterization of the human eukaryotic initiation factor 3 protein complex by mass spectrometry. *Mol. Cell Proteomics* **2007**, *6*, 1135–1146.
30. Zhang, L.; Smit-McBride, Z.; Pan, X.; Rheinhardt, J.; Hershey, J.W. An oncogenic role for the phosphorylated h-subunit of human translation initiation factor eIF3. *J. Biol. Chem.* **2008**, *283*, 24047–24060.
31. LeFebvre, A.K.; Korneeva, N.L.; Trutschl, M.; Cvek, U.; Duzan, R.D.; Bradley, C.A.; Hershey, J. W.; Rhoads, R.E. Translation initiation factor eIF4G-1 binds to eIF3 through the eIF3e subunit. *J. Biol. Chem.* **2006**, *281*, 22917–22932.
32. Roll-Mecak, A.; Alone, P.; Cao, C.; Dever, T.E.; Burley, S.K. X-ray structure of translation initiation factor eIF2γ: Implications for tRNA and eIF2alpha binding. *J. Biol. Chem.* **2004**, *279*, 10634–10642.
33. Yatime, L.; Mechulam, Y.; Blanquet, S.; Schmitt, E. Structural switch of the gamma subunit in an archaeal aIF2 α γ heterodimer. *Structure* **2006**, *14*, 119–128.

34. Alone, P.V.; Cao, C.; Dever, T.E. Translation initiation factor 2γ mutant alters start codon selection independent of Met-tRNA binding. *Mol. Cell. Biol.* **2008**, *28*, 6877–6888.
35. Sokabe, M.; Yao, M.; Sakai, N.; Toya, S.; Tanaka, I. Structure of archaeal translational initiation factor 2βγ-GDP reveals significant conformational change of the β-subunit and switch 1 region. *Proc. Natl. Acad. Sci. USA* **2006**, *103*, 13016–13021.
36. Schmitt, E.; Blanquet, S.; Mechulam, Y. The large subunit of initiation factor aIF2 is a close structural homologue of elongation factors. *EMBO J.* **2002**, *21*, 1821–1832.
37. Schmitt, E.; Naveau, M.; Mechulam, Y. Eukaryotic and archaeal translation initiation factor 2: A heterotrimeric tRNA carrier. *FEBS Lett.* **2010**, *584*, 405–412.
38. Nika, J.; Rippel, S.; Hannig, E.M. Biochemical analysis of the eIF2beta gamma complex reveals a structural function for eIF2alpha in catalyzed nucleotide exchange. *J. Biol. Chem.* **2001**, *276*, 1051–1056.
39. Westermann, P.; Nygard, O.; Bielka, H. The α and γ subunits of initiation factor eIF-2 can be cross-linked to 18S ribosomal RNA within the quaternary initiation complex, eIF-2-Met-tRNAf GDPCP small ribosomal subunit. *Nucleic Acids Res.* **1980**, *8*, 3065–3071.
40. Wu, R.; Haas, W.; Dephoure, N.; Huttlin, E.L.; Zhai, B.; Sowa, M.E.; Gygi, S.P. A large-scale method to measure absolute protein phosphorylation stoichiometries. *Nat. Methods* **2011**, *8*, 677–683.
41. Llorens, F.; Duarri, A.; Sarro, E.; Roher, N.; Plana, M.; Itarte, E. The *N*-terminal domain of the human eIF2β subunit and the CK2 phosphorylation sites are required for its function. *Biochem. J.* **2010**, *394*, 227–237.
42. Heaney, J.D.; Michelson, M.V.; Youngren, K.K.; Lam, M.Y.; Nadeau, J.H. Deletion of eIF2β suppresses testicular cancer incidence and causes recessive lethality in agouti-yellow mice. *Hum. Mol. Genet.* **2009**, *18*, 1395–1404.
43. Siridechadilok, B.; Fraser, C.S.; Hall, R.J.; Doudna, J.A.; Nogales, E. Structural roles for human translation factor eIF3 in initiation of protein synthesis. *Science* **2005**, *310*, 1513–1515.
44. Zhou, M.; Sandercock, A.M.; Fraser, C.S.; Ridlova, G.; Stephens, E.; Schenauer, M.R.; Yokoi-Fong, T.; Barsky, D.; Leary, J.A.; Hershey, J.W.; *et al.* Mass spectrometry reveals modularity and a complete subunit interaction map of the eukaryotic translation factor eIF3. *Proc. Natl. Acad. Sci. USA* **2008**, *105*, 18139–18144.
45. Cai, Q.; Todorovic, A.; Andaya, A.; Gao, J.; Leary, J.A.; Cate, J.H. Distinct regions of human eIF3 are sufficient for binding to the HCV IRES and the 40S ribosomal subunit. *J. Mol. Biol.* **2010**, *403*, 185–196.

8

Recent Advances in Bacteria Identification by Matrix-Assisted Laser Desorption/Ionization Mass Spectrometry Using Nanomaterials as Affinity Probes

Tai-Chia Chiu

Department of Applied Science, National Taitung University, 684 Section 1, Chunghua Road, Taitung 95002, Taiwan; E-Mail: tcchiu@nttu.edu.tw

Abstract: Identifying trace amounts of bacteria rapidly, accurately, selectively, and with high sensitivity is important to ensuring the safety of food and diagnosing infectious bacterial diseases. Microbial diseases constitute the major cause of death in many developing and developed countries of the world. The early detection of pathogenic bacteria is crucial in preventing, treating, and containing the spread of infections, and there is an urgent requirement for sensitive, specific, and accurate diagnostic tests. Matrix-assisted laser desorption/ionization mass spectrometry (MALDI-MS) is an extremely selective and sensitive analytical tool that can be used to characterize different species of pathogenic bacteria. Various functionalized or unmodified nanomaterials can be used as affinity probes to capture and concentrate microorganisms. Recent developments in bacterial detection using nanomaterials-assisted MALDI-MS approaches are highlighted in this article. A comprehensive table listing MALDI-MS approaches for identifying pathogenic bacteria, categorized by the nanomaterials used, is provided.

Keywords: affinity probes; bacteria; matrix-assisted laser desorption/ionization mass spectrometry; nanomaterials

1. Introduction

Worldwide, infectious diseases cause nearly 40% of the total 50 million deaths annually [1]. According to the World Health Organization, microbial hazards are the primary concern [2] because microbial diseases are a major cause of death in many developing and developed countries of the world [3,4]. Therefore, the development of rapid, accurate, and sensitive methods for bacterial identification is important for the clinical diagnosis, efficient treatment and prevention of diseases, environmental monitoring and food safety [5–8]. In clinical laboratories, bacterial identification is typically based on phenotypic tests, including Gram staining, culture and growth characteristics, and biochemical patterns. A number of methods are currently employed to detect and identify pathogenic agents, and these mainly rely on specific microbiological and biochemical identification methods [9–11]. These methods include culturing the microbes and counting the bacterial colonies, immunology-based methods, antigen–antibody interaction methods, and the polymerase chain reaction method, which involves DNA analysis. These methods can be sensitive and inexpensive, and can provide both qualitative and quantitative information about the test microorganisms; however, they are often time-consuming and laborious because each involves a pathogen amplification step. At present, most bacteria can be identified between a few hours to 1–2 days using these methods, with slow-growing microorganisms requiring additional time or supplementary tests [12]. Consequently, there is an urgent requirement for developing a rapid, sensitive, and selective detection method for such pathogens to treat individuals at risk, to improve public health surveillance and epidemiology, which is essential for ensuring the safety of food supplies, and to diagnose infectious bacterial diseases accurately.

There are challenges associated with identifying various types of pathogenic bacteria in a wide range of samples. Matrix-assisted laser desorption/ionization mass spectrometry (MALDI-MS) has been used to analyze various biomolecules, including peptides, proteins, DNA, RNA, oligonucleotides, oligosaccharides, and polymers [13–16]. This approach was first introduced by Tanaka and Karas in the late 1980s [17,18], and is a soft ionization method that provides mass spectra of the analytes with a minimum amount of fragmentation. MALDI-MS has a number of advantages over conventional methods including ease of operation, providing structural information of molecules with high throughput, speed, sensitivity, accuracy, and reproducibility [19,20]. Therefore, it has become a powerful tool for rapid characterization, differentiation, and identification of microorganism species [21–27]. For example, the mass-spectral profiling of whole cells can indicate the presence of unique biomarkers that can serve as the basis for identifying microbes [28,29]. In general, mass spectra of microbes isolated from a sample may contain unique patterns that can be automatically matched with spectra in a well-established reference library of microorganisms that have been characterized using appropriate sample preparation protocols. Matching the spectra allow the microbes to be identified as well as evaluated.

A sufficient number of bacterial cells (typically ~10^4 cells per well) are required to generate detectable MALDI-MS ion signals. However, samples obtained from infectious biological fluids or food poisoning samples are difficult to characterize directly by MALDI-MS because the ions generated from the bacterial cells may be seriously suppressed by the complex sample matrices. This led to the idea of using nanoparticles as affinity probes, to enhance the ability of MALDI-MS to detect bacteria [30,31]. Nanoparticles provide a high surface to volume ratio, giving them high binding and capture efficiencies for bacteria. Affinity separation approaches are methods of selectively concentrating trace amounts of bacteria from complex biological and food samples before they are characterized using MALDI-MS. When inorganic nanoparticles are used in MALDI-MS, instead of organic matrices, the method is called surface-assisted laser desorption and ionization MS (SALDI-MS) [32–36]. SALDI-MS was originally proposed by Sunner and Chen as early as 1995, and graphite particles were originally used as ion emitters [37]. This method was called SALDI-MS to emphasize that the surfaces and surface structures are critical to not only sample preparation but also desorption and ionization processes [38]. Numerous types of nanoparticles such as gold (Au) nanoparticles [39–41], silver (Ag) nanoparticles [42,43], magnetic nanoparticles [44,45], titanium dioxide (TiO_2) nanoparticles [46,47], carbon nanotubes [48,49], carbon nanoparticles [50], nanodiamonds [51], and graphene and graphene oxide [52] have been successfully used as matrices in SALDI-MS. The nanomaterials used in SALDI-MS play similar roles to the organic matrices used in MALDI-MS, absorbing energy from the laser irradiating them and efficiently transferring the energy to the analytes, causing the analytes to be desorbed and ionized [30]. The method provides several advantages including lower background noise in the low mass region, high surface areas, simple sample preparation, flexibility in sample desorption under different conditions, and high UV absorptivity [34]. Nanoparticles can also act as affinity probes, making it easy to concentrate the analytes, and offering good sensitivity and reproducibility [34].

In this review article, we focused on the overview of the recent advancements in the use of nanoparticles as affinity probes to enhance the detection sensitivity and selectivity of bacteria using MALDI-MS. Several examples of successful MALDI-MS approaches for detecting pathogenic bacteria have been provided to illustrate the advantages of this approach with respect to simplicity, sensitivity, and reproducibility. Furthermore, this article also provides some examples for the identification of bacteria in real samples using nanomaterials-assisted MALDI-MS approaches.

2. Bacterial Identification Using MALDI-MS

MALDI-MS is a very sensitive method where a single bacteria colony is sufficient for analysis, while other methods typically require culturing or enrichment of bacteria to obtain sufficient materials. Therefore, in clinical microbiological laboratories, the MALDI-MS is increasingly used for bacterial identification through the determination of the exact molecular masses of numerous peptides and small proteins, many of which are ribosomal. Conventional biochemical differentiation methods [24] have already been replaced by MALDI-MS. Because MALDI-MS is primarily applicable for analyzing clonal isolates, cultivation of the microorganism is still required. Moreover, for accurate identification, MALDI-MS can be used directly on the clinical samples that contain very few bacteria for accurate identification. In 2010, Ferreira *et al.* [53] introduced a MALDI-MS method for direct analysis of urine samples (4 mL) and observed that the inoculum level in the samples must be greater than 10^5 cfu/mL (colony-forming unit/mL). In 2010, a protocol for direct analysis of blood was introduced by Stevenson *et al.* [54], who separate bacteria from the red blood cells and plasma proteins via several centrifugation steps. A total of 212 positive cultures representing 32 genera and 60 species or groups were examined. Besides urine and blood, Barreiro *et al.* [55] inoculated pasteurized and homogenized samples of whole milk with the bacterial loads of 10^3–10^8 cfu. Sepsityper™ Kit (Bruker, Billerica, MA, USA) was used to for the testing milk sample and then analyzed by the Bruker BioTyper database. For a slightly contaminated (10^4 cfu/mL bacteria) milk sample, bacterial identification could be performed after initial incubation at 37 °C for 4 h. The detection limits for bacteria were in the range of 10^6–10^7 cfu/mL.

3. Nanoparticles Used as Affinity Probes

Nanoparticles are clusters of a few hundred to a few thousand atoms, and range from 1 to 100 nm in diameter. The chemical and physical properties of nanoparticles depend on their surfaces; therefore, these properties are highly dependent on the sizes, shapes, and compositions of the nanoparticles [56–58]. Nanoparticles have high surface-to-volume ratios, and those with excellent optical, magnetic, and electronic properties have been employed in sensing, imaging, catalysis, electronics, optics, and optoelectronics applications [59–65]. Nanoparticles can play an important role in determining the sensitivity of MALDI-MS and provide a high surface-to-volume ratio to give a high binding efficiency for bacteria. The affinity separation approach has been used to attempt to selectively concentrate trace amounts of bacteria from biological and food samples. Nanoparticles (functionalized or unmodified) that have been used as affinity probes to increase the sensitivity of MALDI-MS for detecting microbes are summarized in Table 1.

Table 1. Nanomaterials used as affinity probes in MALDI-MS.

Nanomaterials	Functionalized molecule	Pathogen	Application	LOD (cfu/mL)	Ref.
Fe_3O_4 NPs	IgG	*S. aureus*; *S. saprophyticus*		3.0×10^5	[66]
Fe_3O_4 NPs	IgG	*S. saprophyticus*	Urine	3.0×10^7	[66]
Fe_3O_4 NPs	Vancomycin	*S. aureus*; *S. saprophyticus*	Urine	7.8×10^4; 7.4×10^4	[67]
Fe_3O_4 NPs	Vancomycin	*B. cereus*; *E. faecium*; *S. aureus*	Tap water, reservoir water	5.0×10^2	[68]
Ag NPs		*E. coli*; *S. marcescen*		N.D.	[69]
Ag NPs		*B. lactis*; *L. acidophilus*; *S. thermophilus*; *L. bulgaricus*	Yogurt	N.D.	[70]
Ag NPs		*L. acidophilus*; *B. longum*; *L. bulgaricus*; *S. thermophilus*	Yogurt	N.D.	[70]
CdS QDs		*E. coli*		N.D.	[71]
CdS QDs		*S. cerevisiae*; *C. utilis*		N.D.	[72]
CdS QDs	Chitosan	*P. aeruginosa*; *S. aureus*		2.0×10^2; 1.5×10^2	[73]
Pt NPs	Mixed with ionic liquid (1-butyl-3-methylimidazolium hexafluorophosphate)	*E. coli*; *S. marcescens*		10^6	[74]
Pt NPs	IgG	*B. thuringiensis*; *B. subtilis*	Rhizospheric soil and root	N.D.	[75]
Pt NPs	IgG	*S. marcescens*; *E. coli*		10^5	[76]
AuNCs	Lysozyme	*E. coli*; *K. pneumoniae*; *P. aeruginosa*; pandrug-resistant *A. baumannii*; *S. aureus;* *E. faecalis*; vancomycin-resistant *E. faecalis*	Fetal bovine serum; urine	N.D.; 10^6	[77]
Graphene magnetic nanosheets	Chitosan	*P. aeruginosa*; *S. aureus*	Blood	6.0×10^2; 5.0×10^2	[78]

Table 1. *Cont.*

Nanomaterials	Functionalized molecule	Pathogen	Application	LOD (cfu/mL)	Ref.
NiO NPs		*E. coli*	Human nasal passage	10^7	[79]
TiO_2 NPs		*S. aureus*		N.D.	[80]
ZnO NPs		*E. coli*		N.D.	[81]
Ag, Pt, NiO, TiO_2, ZnO NPs		*S. aureus*; *P. saeruginosa*		N.D.	[82]

Ref., Reference; Ag, silver; AuNCs, gold nanoclusters; CdS, cadmium sulfide; IgG, immunoglobulin; LOD, limit of detection; N.D., not determined; NiO, nickel oxide; NPs, nanoparticles; Pt, platinum; QDs, quantum dots; TiO_2, titanum dioxide; ZnO, zinc oxide.

3.1. Magnetic Nanoparticles

Ho *et al.* [66] immobilized human immunoglobulin (IgG) onto the surfaces of magnetic Fe_3O_4 nanoparticles through covalent bonding (Figure 1). The functionalized magnetic nanoparticles were used as affinity probes to selectively concentrate pathogens, such as *Staphylococcus aureus* (*S. aureus*) and *Staphylococcus saprophyticus* (*S. saprophyticus*), from sample solutions. The bacteria were then characterized using MALDI-MS. The lowest bacterial concentration detected in an aqueous sample solution (0.5 mL) was 3×10^5 cfu/mL, for both *S. aureus* and *S. saprophyticus*, and the lowest detectable *S. saprophyticus* concentration in a urine sample was 3×10^7 cfu/mL.

Figure 1. Synthetic route for immobilizing immunoglobulin (IgG) onto the surfaces of Fe_3O_4 magnetic nanoparticles.

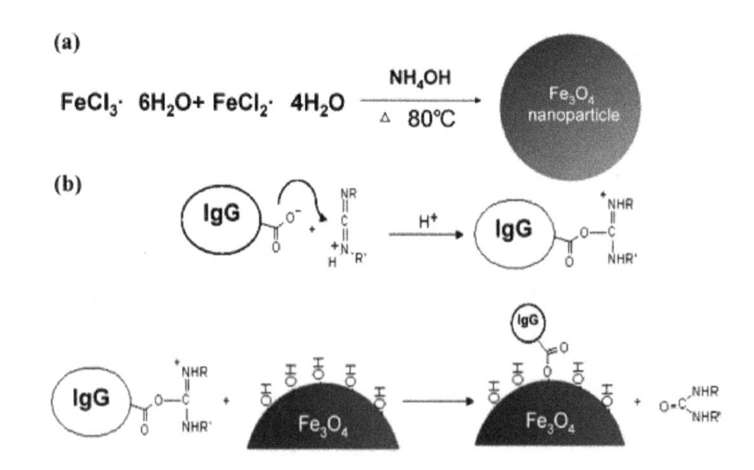

Vancomycin-modified 11 nm magnetic (Fe_3O_4) nanoparticles were used as affinity probes to selectively bind to the surface walls of Gram-positive bacteria (*S. aureus* and *S. saprophyticus*), as shown in Figure 2, allowing the bacteria to then be directly characterized using MALDI-MS [67]. Vancomycin is one of the most potent antibiotics, and has a high specificity for the *D*-Alanine (Ala) (*D*-Ala) moieties on the cell walls of Gram-positive bacteria. The lowest cell concentrations that could be detected in a urine sample (3 mL) were 7.4×10^4 cfu/mL for *S. aureus* and 7.8×10^4 cfu/mL for *S. saprophyticus*.

Figure 2. Cartoon illustrations of the proposed method for anchoring vancomycin-immobilized magnetic nanoparticles onto the surface of a Gram-positive bacterial cell and the binding of vancomycin to the terminal of *D*-Alanine (*D*-Ala)–*D*-Ala units of the peptides on the cell wall of a Gram-positive bacterium.

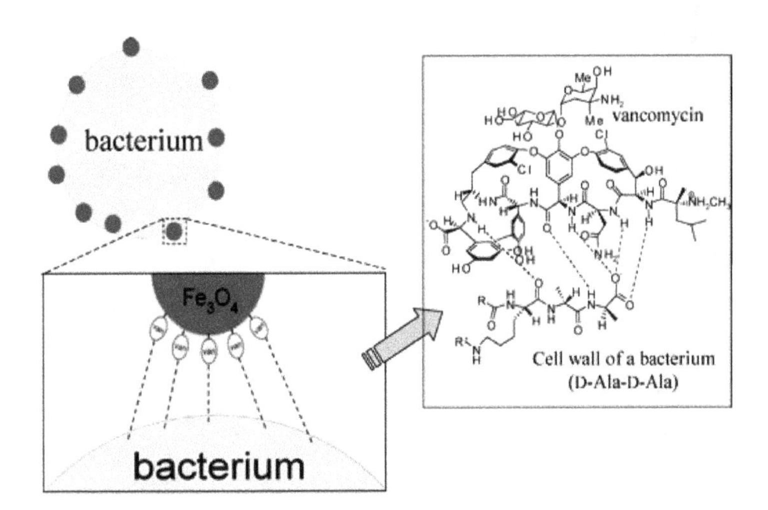

IgG- and vancomycin-modified magnetic nanoparticles have been demonstrated to exhibit effective affinities for selectively concentrating traces of bacteria from the sample solutions. However, because interferences from the urine matrix affect the binding capacity of these nanoprobes, further improvements are required to reduce the matrix effects in the analysis of biological samples.

A combination of membrane filtration and vancomycin-modified magnetic (Fe_3O_4) 15–20 nm nanoparticles has been used to selectively concentrate Gram-positive bacteria from tap water and reservoir water, allowing the bacteria to be rapidly analyzed using whole-cell MALDI-MS [68]. The capture efficiency for Gram-positive bacteria using these vancomycin-modified magnetic nanoparticles was 26.7%–33.3%, and the analysis time was approximately 2 h. This approach enhanced the sensitivity of the method by a factor of approximately 6×10^4, giving a limit of detection of 5×10^2 cfu/mL for *Bacillus cereus* (*B. cereus*), *Enterococcus faecium* (*E. faecium*), and *S. aureus* in water samples (2 L).

3.2. Silver (Ag) Nanoparticles

The bifunctional properties of Ag nanoparticles allowed them to be used as affinity probes for *Escherichia coli* (*E. coli*) and *Serratia marcescens* (*S. marcescens*), by Gopal *et al.* [69], to increase the sensitivity of MALDI-MS when characterizing the bacteria. The critical concentration of affinity probes for Ag nanoparticles was 1 mL/L in the case of *E. coli* and 0.5 mL/L in the case of *S. marcescens*. Ag nanoparticle concentrations higher than these values showed pronounced bactericidal activities.

The same research group also observed that an ionic solution (CrO_4^{2-}) and 0.035 mg of Ag nanoparticles could be used to capture yogurt bacteria (*Bifidobacterium lactis* (*B. lactis*), *Lactobacillus acidophilus* (*L. acidophilus*), *Streptococcus thermophilus* (*S. thermophilus*), and *Lactobacillus bulgaricus* (*L. bulgaricus*) from AB yogurt and *L. acidophilus*, *Bifidobacterium longum* (*B. longum*), *L. bulgaricus*, and *S. thermophilus* from Lin yogurt), improving the sensitivity achieved for detecting bacteria in yogurt samples [70]. This method has demonstrated a rapid, selective and sensitive means of bacterial detection using MALDI-MS for food microbiology.

3.3. Cadmium Sulfide (CdS) Quantum Dots (QDs)

Gopal *et al.* [71] has reported that CdS QDs can degrade the extracellular polysaccharides of *E. coli* cells when using MALDI-MS. Adding 20 μL/L of CdS QDs was observed to enhance the extracellular polymeric substance (EPS) peaks using an incubation time of up to 3 h. The authors confirmed that CdS QDs can function as more than just affinity probes, being able to degrade EPSs. CdS QDs can, therefore, be used to inactivate pathogenic *E. coli* and also inhibit the growth of *E. coli* biofilms.

Manikandan and Wu [72] observed that CdS QDs (10 mg/L) with particle sizes of 1–7 nm performed fungicidal roles and functioned as protein signal enhancement probes in the MALDI-MS analysis of the fungi *Saccharomyces cerevisiae* (*S. cerevisiae*) and *Candida utilis* (*C. utilis*). From their MALDI-MS results, the authors proposed the mechanism involving the CdS QDs interacting with the EPSs and removing small molecules from the EPS layers. The MALDI-MS protein signals were enhanced at all of the CdS QD concentrations that were tested (10–30 mg).

Chitosan-modified CdS QDs have been used as effective bacterial biosensors because of the strong affinities between chitosan molecules and bacterial membranes [73]. In that study, *Pseudomonas aeruginosa* (*P. aeruginosa*) and *S. aureus* cells were detected at low concentrations, 200 and 150 cfu/mL, respectively, after an extremely short time (1 min). MALDI-MS and transmission electron microscopy were used to confirm the interactions and the biocompatibility of the chitosan-modified CdS QDs with bacterial cells.

3.4. Platinum (Pt) Nanoparticles

Ahmad and Wu [74] employed a single drop microextraction technique, using an ionic liquid (1-butyl-3-methylimidazolium hexafluorophosphate) drop mixed with Pt nanoparticles, to extract bacterial proteins from aqueous samples to characterize pathogenic bacteria using MALDI-MS. This approach is based on surface changes in the ionic liquid and the membrane proteins of the bacteria, and it was successfully used to detect *E. coli* and *S. marcescens* at concentrations as low as 10^6 cfu/mL.

A rapid method for detecting bacteria associated with plants by an on-particle ionization and enrichment approach using IgG-functionalized Pt nanoparticle-assisted MALDI-MS was reported by Ahmad *et al.* [75]. The approach was successfully used to detect *Bacillus thuringiensis* (*B. thuringiensis*) and *Bacillus subtilis* (*B. subtilis*) isolated from rhizospheric soil and carrot plant roots. This study proved that bacteria can be directly detected even at low concentrations.

A rapid and sensitive approach to studying interactions between an affinity probe and a bacterial wall was introduced by Ahmad and Wu [76]. IgG was immobilized on Pt nanoparticles and MALDI-MS was used to detect the specific surface proteins of the bacteria *S. marcescens* and *E. coli*. This approach enabled the rapid detection of bacterial proteins, at a high resolution and with good sensitivity, without the need for tedious washing and separation procedures, and can be used to detect approximately 10^5 cfu/mL of *S. marcescens* and *E. coli*.

3.5. Other Nanomaterials

Chan *et al.* [77] demonstrated that pathogenic bacteria, including *E. coli*, *Klebsiella pneumonia* (*K. pneumoniae*), *P. aeruginosa*, pandrug-resistant *Acinetobacter baumannii* (*A. baumannii*), *S. aureus*, *E. faecalis*, and vancomycin-resistant *E. faecalis*, can be concentrated by lysozyme-encapsulated gold nanoclusters (AuNCs) that photoluminesce red, and distinguished by the results combining MALDI-MS and principal component analysis. Figure 3A shows photographs of sample tubes after the lysozyme-AuNCs were used as probes for *E. coli* J96 in urine samples containing different concentrations of the *E. coli*. Photographs of the control experiment sample tubes are shown in Figure 3B for comparison. Figure 3C shows the MALDI-MS spectra of the conjugates containing *E. coli* J96, in which the peaks at *m/z* 6177 and 6237 correspond to *E. coli* J96. The lowest *E. coli* concentration that could be detected using this approach was approximately 10^6 cfu/mL. The advantages of this method include speed (without cell culturing) and simplicity, and it can be used as universal affinity probes for Gram-positive/negative and antibiotic-resistant bacteria.

Figure 3. Photographs obtained by vortex-mixing (**A**) the lysozyme-AuNCs with *E. coli* J96 at different cell concentrations and (**B**) *E. coli* J96 alone for 1 h at different cell concentrations, followed by centrifugation at 3500 rpm for 5 min. The samples were prepared in urine that was diluted 50-fold with PBS solution (pH 7.4) containing BSA (~10 μM). The photographs were taken under illumination of UV light (λ_{max} = 365 nm); (**C**) Examination of the limit of detection. Matrix-assisted laser desorption/ionization mass spectrometry (MALDI-MS) obtained after using the lysozyme-AuNCs (1.36 mg/mL, 0.1 mL) as affinity probes to concentrate target species from a urine sample (0.90 mL) containing *E. coli* J96 (1.59×10^6 cells/mL) for 1 h. The urine sample was diluted 50-fold with PBS solution (pH 7.4) containing BSA (~10 μM) prior to bacterial spiking. Reprinted with permission from [77].

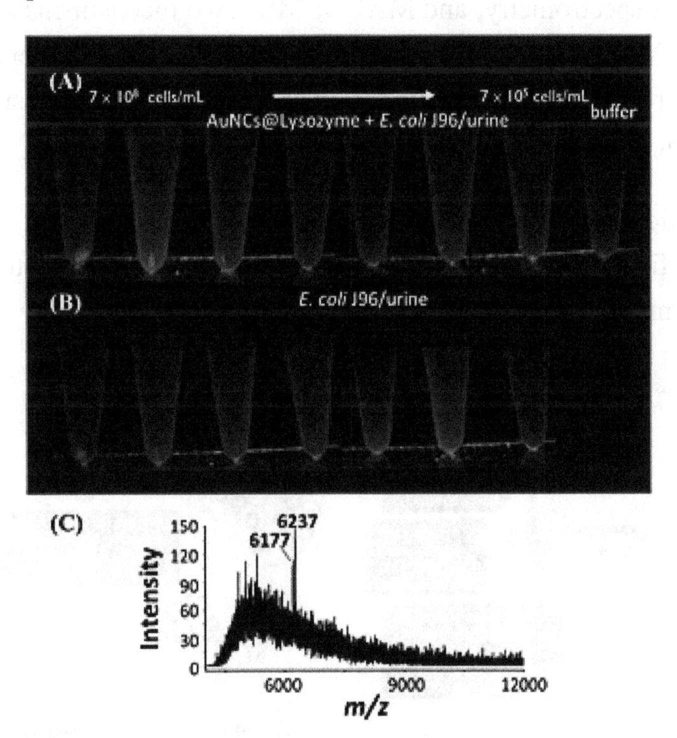

Abdelhamid and Wu [78] demonstrated that multifunctional graphene magnetic nanosheets modified with chitosan (GMCS) can be used in MALDI-MS for the sensitive detection of pathogenic bacteria (*P. aeruginosa* and *S. aureus*). The GMCS were observed to act as efficient separation and preconcentration nanoprobes for SALDI and enhance the ionization of bacterial biomolecules. GMCS have been used in the direct detection of low concentrations of *P. aeruginosa* and *S. aureus* in blood samples, demonstrating their practical applicability. This approach offers many advantages such as robustness, simplicity, and the capability for fluorescence based real-sample monitoring.

The heat stress response of *E. coli* (at 10^7 cfu/mL) at different temperatures has been studied using nickel oxide (NiO) nanoparticle-assisted MALDI-MS by Hasan *et al.* [79]. MALDI-MS was successfully used to detect 10 kDa chaperonin proteins produced by *E. coli* under heat stress at temperatures between 40 and 80 °C in the absence or presence of NiO nanoparticles. Dramatic decreases in the viability of *E. coli* in the presence of NiO were confirmed from the MALDI-MS results. This technique is a rapid, sensitive, and efficient approach for bacterial detection under extremely harsh conditions.

Gopal *et al.* [80] demonstrated that *S. aureus* isolated from the human nasal passage can be directly detected using MALDI-MS assisted by TiO_2 nanoparticles, without any culturing steps or sample pretreatment being required. TiO_2 nanoparticles were used to enhance the bacterial signals in the direct MALDI-MS analysis.

MALDI-MS has been used to evaluate bactericidal activity, by detecting proteins produced because of the inactivation of *E. coli* cells by zinc oxide (ZnO) nanoparticles [81]. The results showed that at concentrations of 1 and 5 g/L ZnO nanoparticles can be used as affinity probes to improve the signal intensities in the MS spectra. The significant differences in the spectral patterns confirmed that MALDI-MS was successfully used to evaluate the bactericidal activity of ZnO nanoparticles.

Gopal *et al.* [82] proposed mechanisms for the interactions between five nanoparticles (Ag, NiO, Pt, TiO_2, and ZnO) and two bacteria (*S. aureus* and *P. aeruginosa*) from studies using transmission electron microscopy, ultra spectrometry, and MALDI-MS. Two mechanisms (Figure 4) were proposed for the interactions: (1) Mechanism A was proposed for Pt and NiO nanoparticles, the function of which is based on their affinities for bacterial walls; and (2) Mechanism B was proposed for bactericidal nanoparticles, such as TiO_2, ZnO, and Ag nanoparticles.

Figure 4. Schematic diagram showing the mechanisms (Mechanism A and Mechanism B) for interactions of five nanoparticles with two pathogenic bacteria postulated in the study. Reprinted with permission from [82].

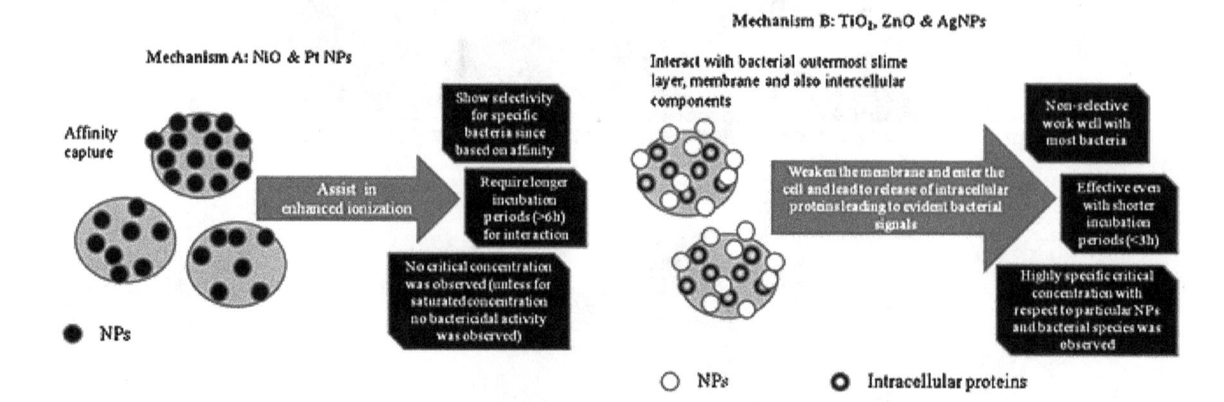

4. Conclusions

MALDI-MS is an emerging analytical tool for detecting and identifying microorganisms. It offers high sensitivity, simple sample preparation processes, low sample consumption volumes, and the possibility of automated and high-throughput analyses. In this review, we have described several MALDI-MS approaches for detecting pathogenic bacteria using nanomaterials (such as AuNCs, Ag, magnetic, and Pt nanoparticles and CdS QDs) as affinity probes. The nanomaterials described here act as concentration probes for the selective capture of unique biomarkers from microorganisms, and as surfaces to absorb energy from the laser irradiation, thereby inducing desorption and ionization of the analytes.

As mentioned above, the most important advantage of the affinity-based nanoparobe methods is their ability to selectively concentrate and purify microorganisms from complex samples, such as urine and blood, and allow the further identification of microorganisms without microbial culturing using MALDI-MS. For the nanomaterials-assisted MALDI-MS, Direct analysis of microorganisms at low microbial levels can be performed using the nanomaterials-based MALDI-MS. Numerous nanomaterials have been demonstrated to be useful as affinity probes for targeting bacteria. However, some nanoparticles such as Ag, TiO_2 and ZnO, also exhibit bactericidal activity, and therefore might not be good affinity probes at higher nanoparticles concentrations. Controlling of the nanoparticle concentration will be a key factor. All these nanomaterials-assisted MALDI-MS methods also encounter challenges with respect to the enrichment of unknown target bacterial species from the urine, blood, and cerebrospinal fluid. Thus, a limiting factor in MALDI-MS analysis is insufficient database entries. The addition of certain species to the database has been demonstrate significantly improve MALDI-MS precision in bacterial identification.

The broad adoption of nanomaterials-assisted MALDI-MS methods for bacterial identification will require a substantial improvement in performance compared with the existing methods, such as conventional MALDI-MS and biochemical tests. Thus, the standardization of terminology is required. Advances in nanomaterials-assisted MALDI-MS methods will support the simple and accurate means of bacterial identification for food safety, environmental monitoring and clinical diagnosis.

Acknowledgments

This work was supported by the Ministry of Science and Technology of Taiwan under contract No. NSC 102-2113-M-143-001.

References

1. Ivnitski, D.; Abdel-Hamid, I.; Atanasov, P.; Wilkins, E. Biosensors for detection of pathogenic bacteria. *Biosens. Bioelectron.* **1999**, *14*, 599–624.
2. WHO Guidelines for Drinking-Water Quality. Available online: http://www.who.int/water_sanitation_health/dwq/guidelines/en/ (accessed on 7 May 2014).
3. Jahid, I.K.; Ha, S.-D. A review of microbial biofilms of produce: Future challenge to food safety. *Food Sci. Biotechnol.* **2012**, *21*, 299–316.
4. Velusamy, V.; Arshak, K.; Korostynska, O.; Oliwa, K.; Adley, C. An overview of foodborne pathogen detection: In the perspective of biosensors. *Biotechnol. Adv.* **2010** *28*, 232–254.

5. Tallury, P.; Malhotra, A.; Byrne, L.M.; Santra, S. Nanobioimaging and sensing of infectious diseases. *Adv. Drug Deliv. Rev.* **2010**, *62*, 424–437.
6. Kim, J.; Yoon, M.-Y. Recent advances in rapid and ultrasensitive biosensors for infectious agents: Lesson from *Bacillus anthracis* diagnostic sensors. *Analyst* **2010**, *135*, 1182–1190.
7. Sekhon, S.S.; Kim, S.-G.; Lee, S.-H.; Jang, A.; Min, J.; Ahn J.-Y.; Kim, Y.-H. Advances in pathogen-associated molecules detection using aptamer based biosensors. *Mol. Cell. Toxicol.* **2013**, *9*, 311–317.
8. Sharma, H.; Mutharasan, R. Review of biosensors for foodborne pathogens and toxins. *Sens. Actuators B* **2013**, *183*, 535–549.
9. Quilliam, R.S.; Williams, A.P.; Avery, L.M.; Malham, S.K.; Jones, D.L. Unearthing human pathogens at the agricultural–environment interface: A review of current methods for the detection of *Escherichia coli* O157 in freshwater ecosystems. *Agric. Ecosyst. Environ.* **2011**, *140*, 354–360.
10. Kirsch, J.; Siltanen, C.; Zhou, Q.; Revzin, A.; Simonian, A. Biosensor technology: Recent advances in threat agent detection and medicine. *Chem. Soc. Rev.* **2013**, *42*, 8733–8768.
11. Bridle, H.; Miller, B.; Desmulliez, M.P.Y. Application of microfluidics in waterborne pathogen monitoring: A review. *Water Res.* **2014**, *55*, 256–271.
12. Biswas, S.; Rolain, J.-M. Use of MALDI-TOF mass spectrometry for identification of bacteria that are difficult to culture. *J. Microbiol. Methods* **2013**, *92*, 14–24.
13. Sandrin, T.R.; Goldstein, J.E.; Schumaker, S. MALDI TOF MS profiling of bacteria at the strain level: A review. *Mass Spectrom. Rev.* **2013**, *32*, 188–217.
14. Bakry, R.; Rainer, M.; Huck, C.W.; Bonn, G.K. Protein profiling for cancer biomarker discovery using matrix-assisted laser desorption/ionization time-of-flight mass spectrometry and infrared imaging: A review. *Anal. Chim. Acta* **2011**, *690*, 26–34.
15. Cho, Y.-T.; Su, H.; Huang, T.-L.; Chen, H.-C.; Wu, W.-J.; Wu, P.-C.; Wu, D.-C.; Shiea, J. Matrix-assisted laser desorption ionization/time-of-flight mass spectrometry for clinical diagnosis. *Clin. Chim. Acta* **2013**, *415*, 266–275.
16. Bergman, N.; Shevchenko, D.; Bergquist, J. Approaches for the analysis of low molecular weight compounds with laser desorption/ionization techniques and mass spectrometry. *Anal. Bioanal. Chem.* **2014**, *406*, 49–61.
17. Karas, M.; Hillenkamp, F. Laser desorption ionization of proteins with molecular masses exceeding 10,000 daltons. *Anal. Chem.* **1988**, *60*, 2299–2301.
18. Tanaka, K.; Waki, H.; Ido, Y.; Akita, S.; Yoshida, Y.; Yoshida, T. Protein and polymer analyses up to *m/z* 100,000 by laser ionization time-of-flight mass spectrometry. *Rapid Commun. Mass Spectrom.* **1988**, *2*, 151–153.
19. Dingle, T.C.; Butler-Wu, S.M. MALDI-TOF mass spectrometry for microorganism identification. *Clin. Lab. Med.* **2013**, *33*, 589–609.
20. Chalupová, J.; Raus, M.; Sedlářová, M.; Šebela, M. Identification of fungal microorganisms by MALDI-TOF mass spectrometry. *Biotechnol. Adv.* **2014**, *32*, 230–241.
21. Ho, Y.-P.; Reddy, P.M. Advances in mass spectrometry for the identification of pathogens. *Mass Spectrom. Rev.* **2011**, *30*, 1203–1224.

22. Wieser, A.; Schneider, L.; Jung, J.; Schubert, S. MALDI-TOF MS in microbiological diagnostics—Identification of microorganisms and beyond (mini review). *Appl. Microbiol. Biotechnol.* **2012**, *93*, 965–974.
23. Krásný, L.; Hynek, R.; Hochel, I. Identification of bacteria using mass spectrometry techniques. *Int. J. Mass Spectrom.* **2013**, *353*, 67–79.
24. Lartigue, M.-F. Matrix-assisted laser desorption ionization time-of-flight mass spectrometry for bacterial strain characterization. *Infect. Genet. Evol.* **2013**, *13*, 230–235.
25. Kostrzewa, M.; Sparbier, K.; Maier, T.; Schubert, S. MALDI-TOF MS: An upcoming tool for rapid detection of antibiotic resistance in microorganisms. *Proteomics Clin. Appl.* **2013**, *7*, 767–778.
26. Havlicek, V.; Lemr, K.; Schug, K.A. Current trends in microbial diagnostics based on mass spectrometry. *Anal. Chem.* **2013**, *85*, 790–797.
27. Del Chierico, F.; Petrucca, A.; Vernocchi, P.; Bracaglia, G.; Fiscarelli, E.; Bernaschi, P.; Muraca, M.; Urbani, A.; Putignani, L. Proteomics boosts translational and clinical microbiology. *J. Proteomics* **2014**, *97*, 69–87.
28. Welker, M.; Moore, E.R.B. Applications of whole-cell matrix-assisted laser-desorption/ionization time-of-flight mass spectrometry in systematic microbiology. *Syst. Appl. Microbiol.* **2011**, *34*, 2–11.
29. Intelicato-Young, J.; Fox, A. Mass spectrometry and tandem mass spectrometry characterization of protein patterns, protein markers and whole proteomes for pathogenic bacteria. *J. Microbiol. Methods* **2013**, *92*, 381–386.
30. Chiu, T.-C.; Huang, L.-S.; Lin, P.-C.; Chen, Y.-C.; Chen, Y.-J.; Lin, C.-C.; Chang, H.-T. Nanomaterials based affinity matrix-assisted laser desorption/ionization mass spectrometry for biomolecules and pathogenic bacteria. *Recent Pat. Nanotechnol.* **2007**, *1*, 99–111.
31. Wu, F.-H.; Gopal, J.; Manikandan, M. Future perspective of nanoparticle interaction-assisted laser desorption/ionization mass spectrometry for rapid, simple, direct and sensitive detection of microorganisms. *J. Mass Spectrom.* **2012**, *47*, 355–363.
32. Chen, W.-T.; Tomalová, I.; Preisler, J.; Chang, H.-T. Analysis of biomolecules through surface-assisted laser desorption/ionization mass spectrometry employing nanomaterials. *J. Chin. Chem. Soc.* **2011**, *58*, 769–778.
33. Rainer, M.; Qureshi, M.N.; Bonn, G.K. Matrix-free and material-enhanced laser desorption/ionization mass spectrometry for the analysis of low molecular weight compounds. *Anal. Bioanal. Chem.* **2011**, *400*, 2281–2288.
34. Chiang, C.-K.; Chen, W.-T.; Chang, H.-T. Nanoparticle-based mass spectrometry for the analysis of biomolecules. *Chem. Soc. Rev.* **2011**, *40*, 1269–1281.
35. Law, K.P.; Larkin, J.R. Recent advances in SALDI-MS techniques and their chemical and bioanalytical applications. *Anal. Bioanal. Chem.* **2011**, *399*, 2597–2622.
36. Lim, A.Y.; Ma, J.; Boey, Y.C.F. Development of nanomaterials for SALDI-MS analysis in forensics. *Adv. Mater.* **2012**, *24*, 4211–4216.
37. Sunner. J.; Dratz, E.; Chen, Y.-C. Graphite surface-assisted laser desorption/ionization time-of-flight mass spectrometry of peptides and proteins from liquid solutions. *Anal. Chem.* **1995**, *67*, 4335–4342.

38. Han. M.; Sunner. J. An activated carbon substrate surface for laser desorption mass spectrometry. *J. Am. Soc. Mass Spectrom.* **2000**, *11*, 644–649.
39. Chiang, C.-K.; Lin, Y.-W.; Chen, W.-T.; Chang, H.-T. Accurate quantitation of glutathione in cell lysates through surface-assisted laser desorption/ionization mass spectrometry using gold nanoaprticles. *Nanomedicine* **2010**, *6*, 530–537.
40. Hsieh, Y.-T.; Chen, W.-T.; Tomalová, I.; Preisler, J.; Chang, H.-T. Detection of melamine in infant formula and grain powder by surface-assisted laser desorption/ionization mass spectrometry. *Rapid Commun. Mass Spectrom.* **2012**, *26*, 1393–1398.
41. Pilolli, R.; Ditaranto, N.; di Franco, C.; Palmisano, F.; Cioffi, N. Thermally annealed gold nanoparticles for surface-assisted laser desorption ionisation–mass spectrometry of low molecular weight analytes. *Anal. Bioanal. Chem.* **2012**, *404*, 1703–1711.
42. Chiu, T.-C.; Chang, L.-C.; Chiang, C.-K.; Chang, H.-T. Determining estrogens using surface-assisted laser desorption/ionization mass spectrometry with silver nanoparticles as the matrix. *J. Am. Soc. Mass Spectrom.* **2008**, *19*, 1343–1346.
43. Nizioł, J.; Rode, W.; Zieliński, Z.; Ruman, T. Matrix-free laser desorption–ionization with silver nanoparticle-enhanced steel targets. *Int. J. Mass Spectrom.* **2013**, *335*, 22–32.
44. Lin, P.-C.; Yu, C.-C.; Wu, H.-T.; Lu, Y.-W.; Han, C.-L.; Su, A.-K.; Chen, Y.-J.; Lin, C.-C. A chemically functionalized magnetic nanoplatform for rapid and specific biomolecular recognition and separation. *Biomacromolecules* **2013**, *14*, 160–168.
45. Huang, S.-Y.; Chen, Y.-C. Magnetic nanoparticle-based platform for characterization of histidine-rich proteins and peptides. *Anal. Chem.* **2013**, *85*, 3347–3354.
46. Chiu, T.-C. Steroid hormones analysis with surface-assisted laser desorption/ionization mass spectrometry using catechin-modified titanium dioxide nanoparticles. *Talanta* **2011**, *86*, 415–420.
47. Radisavljević, M.; Kamčeva, T.; Vukićević, I.; Radoičić, M.; Šaponjić, Z.; Petković, M. Colloidal TiO_2 nanoparticles as substrates for M(S)ALDI mass spectrometry of transition metal complexes. *Rapid Commun. Mass Spectrom.* **2012**, *26*, 2041–2050.
48. Hsu, W.-Y.; Lin, W.-D.; Hwu, W.-L.; Lai, C.-C.; Tsai, F.-J. Screening assay of very long chain fatty acids in human plasma with multiwalled carbon nanotube-based surface-assisted laser desorption/ionization mass spectrometry. *Anal. Chem.* **2010**, *82*, 6814–6820.
49. Cegłowski, M.; Schroeder, G. Laser desorption/ionization mass spectrometric analysis of surfactants on functionalized carbon nanotubes. *Rapid Commun. Mass Spectrom.* **2012**, *27*, 258–264.
50. Ng, E.W.Y.; Lam, H.S.; Ng, P.C.; Poon, T.C.W. Quantification of citrulline by parallel fragmentation monitoring—A novel method using graphitized carbon nanoparticles and MALDI-TOF/TOF mass spectrometry. *Clin. Chim. Acta* **2013**, *420*, 121–127.
51. Wei, L.-M.; Shen, Q.; Lu, H.-J.; Yang, P.-Y. Pretreatment of low-abundance peptides on detonation nanodiamond for direct analysis by matrix-assisted laser desorption/ionization time-of-flight mass spectrometry. *J. Chromatogr. B* **2009**, *877*, 3631–3637.
52. Liu, Y.; Liu, J.; Deng, C.; Zhang, X. Graphene and graphene oxide: Two ideal choices for the enrichment and ionization of long-chain fatty acids free from matrix-assisted laser desorption ionization matrix interference. *Rapid Commun. Mass Spectrom.* **2011**, *25*, 3223–3234.

53. Ferreira, L.; Sánchez-Juanes, F.; González-Ávila, M.; Cembrero-Fuciños, D.; Herrero-Hernández, A.; González-Buitrago, J.M.; Muñoz-Bellido, J.L. Direct identification of urinary tract pathogens from urine samples by matrix-assisted laser desorption ionization-time of flight mass spectrometry. *J. Clin. Microbiol.* **2010**, *48*, 2110–2115.
54. Stevenson, L.G.; Drake, S.K.; Murray, P.R. Rapid identification of bacteria in positive blood culture broths by matrix-assisted laser desorption ionization-time of flight mass spectrometry. *J. Clin. Microbiol.* **2010**, *48*, 444–447.
55. Barreiro, J.R.; Braga, P.A.C.; Ferreira, C.R.; Kostrzewa, M.; Maier, T.; Wegemann, B.; Böettcher, V.; Eberlin, M.N.; dos Santos, M.V. Nonculture-based identification of bacteria in milk by protein fingerprinting. *Proteomics* **2012**, *12*, 2739–2745.
56. Burda, C.; Chen, X.; Narayana, R.; El-Sayed, M.A. Chemistry and properties of nanocrystals of different shapes. *Chem. Rev.* **2005**, *105*, 1025–1102.
57. An, K.; Somorjai, G.A. Size and shape control of metal nanoparticles for reaction selectivity in catalysis. *ChemCatChem* **2012**, *4*, 1512–1524.
58. Lohse, S.E.; Murphy, C.A. Applications of colloidal inorganic nanoparticles: From medicine to energy. *J. Am. Chem. Soc.* **2012**, *134*, 15607–15620.
59. Sanvicens, N.; Pastells, C.; Pascual, N.; Marco, M.-P. Nanoparticle-based biosensors for detection of pathogenic bacteria. *Trends Anal. Chem.* **2009**, *28*, 1243–1252.
60. Coto-García, A.M.; Sotelo-González, E.; Fernández-Argüelles, M.T.; Pereiro, R.; Costa-Fernández, J.M.; Sanz-Medel, A. Nanoparticles as fluorescent labels for optical imaging and sensing in genomics and proteomics. *Anal. Bioanal. Chem.* **2011**, *399*, 29–42.
61. Veerapandian, M.; Yun, K. Functionalization of biomolecules on nanoparticles: Specialized for antibacterial applications. *Appl. Microbiol. Biotechnol.* **2011**, *90*, 1655–1667.
62. Gilmartin, N.; O'Kennedy, R. Nanobiotechnologies for the detection and reduction of pathogens. *Enzym. Microb. Technol.* **2012**, *50*, 87–95.
63. Burris, K.P.; Stewart, C.N., Jr. Fluorescent nanoparticles: Sensing pathogens and toxins in foods and crops. *Trends Food Sci. Technol.* **2012**, *28*, 143–152.
64. Shinde, S.B.; Fernandes, C.B.; Patravale, V.B. Recent trends in *in vitro* nanodiagnostics for detection of pathogens. *J. Control. Release* **2012**, *159*, 164–180.
65. Zamborini, F.P.; Bao, L.; Dasari, R. Nanoparticles in measurement science. *Anal. Chem.* **2012**, *84*, 541–576.
66. Ho, K.-C.; Tsai, P.-J.; Lin, Y.-S.; Chen, Y.-C. Using biofunctionalized nanoparticles to probe pathogenic bacteria. *Anal. Chem.* **2004**, *76*, 7162–7168.
67. Lin, Y.-S.; Tsai, P.-J.; Weng, M.-F.; Chen, Y.-C. Affinity capture using vancomycin-bound magnetic nanoparticles for the MALDI-MS analysis of bacteria. *Anal. Chem.* **2005**, *77*, 1753–1760.
68. Li, S.; Guo, Z.; Wu, H.-F.; Liu, Y.; Yang, Z.; Woo, C.H. Rapid analysis of Gram-positive bacteria in water via membrane filtration coupled with nanoprobe-based MALDI-MS. *Anal. Bioanal. Chem.* **2010**, *397*, 2465–2476.
69. Gopal, J.; Wu, H.-F.; Lee, C.-H. The biofunctional role of Ag nanoparticles on bacteria—A MALDI-MS perspective. *Analyst* **2011**, *136*, 5077–5083.

70. Lee, C.-H.; Gopal, J.; Wu, F.-H. Ionic solution and nanoparticle assisted MALDI-MS as bacterial biosensors for rapid analysis of yogurt. *Biosens. Bioelectron.* **2012**, *31*, 77–83.
71. Gopal, J.; Wu, F.-H.; Gangaraju, G. Quantifying the degradation of extracellular polysaccharides of *Escherichia coli* by CdS quantum dots. *J. Mater. Chem.* **2011**, *21*, 13445–13451.
72. Manikandan, M.; Wu, H.-F. Probing the fungicidal properties of CdS quantum dots on *Saccharomyces cerevisiae* and *Candida utilis* using MALDI-MS. *J. Nanopart. Res.* **2013**, *15*, 1728.
73. Abdelhamid, H.N.; Wu, F.-H. Probing the interactions of chitosan capped CdS quantum dots with pathogenic bacteria and their biosensing application. *J. Mater. Chem. B* **2013**, *1*, 6094–6106.
74. Ahmad, F.; Wu, F.-H. Characterization of pathogenic bacteria using ionic liquid via single drop microextraction combined with MALDI-TOF MS. *Analyst* **2011**, *136*, 4020–4027.
75. Amhad, F.; Siddiqui, M.A.; Babalola, O.O.; Wu, F.-H. Biofunctionalization of nanoparticle assisted mass spectrometry as biosensors for rapid detection of plant associated bacteria. *Biosens. Bioelectron.* **2012**, *35*, 235–242.
76. Ahmad, F.; Wu, F.-H. Rapid and sensitive detection of bacteria via platinum-labeled antibodies and on-particle ionization and enrichment prior to MALDI-TOF mass spectrometry. *Microchim. Acta* **2013**, *180*, 485–492.
77. Chan, P.-H.; Wong, S.-Y.; Lin, S.-H.; Chen, Y.-C. Lysozyme-encapsulated gold nanocluster-based affinity mass spectrometry for pathogenic bacteria. *Rapid Commun. Mass Spectrom.* **2013**, *27*, 2143–2148.
78. Abdelhamid, H.N.; Wu, H.-F. Multifunctional graphene magnetic nanosheet decorated with chitosan for highly sensitive detection of pathogenic bacteria. *J. Mater. Chem. B* **2013**, *1*, 3950–3961.
79. Hasan, N.; Ahmad, F.; Wu, F.-H. Monitoring the heat stress response of *Escherichia coli* via NiO nanoparticle assisted MALDI–TOF mass spectrometry. *Talanta* **2013**, *103*, 38–46.
80. Gopal, J.; Narayana, J.L.; Wu, F.-H. TiO_2 nanoparticle assisted mass spectrometry as biosensors of *Staphylococcus aureus*, key pathogen in nosocomial infections from air, skin surface and human nasal passage. *Biosens. Bioelectron.* **2011**, *27*, 201–206.
81. Gopal, J.; Wu, F.-H.; Lee, Y.-H. Matrix-assisted laser desorption ionization-time-of-flight mass spectrometry as a rapid and reliable technique for directly evaluating bactericidal activity: Probing the critical concentration of ZnO nanoparticles as affinity probes. *Anal. Chem.* **2010**, *82*, 9617–9621.
82. Gopal, J.; Manikandan, M.; Hasan, N.; Lee, C.-H.; Wu, H.-F. A comparative study on the mode of interaction of different nanoparticles during MALDI-MS of bacterial cells. *J. Mass Spectrom.* **2013**, *48*, 119–127.

9

Quantitative Proteomics to Characterize Specific Histone H2A Proteolysis in Chronic Lymphocytic Leukemia and the Myeloid THP-1 Cell Line

Pieter Glibert [1], Liesbeth Vossaert [1], Katleen Van Steendam [1], Stijn Lambrecht [2], Filip Van Nieuwerburgh [1], Fritz Offner [3], Thomas Kipps [4], Maarten Dhaenens [1,†] and Dieter Deforce [1,†,*]

[1] Laboratory of Pharmaceutical Biotechnology, Ghent University, 72 Harelbekestraat, B-9000 Ghent, Belgium; E-Mails: pieter.glibert@ugent.be (P.G.); liesbeth.vossaert@ugent.be (L.V.); katleen.vansteendam@ugent.be (K.V.S.); filip.vannieuwerburgh@ugent.be (F.V.N.); maarten.dhaenens@ugent.be (M.D.)

[2] Department of Rheumatology, Ghent University Hospital, 185 1P7 De Pintelaan, B-9000 Ghent, Belgium; E-Mail: stijn.lambrecht@hotmail.com

[3] Department of Hematology, Ghent University Hospital, 185 1P7 De Pintelaan, B-9000 Ghent, Belgium; E-Mail: fritz.offner@ugent.be

[4] Department of Medicine, Moores Cancer Center, University of California at San Diego (UCSD), 3855 Health Sciences Drive, La Jolla, CA 92093, USA; E-Mail: tkipps@ucsd.edu

† These authors contributed equally to this work.

* Author to whom correspondence should be addressed; E-Mail: dieter.deforce@ugent.be

---

Abstract: Proteome studies on hematological malignancies contribute to the understanding of the disease mechanism and to the identification of new biomarker candidates. With the isobaric tag for relative and absolute quantitation (iTRAQ) method we analyzed the protein expression between B-cells of healthy people and chronic lymphocytic leukemia (CLL) B-cells. CLL is the most common lymphoid cancer of the blood and is characterized by a variable clinical course. By comparing samples of patients with an aggressive *vs.* indolent disease, we identified a limited list of differentially regulated proteins. The enhanced sensitivity attributed to the iTRAQ labels led to the discovery of a previously reported but still not clarified proteolytic product of histone H2A (cH2A) which we further investigated in light of the suggested functional properties of this modification. In the exploratory proteome study the Histone H2A peptide was up-regulated in CLL samples but a more specific and sensitive screening of a larger patient cohort indicated that cH2A is of myeloid origin. Our subsequent quantitative analysis led to a more profound characterization of the clipping in acute monocytic leukemia THP-1 cells subjected to induced differentiation.

Keywords: histone H2A; proteolysis; histone clipping; chronic lymphocytic leukemia; THP-1 cells; quantitative proteomics

---

1. Introduction

Proteomics approaches are often trailing genetic studies but are essential in the multi-disciplinary field of hematological research. As opposed to, e.g., mRNA microarray data, there is a better understanding of which proteins are actually expressed, although seeing the forest for the trees in long lists of protein identifications remains challenging [1,2]. In neoplastic hematology, protein studies have contributed to the elucidation of the disease mechanism, defined prognostic or therapeutic biomarkers and clarified previously reported uncharacterized phenomena [2]. By analyzing body fluids, cell lines, and tissues with quantitative high throughput mass spectrometry techniques complementary biological insights in hematopoietic malignancies can be generated. Uncovering relevant posttranslational modifications (PTMs), such as phosphorylations and proteolytical cleavages, might be associated with specific disease stages and could, hence, be informative on the biology of the disease [3].

The most common adult hematopoietic malignancy is chronic lymphocytic leukemia (CLL), a disease characterized by a widely variable median survival. After the initial staging of the patient, the risk of progression is defined by a set of genetic and protein based laboratory assays. A well-established prognostic marker is the mutational status of the immunoglobulin heavy chain (*IGVH*) genes encoding for the B-cell antigen-binding domain: CLL patients who have B-cells with unmutated (UM, >98% germ line identity) *IGVH* genes have an unfavorable outcome, whereas mutated (M) *IGHV* genes predict a more indolent course [4]. Surrogate markers on the protein level, such as the Zeta-chain-associated protein kinase (ZAP) 70 and CD38, are more easily applicable in the clinical practice although CD38 is considered to have less predictive value since discordancy with gene status is commonly observed and problems with standardization occur. ZAP70 expression however, is still used in the research on CLL pathogenesis and is considered as an independent biomarker [5]. Hence, for determining disease progression and survival of CLL patients, specific genetic markers, e.g., chromosomal aberrations, such as 13q14 deletion, are of increasing importance [6]. More recently, also genome sequencing, miRNA expression profiling and methylome studies are starting to offer new insights in the disease onset and progression. By the same token, specific epigenetic modifications together with protein alterations became valuable targets in leukemia research due to their reversible character and thus potential in therapy [7].

To extend the knowledge on CLL pathology and to identify new biomarker candidates, we applied quantitative mass spectrometry strategies to target the lower abundant proteins and peptides on patient and control samples [3]. The expressional differences between isolated age-matched healthy B-cells and CLL B-cells clearly showed that the morphological differences inherent to cancerous cells challenge disease marker discovery. Our comparative proteome analysis of UM and M CLL B-cells however, revealed that remarkably, only a limited amount of the identified proteins was differentially expressed between patients with a different outcome. For both the UM and M patient group, known up-regulations of proteins contributing to cell proliferation were corroborated [8,9].

Analysis of the iTRAQ (isobaric tag for relative and absolute quantitation) data at the peptide level surfaced an interesting aberrant proteolytic product of a histone protein: clipping of the histone H2A *C*-tail. The specific clipping of histone H2A after V_{114} (cH2A) was previously reported in leukemia and leukemia cell lines and is catalyzed by the so-called "H2A specific protease" (H2Asp) [10–14]. Recently, we identified the enzyme Neutrophil Elastase (NE) as an important candidate for the identity of the H2Asp [15]. Even though proteolysis is often not considered as a regulated PTM and the importance of protein degradation in biological functions is frequently unclear, new technologies have started to unravel the critical role of clipping in cellular homeostasis and disease [16,17]. In some reports, histone clipping has even been suggested to be a functional modification with epigenetic potential [18,19]. More specifically, cH2A caught our attention as H2A is the only histone with a *C*-terminus protruding out of the nucleosomal core and the clipping site is localized at the entry and exit points of the DNA [20]. In general, histone modifications help in determining the heritable transcriptional state and lineage commitment development in normal B-cells [21]. Consequently, we persisted in the investigation of the H2A clipping in CLL as disruption of the histone code is suggested to drive hematopoietic cells in lymphomagenesis [22]. Here, we initially observed that H2A clipping was more abundant in CLL patients compared to healthy controls while no differences were found between M and UM. However, we showed this was not due to the disease itself, since this clipping seemed to be associated with the amount of myeloid cells present in the predominantly lymphoid samples. To further unfold the actual role of histone H2A proteolysis, we examined this cH2A clipping during induced differentiation of myeloid THP-1 cells into macrophages through quantitative mass spectrometry [12]. We concluded that synchronization of the THP-1 cells before the stimulation abrogates the temporally uprise of H2A clipping which has initially been observed at the onset of the differentiation.

2. Results

2.1. Differential Protein Expression between Healthy B-Cells and Chronic Lymphocytic Leukemia (CLL) B-Cells of M^- and UM^+ Staged Patients

The difference in protein expression between the B-cells of patients with indolent mutated $ZAP70^-$ (M^-) CLL, the aggressive unmutated $ZAP70^+$ (UM^+) CLL and healthy 50+ donors was quantified using iTRAQ. The six runs were merged into a dataset which comprised 18,014 MS/MS (mass spectrometry) queries, yielding 7473 annotated peptides which were derived from 536 unique proteins (Table S1a). For the ratios included in the analysis, in total 107 proteins differed significantly between groups (Table S1b). Among the leukemia samples, only 22 proteins were distinct between M^- and UM^+ as opposed to 70 between healthy and both CLL samples. One label in each run comprised the same pool of all CLL samples, which was included to simplify inter-run comparison and to increase sensitivity.

Functional clustering annotation of the protein lists from differentially expressed proteins between healthy and leukemia B-cells, showed that most up-regulated proteins in the leukemia samples (both UM^+ and M^-) are involved in mRNA processing, implying an increased transcriptional activity in cancer cells (Table S1c). Compared to healthy, most of the proteins that are down-regulated in the leukemia cells are predominantly connected to actin binding and cellular localization [23]. A cellular component analysis confirmed that 64% of the proteins up-regulated in CLL are categorized as nuclear whereas down-regulated proteins are primarily located in the cytosol (60%) (Table S1c) [24]. These results suggest morphological differences between the healthy and cancerous cells rather than true molecular aberrancies.

In the limited list of proteins specifically overexpressed in the aggressive UM^+ compared to M^- two proteins are involved in ATP-binding, one is a ribosomal subunit and the H2B and Prohibitin proteins are involved in chromosomal organization (Table 1). For proteins significantly down-regulated in UM^+/M^-, the Interleukin enhancer-binding factor 3 and Bcl-2-associated transcription factor 1 are associated with DNA but most proteins are involved in metabolic mechanisms or are cytoskeletal.

2.2. Quantitative Mass Spectrometry and Western Blot Analysis on CLL Samples Endorsed the Myeloid Origin of H2A Clipping

Additional in-depth analysis at the peptide level, surfaced one remarkably aberrant modification in all the samples of the described iTRAQ analysis: clipping of histone H2A after V_{114} (Figure 1). The peptide VTIAQGGVLPNIQAV (*m*/*z* 740.4, charge 2+) was the only semi-tryptic peptide out of the >7400 annotated MS/MS spectra that was identified in all six runs. The annotation of the peptide was confirmed by *de novo* sequencing on the MS/MS spectrum (Figure S1). These results highlight how iTRAQ chemistry contributes to a better annotation of semi-tryptic peptides by enhancing the sensitivity due to the multiplexing and increased *b*-ion formation.

Figure 1. Sequence of histone H2A. Sequence from Uniprot [25]. The VTIAQGGVLPNIQAV peptide (*m*/*z* 740.4, charge 2+) was the only semi-tryptic peptide identified in all six runs. * indicates the clipping site after V_{114} (here described as amino acid 115 since methionine, generated by the start codon, is the first amino acid in the Uniprot sequence); Bold underlined sequences are the same as the isotopically labeled absolute quantification (AQUA) peptides, used for the subsequent specific cH2A quantitation. The bold double underlined sequence is the Western blot epitope.

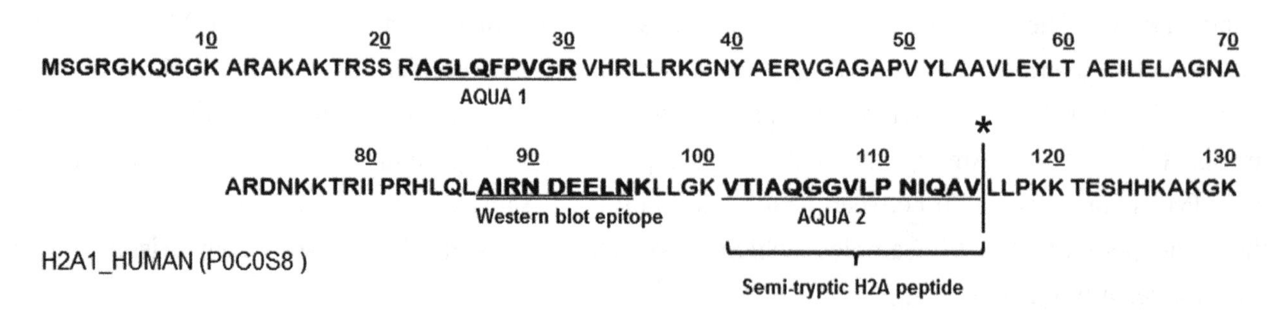

Table 1. Proteins which are differentially expressed between UM^+ and M^- Chronic Lymphocytic Leukemia (CLL) B-cells. The table lists the protein name, Uniprot_ID, average UM^+/M^- ratio, the number of iTRAQ samples where the protein was identified (*n*), the *p*-value of the performed *t*-test and the protein function (derived from UniProt GO).

UM^+ *vs.* M^-	Protein Name	Identified Protein	UM^+/M^- Average Ratio	*n*	*p*-Value	Function (GO)
Down-regulated in UM^+/M^-	Serine/arginine-rich splicing factor	SRSF1_HUMAN	0.61	4	2.77×10^{-4}	RNA binding
		SRSF3_HUMAN	0.77	5	1.30×10^{-2}	RNA binding
		SRSF7_HUMAN	0.90	3	2.70×10^{-2}	RNA binding
	Bcl-2-associated transcription factor	BCLF1_HUMAN	0.69	3	3.84×10^{-2}	DNA binding, induction of apoptosis
	40S ribosomal protein	RS21_HUMAN	0.70	3	3.64×10^{-2}	Ribosomal subunit
		RS28_HUMAN	0.67	4	1.47×10^{-2}	Ribosomal subunit
	Stress-induced-phosphoprotein	STIP1_HUMAN	0.73	3	4.68×10^{-2}	Golgi apparatus
	Protein S100-A6	S10A6_HUMAN	0.75	5	2.24×10^{-2}	Calcium ion binding
	Heat shock cognate 71 kDa protein	HSP7C_HUMAN	0.79	5	1.44×10^{-2}	ATP binding
	Matrin-3	MATR3_HUMAN	0.80	3	1.85×10^{-2}	RNA binding
	Ezrin	EZRI_HUMAN	0.84	4	3.32×10^{-2}	Actin filament
	Transitional endoplasmic reticulum ATPase	TERA_HUMAN	0.84	5	1.13×10^{-2}	ATP binding
	Interleukin enhancer-binding factor	ILF3_HUMAN	0.87	3	5.81×10^{-3}	DNA binding
	Heterogeneous nuclear ribonucleoprotein	HNRPK_HUMAN	0.88	6	1.38×10^{-3}	ATP binding, RNA binding
	Heterogeneous nuclear ribonucleoprotein	ROA1_HUMAN	0.89	5	1.15×10^{-2}	mRNA splicing
Up-regulated in UM^+/M^-	40S ribosomal protein	RS13_HUMAN	1.27	5	4.81×10^{-2}	Ribosomal subunit
	Prohibitin	PHB_HUMAN	1.35	4	3.72×10^{-2}	DNA replication
	Endoplasmin	ENPL_HUMAN	1.80	3	2.05×10^{-2}	ATP binding
	ATP synthase subunit beta, mitochondrial	ATPB_HUMAN	1.93	3	3.72×10^{-2}	ATP binding
	Histone H2B	H2B1C_HUMAN	2.59	3	1.24×10^{-2}	DNA binding
		H2B1D_HUMAN	1.10	3	1.84×10^{-2}	DNA binding
		H2B1O_HUMAN	1.18	3	2.97×10^{-2}	DNA binding

For quantitation of this fragment, we compensated for abovementioned morphological differences by normalizing the reporter intensities of the clipping fragments to the average of the whole H2A protein. The relative amount of H2A that is cleaved after V_{114} was found to be on average higher in leukemia cells compared to healthy B-cells as seen by the log-transformed ratios. Particularly, all six ratios of the pool of UM^+ and M^- samples were consistently positive opposed to healthy (Figure 2A).

Figure 2. Quantitative mass spectrometry and Western blot analysis on CLL samples revealed that histone H2A clipping is from myeloid origin. (**A**) The iTRAQ ratios of the cH2A peptide (normalized to the average of the H2A protein to compensate for morphological differences) hint towards an increased abundance in CLL compared to healthy B-cells. The individual log ratios of all six runs are presented as dots, the average ratios as the horizontal bars. Histone H2A (cH2A) is up-regulated in the samples of the leukemia pool compared to the samples of the healthy B-cells in all six runs; (**B**) The histone extracts from 12 of the 36 samples (From left to right: patient samples UM^+ 8, 9, 10 and 11; M^- 9, 10, 11, 12, 13 and 14; M^+ 5 and UM^- 6). Western blot with an H2A antibody against the epitope depicted in Figure 1 detects the H2A variants H2Ax, Ubiquitinated H2A and macroH2A. The band under ubiquitinated H2A could not be identified. cH2A was only faintly detected, except for one sample (*); (**C**) Specific cH2A screening with AQUA peptides and flow cytometry data revealed the myeloid characteristic of the clipping. **Left panel**: %cH2A differs significantly ($p = 0.049$) between CLL patients with a distinct mutational status (UM: 11× UM^+ & 6× UM^-, M: 14× M^- & 5× M^+); **Middle panel**: Although not significant ($p = 0.15$), the %CD66b suggested a similar correlation with the mutational status; **Right panel**: Relation between %cH2a and %CD66b. Spearman's Rho correlation between $CD66b^+$ and %cH2A was significant at the 0.01 level (Spearman's Rho correlation coefficient: 0.439; $p = 0.007$; Data: Table S2b).

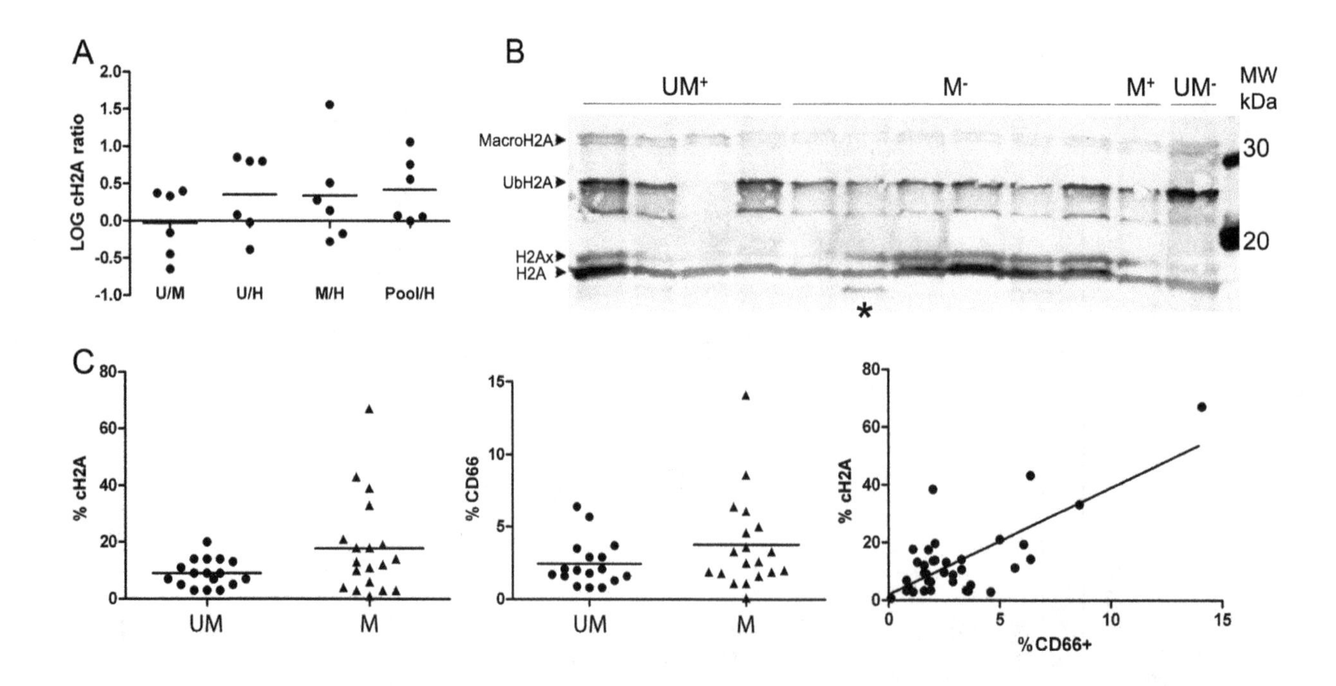

The consequences of proteolytic PTMs are ubiquitously underappreciated so we persisted in the more specific investigation of cH2A in a larger patient population due to the biomarker potential of this previously reported modification (Scheme of workflow: Figure S2). Immunodetection of H2A on histone extracts visualized cH2A only in one sample (Figure 2B, SYPRO staining: Figure S3). The flow cytometry data showed that this particular sample had the highest amount of granulocytes (expressed as %CD66b^{+} cells) in the screened patient population. In line with this, direct comparison of %cH2A with the more sensitive and specific AQUA (absolute quantification) approach suggested higher clipping in the M compared to UM CLL B-cells (Figure 2C, left panel) where a similar relationship between the amount of granulocytes cells and the respective mutational status was observed (Figure 3C, middle panel). Indeed, a statistical analysis of the possible correlation between the amounts of V_{114} clipped H2A and all cellular markers investigated for each sample confirmed a significant correlation between the amount of CD66b^{+} cells present in the samples and %cH2A (Figure 3C, right panel) (Details patient screening are listed in Table S2). Even though all steps were performed at 4 °C in the presence of protease inhibitors, we observed that H2A clipping can actually be caused *in vitro* to some extent as we observed some residual clipping activity on spiked-in biotinylated H2A (data not shown).

2.3. Characterization of cH2A in THP-1 Cells

To further corroborate the myeloid nature of cH2A we cultured human leukemia cell lines of different origin and prepared histone extracts when a density of approximately 1×10^{6}/mL was achieved. In all the investigated lymphatic cell lines, no clipping of H2A was detected. For the myeloid cells, ±10% cH2A was measured in the monocyte-like THP-1 and U-937 cell lines and, although not unambiguously, ±2% in the promyelocytic HL-60 cells (Figure 3A).

Next, we applied our sensitive and specific AQUA approach to specifically quantify H2AV_{114} clipping in a previously reported model in which Ohkawa *et al.* describe a temporal increase of truncated H2A proteins during stimulated differentiation of acute monocytic leukemia THP-1 cells into macrophages using both phorbol 12-myristate 13-acetate (PMA) and retinoic acid (RA) [13]. For three biological replicates, %cH2A was specifically quantified in histone extracts, isolated at different time-points shortly after PMA supplementation (Figure 3B). The clipping at T_{PMA0} was significantly different than at $T_{PMA\,10}$ ($p = 0.045$) and T_{PMA60} ($p = 0.011$). Equally as to T_{PMA60}, %cH2A was significantly different than at $T_{PMA\,5}$ ($p = 0.041$), $T_{PMA\,10}$ ($p = 0.010$) and $T_{PMA\,30}$ ($p = 0.0075$). These results confirm that H2A clipping at V_{114} indeed ascends and sequentially decreases upon THP-1 differentiation.

Finally, since we hypothesized that the variation might be due to differences in cell cycle synchronization, THP-1 cells were synchronized by double thymidine block, as the more open chromatin structure during cell division could very well be more susceptible to proteolytical activity during cell lysis. After synchronization, cells were either left untreated or were subjected to differentiation and cH2A was subsequently quantified by AQUA screening on histone extracts, obtained from different time points (Figure 3B). %cH2A was significantly lower in both the control ($p = 6.09 \times 10^{-4}$) and stimulated ($p = 9.18 \times 10^{-5}$) THP-1 cells compared to the unsynchronized cells. Further, no difference was observed between the synchronized stimulated and control cells ($p = 0.22$). Sampling 48 h after the start of the differentiation indicated that the amount of clipped H2A was further reduced in both cell lines (data not shown).

Figure 3. Specific cH2A quantitation in myeloid cell lines. (**A**) Example of an MS spectrum of the AQUA 2 peptide if H2A V_{114} clipping is absent (**left**) or present (**right**). The analysis of different cell lines confirms the myeloid characteristic of cH2A in the U-937, THP-1 and HL-60 cells. H2A clipping is not present in the investigated lymphatic cell lines and could not be unambiguously detected in the promyelocytic HL-60 cell line; (**B**) THP-1 cells stimulated with PMA show a transient H2A clipping pattern. Three biological replicates display the high variance. *1: significant differences of %cH2A between T_{PMA0} and the other data points; *2: the equivalent for T_{PMA60}. The %cH2A of synchronized stimulated and control THP-1 cells, presented respectively as squares and triangles, is lower than the %cH2A of non-synchronized cells.

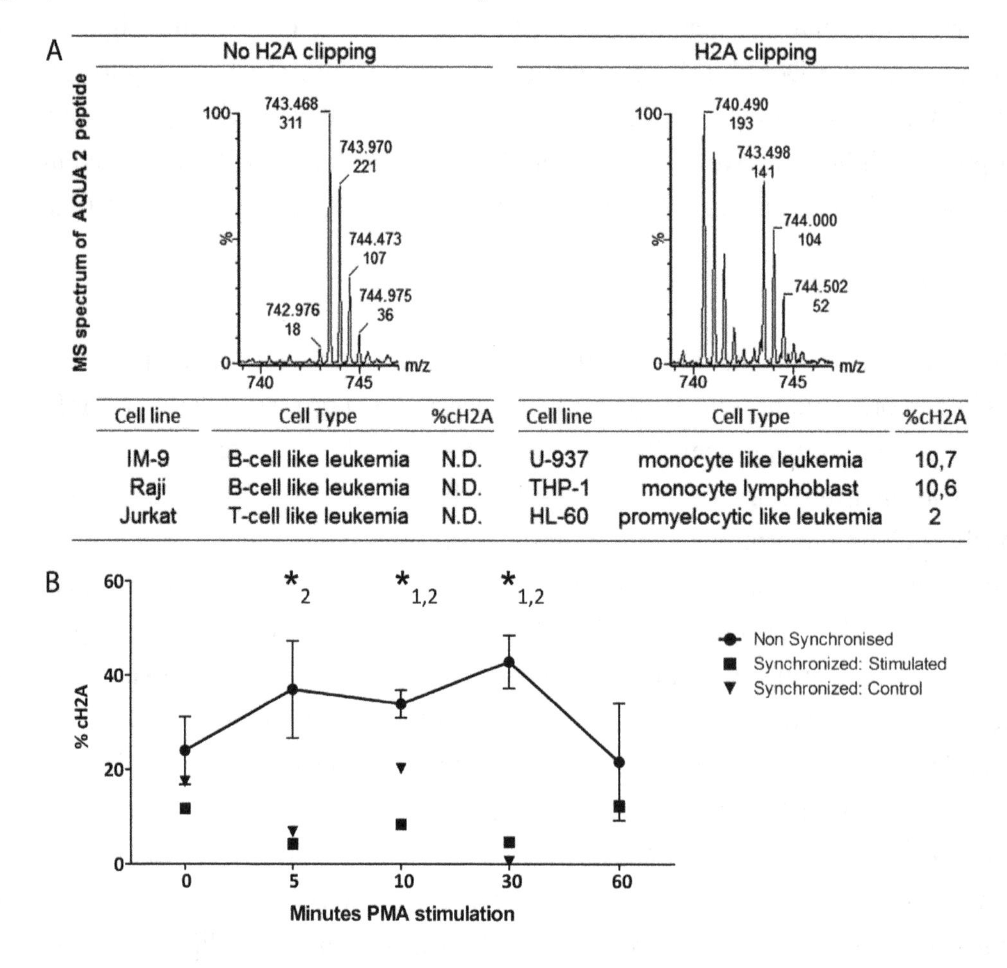

Cell line	Cell Type	%cH2A	Cell line	Cell Type	%cH2A
IM-9	B-cell like leukemia	N.D.	U-937	monocyte like leukemia	10,7
Raji	B-cell like leukemia	N.D.	THP-1	monocyte lymphoblast	10,6
Jurkat	T-cell like leukemia	N.D.	HL-60	promyelocytic like leukemia	2

3. Discussion

Quantitative proteomics on patient samples and on leukemia cell lines can help to define new biomarkers and insights into the pathogenesis of lymphoid and hematopoietic neoplasms [2,3]. In our comparative proteome study between healthy and CLL B-cells of patients with different prognosis, we implemented two quantitative label-based mass spectrometry methods: iTRAQ and AQUA. The isobaric character of the iTRAQ labels allows multiplexing different samples, resulting in an increased signal and lower sample complexity [26]. We included both CLL samples of patients with a different disease prognosis and healthy B-cells in the analysis to obtain deeper insights in disease biology. A substantial part of the observed significant differences in relative protein expression between healthy and CLL B-cells is primarily due to differences in morphology between healthy and neoplastic cells [27]. Cancerous B-cells are indeed characterized by a larger nucleus and a denser cytoplasm, which we affirmed here by functional grouping and cellular component analysis of the up- and down-regulated proteins. We caution for the interpretation of proteomics data when healthy and malignant cells differ strongly in their morphology, an important restriction that is not always validated prior to relative comparison in proteomics approaches. From a clinical perspective however, aberrant ratios between patient groups with UM^+, who have a bad prognosis and poor overall survival range and the less aggressive M^-, are more relevant. Notably, less than 5% of the identified proteins were significantly different between these two groups. In this list of proteins with a log ratio significantly different from zero, defining distinctions between M^- and UM^+, one of the most remarkable candidates is the Bcl-2-associated transcription factor 1 ($p = 3.84 \times 10^{-2}$). The Bcl-2 family regulates apoptosis and is an established hallmark in CLL as aberrant expression of Bcl-2 proteins causes apoptosis resistance of CLL B-lymphocytes. Although not all the Bcl-2 proteins correlate consistently with known CLL biomarkers, several Bcl-2-anatagonists are in clinical trials for CLL treatment [8].

The use of the iTRAQ label and the subsequent manual analysis of the results at the peptide level together surfaced another fascinating finding: a semi-tryptic peptide derived from histone H2A clipped at V_{114}. Of all the trypsin-based mass spectrometry experiments uploaded in the PRIDE database, this fragment was only annotated once (accession: 10,528) [28]. Remarkably, we identified the peptide in all six runs. This could be explained by the contribution of the iTRAQ labels which are covalently bound to the peptide *N*-termini, generating intense b-series and consequently contributing to an in increased peptide score [26]. cH2A is generated by the removal of fifteen amino acids from the carboxy-terminal end of the intact H2A molecule after V_{114}, coordinately removing K119 which is an important site of mono-ubiquitination [29–33]. cH2A had been described over 35 years ago as a product of H2A specific protease (H2Asp) activity [14] and shortly afterward, other groups observed similar cleavage patterns in extracts from both myeloid and lymphatic leukemia's [10,11]. Since then, this clipping event was only referenced on a few occasions.

Although the biological function of protein degradation is largely unknown, proteolysis is an important category of PTMs, e.g., 5%–10% of all drug targets are proteases [16]. Truncation of histone tails has already been linked to cell differentiation and the *C*-tail of H2A is known to be important in cellular homeostasis and chromatin biology [18,20]. We recently identified the myeloid enzyme NE as being a prime candidate to fulfill the reaction mediated by the H2Asp but could not clarify if cH2A formation is involved in an epigenetic process or is rather a consequence of NET formation [15]. In healthy hematopoietic cells we only observed H2A clipping in cells of myeloid origin but the references to clipped H2A found in literature are mainly in the context of leukemia [10–13]. We thus persisted in a more detailed analysis of this modification and examined if cH2A formation is an epigenetic hallmark of CLL.

The preliminary iTRAQ results indicated a greater average abundance of the cH2A peptide in leukemia samples compared to the healthy controls. As validation is required, we specifically quantified cH2A in histone extracts derived from CLL B-cells of 36 clinically staged patients by applying the isotopic synthesized AQUA peptides that allow to target specific known H2A peptides, present in low concentrations [26]. We could this time not define any direct connection between %cH2A and any known disease marker. However, the amount of granulocytes in the samples did correlate with %cH2A, which corroborates the myeloid character of cH2A and is in line with the identification of NE. As complete inhibition of proteases is known to be very challenging, as also seen by us in a spike-in experiment of biotinylated H2A (not shown), it is difficult to define how much, if any, cH2A is endogenously generated [34].

On the other hand, the reported transient clipping of the H2A *C*-tail during the induced differentiation of THP-1 promonocytes, implies a potential biological role of H2A processing in the hematopoietic development of cells from myeloid origin [13]. Cells may for instance have mechanisms to control histone degradation for re-establishing the epigenetic marks on their tails in the proliferating state [35]. However, although our AQUA results demonstrate a brief uprise in specific H2A V_{114} clipping upon PMA or RA stimulation, synchronization of the THP-1 cells before the stimulation abrogated such trend. Histone clipping seems to correlate with the myeloid cell cycle and as our results suggest, cH2A fluctuation is more likely caused by possible differences in cell cycle stage synchronization, rather than PMA or RA induced differentiation. Instead of being a regulated mechanism, we hypothesize that the high degree of variation found in these experiments probably is due to the more open chromatin structure during cell division rendering histones more susceptible to proteolytical activity.

4. Experimental Section

4.1. Cells and Reagents

Phosphate buffered saline (PBS), media, L-glutamine, Fetal bovine serum (FBS), penicillin/streptomycin, Dynabeads and SYPRO Ruby were from Life Technologies (San Diego, CA, USA), ammonium bicarbonate (ABC), sodium dodecyl sulfate (SDS), *N*-cyclohexyl-3-aminopropanesulfonic acid (CAPS) and Tween-20 from Millipore (Billerica, MA, USA). ReadyPrep sequential extraction kit was from BioRad (Hercules, CA, USA) and Vivaspin-2 columns from Sartorius (Göttingen, Germany). The Recombinant human H2A (M2502S) was obtained from New England Biolabs (Ipswich, MA, USA) and bovine histone extract (cat. no. 223565) from Roche (Basel, Switzerland). All other reagents were purchased from Sigma Aldrich (St. Louis, MO, USA) unless described otherwise. Raji, Jurkat and HL-60 cells were cultured in Dulbecco's Modified Eagle Medium and IM-9, U-937 and THP-1 cells in RPMI-1640 medium, both enriched with 2% (*w*/*v*) L-glutamine, 10% (*w*/*v*) FBS and 50 IU/mL penicillin/streptomycin. To achieve synchronization at the late G1− early S phase, 2 mM thymidine was added over two intervals of 12–16 h, with an incubation in non-thymidine containing RMPI-1640 medium for 8 h in between 50 ng/mL PMA was added to the THP-1 medium for the differentiation of the non-synchronized cells and 1 μM RA for the differentiation of the synchronized cells.

4.2. Patient Samples

For the iTRAQ analysis, whole blood of six UM^+ and six M^- clinically staged CLL patients was obtained from Ghent University Hospital, Department of Hematology. Informed consent was given according to the requirements of the Ethics Committee of the Ghent University Hospital. PBMCs were isolated from a Ficoll-Paque (GE Healthcare, Waukesha, WI, USA) density gradient, B-cells were purified using Dynabeads Untouched Human B (Life Technologies, San Diego, CA, USA). Correspondingly, control samples were isolated from healthy volunteers aged 50+. For the high throughput screening of cH2A, samples were obtained from the UCSD CLL Research Consortium (CRC). Immediately after thawing, cells were washed twice with cold PBS containing 1 mM phenylmethanesulfonyl fluoride (PMSF) and protease inhibitor cocktail (Roche, Basel, Switzerland). $CD5^{+}CD19^{+}$ cells consistently made out more than 85% of the lymphocyte population as seen by flow cytometry.

4.3. Flow Cytometry

For each measurement 2×10^5 cells were washed twice at 4 °C with PBS 1% Bovine Serum Albumin (BSA) and analyzed using a Cytomics FC500 flow cytometer (Becton Dickinson Immunocytometry Systems, San Jose, CA, USA) with monoclonal antibodies (mAbs) antibodies from BD-Biosciences (Franklin Lakes, NJ, USA): Isotype controls, anti-CD5 (PECy5), anti-CD19 (FITC), anti-CD33 (PECy5), anti-CD66b (FITC) and anti-Annexin V (FITC). The synchronization of THP-1 cells was monitored with Propidium Iodide staining.

4.4. Cell Lysis and Histone Isolation

All steps were performed at 4 °C. To obtain a complete cell lysate for the iTRAQ analysis, cells were washed twice with PBS, pelleted and resuspended in the ReadyPrep sequential extraction buffer 1 at 5×10^6 cells/mL, supplemented with protease inhibitor cocktail (Roche), 1 mM PMSF, 10 µL 200 mM Tributylphosphine (Biorad), 20 µL phosphatase inhibitor cocktail 1 and 2 and 1 µL 250-units/µL benzonase. After sonication and centrifugation at 1500 rpm for 5 min, the proteins in the supernatant were transferred to a new eppendorf. The obtained pellet was resuspended in buffer 3 from the extraction kit and sonicated for 10 min. After centrifugation at 1500 rpm for 5 min the supernatant was pooled with the previous extract. Detergents, inhibitors and urea were removed by washing twice with Milli-Q water on a Vivaspin-2 column.

For the histone extracts, harvested cells were washed twice in PBS containing 1 mM PMSF, and protease inhibitor cocktail. 10^7 cells/mL were resuspended in Triton extraction buffer (PBS containing 0.5% (*v*/*v*) Triton 100×, 1 mM PMSF and protease inhibitor cocktail) and lysed by gentle stirring. Pelleted nuclei were subsequently washed in PBS containing 1 mM PMSF and proteinase inhibitor cocktail. Histones were extracted overnight after benzonase treatment of the sonicated nuclei by acid extraction: incubation in 250 µL 0.2 M HCl at 4 °C with gentle stirring. Precipitated proteins were pelleted and the supernatant containing the histones was dried and stored at −20 °C until further use. The Bradford Coomassie Assay determined the protein content of all samples.

4.5. Western Blot Analysis

An amount of 3 µg of a dried histone extract was suspended in Laemmli buffer and run on a 15% PreCast Gel (Biorad) for 30 min at 150 V and 60 min at 200 V and subsequently transferred to a nitrocellulose membrane in a 10 mM CAPS buffer with 20% MeOH (Merck, New York, NY, USA). The remaining proteins in the gel were visualized after overnight Sypro Ruby staining. For Western blot, all steps were performed at room temperature with intermediate washing steps in 0.3% Tween-20 in PBS. The Histone H2A (LS-C24265, LifeSpan BioScience, Seattle, WA, USA) antibody incubation was performed overnight in a 1:1000 dilution in PBS 1% BSA, followed by 1 h incubation in the same buffer, with a stabilized horseradish-peroxidase (HRP)-conjugated goat anti-rabbit immunoglobulin G

(Pierce, Rockford, IL, USA). The Supersignal West Dura Extended Duration Substrate (Pierce) was applied to perform the chemiluminescence and both the gel and Western blot membrane were visualized with a VersaDoc Imaging System.

4.6. Quantitative Mass Spectrometry Analysis

All digests were performed according to the iTRAQ (ABSciex, Foster City, CA, USA) reagent Kit guidelines, as was the given labeling itself. For the iTRAQ analysis, six 4plex runs each encompassed 4 × 100 μg of the protein lysates from CLL B-cells from UM^+, M^- and B-cells from healthy donors aged 50^+. The fourth label was used for an additional pool of all 12 CLL samples for inter-run comparison and to increase the number of identifications. Technical variation was also minimized by reversing the labeling order. All six samples were first fractionated off line into 12 fractions on a Poros 10S strong-cation exchange (SCX) column (300 μm i.d. × 15 cm, ABSciex, Foster City, CA, USA) for subsequent nano reversed phase liquid chromatography (Dionex U3000, Dionex, Chelmsford, MA, USA) separation on a gradient specifically optimized for sample content of each fraction [36]. All other samples were separated on a 70 min organic gradient from 4%–100% buffer B (80% (*v*/*v*) acetonitrile (Millipore, Billerica, MA, USA) in 0.1% (*v*/*v*) FA). All samples were analyzed on an Electrospray ionization—Q-TOF Premier (Waters, Wilmslow, UK). Data was searched against the SwissProt (541,954 sequence entries) database using Mascot 2.3 and additionally manually interpreted with Mascot Distiller (Matrix Science, London, UK). Proteins were included for the analysis if a proteotypic peptide had a Mascot score above 45 in at least 3 runs. For iTRAQ quantitation, ratios were normalized based on summed intensities and only proteins recurring in at least three out of six different runs were withheld by an automated in-house approach. Gene ontology (GO) analysis was performed on the significantly up- or down-regulated proteins with Uniprot, DAVID and the WEB-based GEne SeT AnaLysis Toolkit (WEBGESTAT) [23–25].

For specific cH2A screening, a mix of the isobaric peptides (AQUA1, Thermo, Waltham, MA, USA) AQUA2, Sigma Aldrich (St. Louis, MO, USA) was added right before MS analysis. The tryptic H2A *N*-terminal peptide R.AGLQFPVGR.V (*m*/*z* 475.7 was used to quantify the total amount of histone H2A present in the sample (AQUA 1). Specific clipping is quantified by means of the semi-tryptic peptide K.VTIAQGGVLPNIQAV.L (*m*/*z* 743.4) (AQUA2). This is the same sequence as was found during the iTRAQ analysis. To 1 μg of HE, 10 pmol of AQUA1 and 1 pmol of AQUA 2 was spiked right before the MS analysis [15]. For the quantitation, the total ion current (TIC) from both AQUA peptides was obtained from a single MS-scan acquired on the top of the two extracted-ion chromatograms. An example of the MS spectra of the two AQUA peptides is given at the right site of the accolade in Figure S2. To confirm the efficiency of the histone extraction, the raw data of the data-directed analysis was equally searched against the SwissProt database using Mascot 2.3 (Matrix Sciences, London, UK).

4.7. Statistical Analysis

For the iTRAQ analysis, ratios were log transformed and averaged. Statistical analysis was performed by means of a homoscedastic two-tailed *t*-test to inspect which log ratios were significantly different from zero (p = 0.05). The average ratio of a protein indicated whether a protein is up- or down-regulated between two samples. The same *t*-tests were performed to test if %cH2A and %CD66b was different between the M and UM patients. For the comprehensive analysis of the AQUA screening, the relationship between cH2A and the other variables was determined by the nonparametric Spearman's Rho correlation test wherein the ZAP and the mutational status were analyzed as dichotomized values (+ or −) with the SPSS Statistics 20 software (Endicott, NY, USA). To corroborate if the clipping temporally increased during the differentiation of the THP-1 cells, one-tailed *t*-tests were applied to compare each data point with the %cH2A at $T_{PMA\ 0}$ and $T_{PMA\ 60}$. With the same test, all the data points of the non-synchronized experiment were examined against all the data points of the synchronized (stimulated or control) experiment. Similarly, a *t*-test was used to validate if the %cH2A between the synchronized stimulated and control cells was analogous.

5. Conclusions

We particularly emphasize both the pitfalls and the benefits of applying quantitative proteomic strategies in hematological research. Although morphological differences hamper proteome-wide comparison of healthy and leukemia B-cells, we resurfaced a previously described clipping product of the H2A *C*-tail. A more profound characterization in CLL and THP-1 samples indicates that H2A V_{114} clipping occurs in hematopoietic cells of the myeloid lineage. Although the process of histone clipping has considerable epigenetic potential, many questions about the relevance of specific histone proteolysis remain.

Acknowledgments

We thank Koji Takada from the Jikei University School of Medicine for the vials with the THP-1 cells.

Author Contributions

P.G. wrote the manuscript and performed the bulk of the described laboratory and mass spectrometry experiments together with the subsequent data analysis. L.V. performed the THP-1 synchronization and associated flow monitoring and shared her expertise on cell culturing and overall histone research. K.V.S. helped with/optimized the electrophoresis and Western blot analyses and reviewed the writing. S.L. assisted with the design of the iTRAQ analysis and performed the statistics of the CLL screening. F.V.N. wrote the semi-automated data analysis tool and contributed to the interpretation of the data. F.O. provided the healthy and CLL patient samples for the iTRAQ analysis and supervised the determination of the mutational status of the IGVH genes and the ZAP70 expression. T.K. provided the CLL patient samples for the AQUA analysis and supervised the determination of the mutational status of the IGVH genes and the ZAP70 expression. M.D. performed the iTRAQ analysis and assisted with the subsequent experiments, data analysis and the writing of the manuscript. D.D. supervised this work.

References

1. Unwin, R.D.; Whetton, A.D. How will haematologists use proteomics? *Blood Rev.* **2007**, *21*, 315–326.
2. Boyd, R.S.; Dyer, M.J.; Cain, K. Proteomic analysis of B-cell malignancies. *J. Proteomics* **2010**, *73*, 1804–1822.
3. Zamo, A.; Cecconi, D. Proteomic analysis of lymphoid and haematopoietic neoplasms: There's more than biomarker discovery. *J. Proteomics* **2010**, *73*, 508–520.
4. Chiorazzi, N.; Rai, K.R.; Ferrarini, M. Chronic lymphocytic leukemia. *N. Engl. J. Med.* **2005**, *352*, 804–815.
5. Chiorazzi, N. Implications of new prognostic markers in chronic lymphocytic leukemia. *Hematol. Am. Soc. Hematol. Educ. Program* **2012**, *2012*, 76–87.
6. Smolewski, P.; Witkowska, M.; Korycka-Wolowiec, A. New insights into biology, prognostic factors, and current therapeutic strategies in chronic lymphocytic leukemia. *ISRN Oncol.* **2013**, *2013*, 740615.
7. Rodriguez-Vicente, A.E.; Diaz, M.G.; Hernandez-Rivas, J.M. Chronic lymphocytic leukemia: A clinical and molecular heterogenous disease. *Cancer Genet.* **2013**, *206*, 49–62.
8. Scarfo, L.; Ghia, P. Reprogramming cell death: BCL2 family inhibition in hematological malignancies. *Immunol. Lett.* **2013**, *155*, 36–39.
9. Woodlock, T.J.; Bethlendy, G.; Segel, G.B. Prohibitin expression is increased in phorbol ester-treated chronic leukemic B-lymphocytes. *Blood Cells Mol. Dis.* **2001**, *27*, 27–34.
10. Simpkins, H.; Mahon, K. The histone content of chromatin preparations from leukaemic cells. *Br. J. Haematol.* **1977**, *37*, 467–473.
11. Pantazis, P.; Sarin, P.S.; Gallo, R.C. Detection of the histone-2A related polypeptide in differentiated human myeloid cells (HL-60) and its distribution in human acute leukemia. *Int. J. Cancer* **1981**, *27*, 585–592.
12. Okawa, Y.; Takada, K.; Minami, J.; Aoki, K.; Shibayama, H.; Ohkawa, K. Purification of *N*-terminally truncated histone H2A-monoubiquitin conjugates from leukemic cell nuclei: Probable proteolytic products of ubiquitinated H2A. *Int. J. Biochem. Cell Biol.* **2003**, *35*, 1588–1600.
13. Minami, J.; Takada, K.; Aoki, K.; Shimada, Y.; Okawa, Y.; Usui, N.; Ohkawa, K. Purification and characterization of *C*-terminal truncated forms of histone H2A in monocytic THP-1 cells. *Int. J. Biochem. Cell Biol.* **2007**, *39*, 171–180.
14. Eickbush, T.H.; Watson, D.K.; Moudrianakis, E.N. A chromatin-bound proteolytic activity with unique specificity for histone H2A. *Cell* **1976**, *9*, 785–792.
15. Dhaenens, M.; Glibert, P.; Lambrecht, S.; Vossaert, L.; van Steendam, K.; Elewaut, D.; Deforce, D. Neutrophil Elastase in the capacity of the "H2A-specific protease". *Int. J. Biochem. Cell Biol.* **2014**, *51C*, 39–44.
16. Doucet, A.; Butler, G.S.; Rodriguez, D.; Prudova, A.; Overall, C.M. Metadegradomics: Toward *in vivo* quantitative degradomics of proteolytic post-translational modifications of the cancer proteome. *Mol. Cell. Proteomics* **2008**, *7*, 1925–1951.
17. Rogers, L.; Overall, C.M. Proteolytic post translational modification ofproteins: Proteomic tools and methodology. *Mol. Cell. Proteomics* **2013**, *12*, 3532–3542.
18. Osley, M.A. Epigenetics: How to lose a tail. *Nature* **2008**, *456*, 885–886.

19. Azad, G.K.; Tomar, R.S. Proteolytic clipping of histone tails: The emerging role of histone proteases in regulation of various biological processes. *Mol. Biol. Rep.* **2014**, *41*, 2717–2730.
20. Vogler, C.; Huber, C.; Waldmann, T.; Ettig, R.; Braun, L.; Izzo, A.; Daujat, S.; Chassignet, I.; Lopez-Contreras, A.J.; Fernandez-Capetillo, O.; *et al.* Histone H2A *C*-terminus regulates chromatin dynamics, remodeling, and histone H1 binding. *PLoS Genet.* **2010**, *6*, e1001234.
21. Parra, M. Epigenetic events during B lymphocyte development. *Epigenetics* **2009**, *4*, 462–468.
22. Taylor, K.H.; Briley, A.; Wang, Z.; Cheng, J.; Shi, H.; Caldwell, C.W. Aberrant epigenetic gene regulation in lymphoid malignancies. *Semin. Hematol.* **2013**, *50*, 38–47.
23. Huang, W.; Sherman, B.T.; Lempicki, R.A. Systematic and integrative analysis of large gene lists using DAVID bioinformatics resources. *Nat. Protoc.* **2009**, *4*, 44–57.
24. Wang, J.; Duncan, D.; Shi, Z.; Zhang, B. WEB-based GEne SeT anaLysis toolkit (WebGestalt): update 2013. *Nucleic Acids Res.* **2013**, *41*, W77–W83.
25. Magrane, M.; Consortium, U. UniProt knowledgebase: A hub of integrated protein data. *Database* **2011**, *2011*, bar009.
26. Vaudel, M.; Sickmann, A.; Martens, L. Peptide and protein quantification: A map of the minefield. *Proteomics* **2010**, *10*, 650–670.
27. Baba A.I.; Câtoi, C. *Tumor Cell Morphology*; The Publishing House of the Romanian Academy: Bucuresti, Romania, 2007.
28. Vizcaino, J.A.; Cote, R.G.; Csordas, A.; Dianes, J.A.; Fabregat, A.; Foster, J.M.; Griss, J.; Alpi, E.; Birim, M.; Contell, J.; *et al.* The PRoteomics IDEntifications (PRIDE) database and associated tools: Status in 2013. *Nucleic Acids Res.* **2013**, *41*, D1063–D1069.
29. Eickbush, T.H.; Moudrianakis, E.N. The histone core complex: An octamer assembled by two sets of protein–protein interactions. *Biochemistry* **1978**, *17*, 4955–4964.
30. Eickbush, T.H.; Godfrey, J.E.; Elia, M.C.; Moudrianakis, E.N. H2a-specific proteolysis as a unique probe in the analysis of the histone octamer. *J. Biol. Chem.* **1988**, *263*, 18972–18978.
31. Davie, J.R.; Numerow, L.; Delcuve, G.P. The nonhistone chromosomal protein, H2A-specific protease, is selectively associated with nucleosomes containing histone H1. *J. Biol. Chem.* **1986**, *261*, 10410–10416.
32. Elia, M.C.; Moudrianakis, E.N. Regulation of H2A-specific proteolysis by the histone H3:H4 tetramer. *J. Biol. Chem.* **1988**, *263*, 9958–9964.
33. Watson, D.K.; Moudrianakis, E.N. Histone-dependent reconstitution and nucleosomal localization of a nonhistone chromosomal protein: The H2A-specific protease. *Biochemistry* **1982**, *21*, 248–256.
34. Groth, I.; Alban, S. Elastase inhibition assay with peptide substrates—An example for the limited comparability of *in vitro* results. *Planta Med.* **2008**, *74*, 852–858.
35. Gunjan, A.; Paik, J.; Verreault, A. The emergence of regulated histone proteolysis. *Curr. Opin. Genet. Dev.* **2006**, *16*, 112–118.
36. Lambrecht, S.; Dhaenens, M.; Almqvist, F.; Verdonk, P.; Verbruggen, G.; Deforce, D.; Elewaut, D. Proteome characterization of human articular chondrocytes leads to novel insights in the function of small heat-shock proteins in chondrocyte homeostasis. *Osteoarthr. Cartil.* **2010**, *18*, 440–446.

10

Investigation into Variation of Endogenous Metabolites in Bone Marrow Cells and Plasma in C3H/He Mice Exposed to Benzene

Rongli Sun [†], **Juan Zhang** [†,*], **Lihong Yin and Yuepu Pu** *

Key Laboratory of Environmental Medicine Engineering, Ministry of Education,
School of Public Health, Southeast University, Nanjing 210009, Jiangsu, China;
E-Mails: sunrongli5318@gmail.com (R.S.); lhyin@seu.edu.cn (L.Y.)

† These authors contributed equally to this work.

* Authors to whom correspondence should be addressed; E-Mails: 101011288@seu.edu.cn (J.Z.); yppu@seu.edu.cn (Y.P.)

Abstract: Benzene is identified as a carcinogen. Continued exposure of benzene may eventually lead to damage to the bone marrow, accompanied by pancytopenia, aplastic anemia or leukemia. This paper explores the variations of endogenous metabolites to provide possible clues for the molecular mechanism of benzene-induced hematotoxicity. Liquid chromatography coupled with time of flight-mass spectrometry (LC-TOF-MS) and principal component analysis (PCA) was applied to investigate the variation of endogenous metabolites in bone marrow cells and plasma of male C3H/He mice. The mice were injected subcutaneously with benzene (0, 300, 600 mg/day) once daily for seven days. The body weights, relative organ weights, blood parameters and bone marrow smears were also analyzed. The results indicated that benzene caused disturbances in the metabolism of oxidation of fatty acids and essential amino acids (lysine, phenylalanine and tyrosine) in bone marrow cells. Moreover, fatty acid oxidation was also disturbed in plasma and thus might be a common disturbed metabolic pathway induced by benzene in multiple organs. This study aims to investigate the underlying molecular mechanisms involved in benzene hematotoxicity, especially in bone marrow cells.

Keywords: benzene; endogenous metabolites; bone marrow; plasma; HPLC-TOF-MS

---

1. Introduction

Benzene is an important industrial chemical widely used in the production of many products and is also a component of cigarette smoke, gasoline, crude oil and automobile emissions [1]. The hematopoietic system is the most critically affected target tissue following exposure to benzene in humans and animals. Benzene is an established cause of acute myeloid leukemia (AML), myelodysplastic syndromes (MDS), and very likely also lymphocytic leukemias and non-Hodgkin lymphoma (NHL) in humans [2–4]. Benzene was identified as an environmental carcinogen in 1982 [5] and placed in the Group 1 human carcinogen category in 1987 by the International Agency for Research on Cancer [6].

The major adverse health effect from exposure to benzene is hematotoxicity. Benzene can cause a decrease in the three major circulating cell types: platelets (thrombocytopenia), red blood cells (anemia) and white blood cells (leukopenia); and an increase in red cell mean corpuscular volume [7,8]. Sustained exposure may result in continued marrow depression involving multiple cell lineages. This multi-lineage depression of blood counts is also known as pancytopenia [9]. Continued exposure may eventually lead to damage to the bone marrow, accompanied by pancytopenia or aplastic anemia. Benzene metabolism is inherently complex [10], and its secondary metabolism occurs in the bone marrow [11,12]. It is clear that bone marrow is the most critical target organ for benzene metabolites, and both progenitor cells and stromal cells in bone marrow have been considered to be potential targets of benzene hematotoxicity. There are a number of mechanistic studies in the literature that can help us to understand the primary mode of action for benzene, and significant progress has been made in this area. These studies describe effects of benzene such as chromosomal aberrations [13], covalent binding [14], and gene mutations [15]; as well as newly identified mechanisms that include alterations in gene expression [16], oxidative stress [17], epigenetic regulation [18], and immune suppression [19]. In addition, a number of biomonitoring studies have estimated internal benzene exposure for humans, and these have identified and quantified benzene or its biological adducts in blood, urine or expired air; urinary biomarkers include S-phenylmercapturic acid (SPMA), trans, trans-muconic acid (ttMA), phenol, catechol and hydroquinone. Although there is a wealth of epidemiologic data regarding benzene in humans and animals, exposure and toxicology data on benzene, and the mechanism of action for the hematotoxic effects of benzene are not completely understood; and assessment of internally based metabolites responsible for these effects are not currently available.

A systems biology approach to disease-related biology is revolutionizing our knowledge of the cellular pathways and gene networks that underlie the onset and progression of disease, and their associated pharmacological treatments [20]. The study of metabonomics depends upon the production of global metabolite profiles that enable diagnostic changes in the concentrations, or proportions, of low-molecular-weight organic metabolites in samples (such as biofluids and organ extracts) to be assayed. Such investigations thus generate metabolic phenotypes (metabotypes [21]), and, by studying these, it may be possible to identify target organ response to a specific toxicant [22], to assess the toxicity of candidate chemical agents, and to gain new insights into the mechanisms of toxicity of xenobiotics [23]. Kristin *et al.* carried out metabolomic analyses of stem cell samples from peripheral blood collected from a cohort of patients before hematopoietic cell transplantation, and the results suggested that the development of therapy-related myelodysplasia syndrome (t-MDS) was associated with dysfunctions in cellular metabolic pathways, including those involved in the metabolism of alanine, aspartate, glyoxylate, dicarboxylate, and phenylalanine; the citrate acid cycle; and aminoacyl-t-RNA biosynthesis [24]. A recent study suggested that the acridone derivative, 2-aminoacetamido-10-(3,5-dimethoxy)-benzyl-9(10*H*)-acridone hydrochloride, altered metabolism of fatty acids, nucleosides, amino acids, glycerophospholipid, and glutathione; it also induced oxidative stress-mediated apoptosis in CCRF-CEM leukemia cells [25]. Metabonomic approaches also enable identification of predictive markers and biomarkers of disease progression. Huang *et al.* used metabonomic profiling to identify a putative specific biomarker pattern in urine as a noninvasive bladder cancer (BC) detection strategy, and found carnitine C9:1 and component I (in a combined biomarker pattern) with a high sensitivity and specificity that allowed discrimination of bladder cancer patients [26]. One of the major analytical techniques used for global metabolic profiling at this time is mass spectrometry. Mass spectrometry (MS) occupies a major role in holistic metabolite profiling due to its sensitivity and widespread availability. Liquid chromatography (LC-MS) is currently the most widely used mass spectrometry technology, especially in the life science and bioanalytical sectors, due to its ability to separate and detect a wide range of molecules [27].

In our previous study [28], the results indicated that pathways of purine, spermidine, fatty acids, tryptophan, and peptide metabolism were disturbed in benzene-exposed mice; but this study was restricted to urine, which meant that relevant information regarding the hematopoietic system and interactions among compartments was lost. Compared with biofluids, bone marrow (as a target organ) more directly reflects the pathophysiologic state of disease processes induced by benzene. Endogenous metabolites in plasma reflects systemic metabolic effects associated with benzene exposure, while bone marrow cell metabolites analyses enable a more precise investigation of local metabolic changes. However, there are presently no studies reported regarding benzene-induced metabolic changes in bone marrow cells or plasma.

In the present study, we used an integrated approach that entailed metabonomic analyses (based upon HPLC-TOF-MS) of bone marrow cells and plasma to discern changes in the respective metabolomes and the interactions between the two compartments. Our aims were to study the variation of endogenous metabolites involved in benzene toxicity from metabonomic information derived from bone marrow, and to identify specific endogenous metabolites in plasma as potential biomarkers of benzene's toxic hematopoietic effects.

2. Results and Discussion

2.1. Body Weights and Relative Organ Weights

The mice in benzene 2 group (receiving 600 mg/kg b.w.) manifested some irritability and lethargy after benzene exposure for seven consecutive days. There were no significant differences in the body weights of mice at any of the time intervals analyzed (Figure 1) ($p > 0.05$), suggesting that the toxicity induced by benzene was insufficient to cause observable body weight changes. The relative organ (liver, spleen, lung, and kidney) weights of mice are presented in Table 1. There was a significant decrease in relative lung weights in benzene 2 group mice on the seventh day of benzene exposure. In addition, there was a significant decrease in relative spleen weights in both benzene 1 and 2 groups.

Figure 1. Body weight of mice at each time point and dose. Each bar represents means ± standard deviation (SD) from one-way ANOVA.

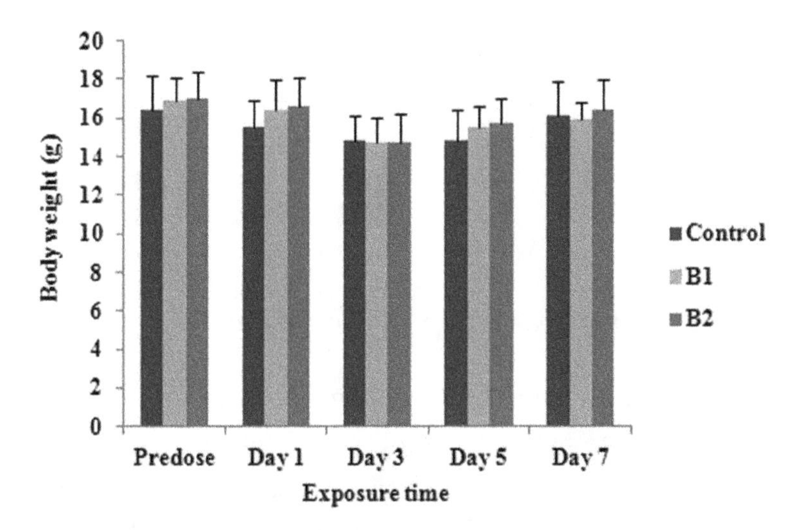

Table 1. Relative organ weights of male C3H/He mice on benzene exposure day 7.

Group	Relative liver weight	Relative spleen weight	Relative lung weight	Relative kidney weight
Control	6.01 ± 0.32	0.36 ± 0.05	0.69 ± 0.10	1.63 ± 0.14
Benzene 1	6.38 ± 0.31	0.27 ± 0.09 *	0.67 ± 0.08	1.65 ± 0.15
Benzene 2	6.54 ± 0.48	0.25 ± 0.07 *	0.56 ± 0.02 *	1.69 ± 0.06

* significant difference compared with control group ($p < 0.05$).

2.2. Blood Parameters and Bone Marrow Smear

The parameters of peripheral blood and bone marrow smears were investigated to assess the hematotoxicity of benzene. As shown in Table 2, a significant decrease in RBC number and hemoglobin (Hgb) concentration occurred in both groups of mice following exposure to benzene for seven days. No effect on WBC number was found, which may be due to the relatively short time of exposure. The platelet (pit) was reduced in the benzene-exposed groups, but not to a statistically significant extent. The bone marrow smears were made to observe the extent of nucleated cell proliferation and cell morphology (Figure 2). The results showed significant myeloid hyperplasia and a marked reduction of erythroidin benzene groups, while no significant difference was observed in the ratio of immature cells in benzene exposure groups. These observations suggest that benzene exposure leads to hematotoxicity.

Table 2. Blood parameters in male C3H/He mice on benzene exposure day 7.

Group	WBC (10^9/L)	RBC (10^{12}/L)	Hgb (g/L)	Pit (10^9/L)
Control	4.05 ± 0.65	7.98 ± 0.39	137.20 ± 5.76	364.25 ± 60.50
Benzene 1	3.87 ± 1.06	7.30 ± 0.14 *	126.67 ± 3.50 *	322.00 ± 107.53
Benzene 2	4.48 ± 0.96	7.32 ± 0.42 *	127.60 ± 7.50 *	282.2 ± 50.01

* significant difference compared with control group ($p < 0.05$).

Figure 2. Bone marrow smear examination in male C3H/He mice following 7 days of benzene exposure. * significant difference compared with control group ($p < 0.05$).

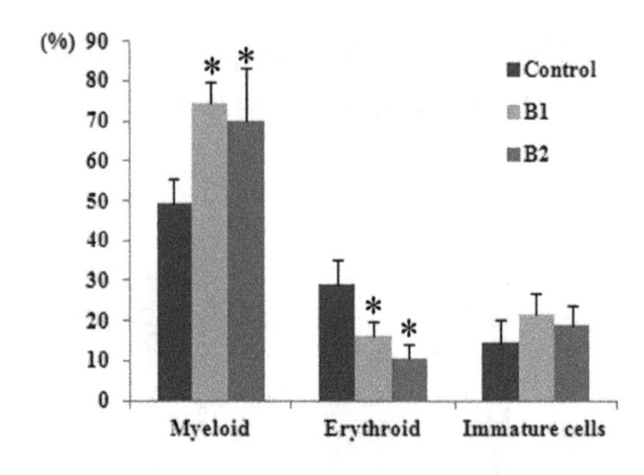

2.3. LC-MS Fingerprinting of Mouse Bone Marrow Cells and Plasma

Typical HPLC-MS total ion current (TIC) chromatograms of mouse bone marrow cell and plasma samples on day seven taken from the control and benzene-exposed groups are shown in Figures 3 and 4. As shown in Figure 3, there were significant visual differences in the TIC among the groups, especially from 1 to 3 min. The difference between the control group and benzene-exposed groups was more apparent than that between the two dosed groups. A similar metabonomic profile difference was also observed in the TIC of plasma (Figure 4).

Figure 3. Total ion chromatograms (TICS) of bone marrow cell samples obtained from the control group (**C**), and benzene 1 (**B1**) and benzene 2 groups (**B2**) of male C3H/He mice following 7 days of benzene exposure, using LC/MS (positive mode).

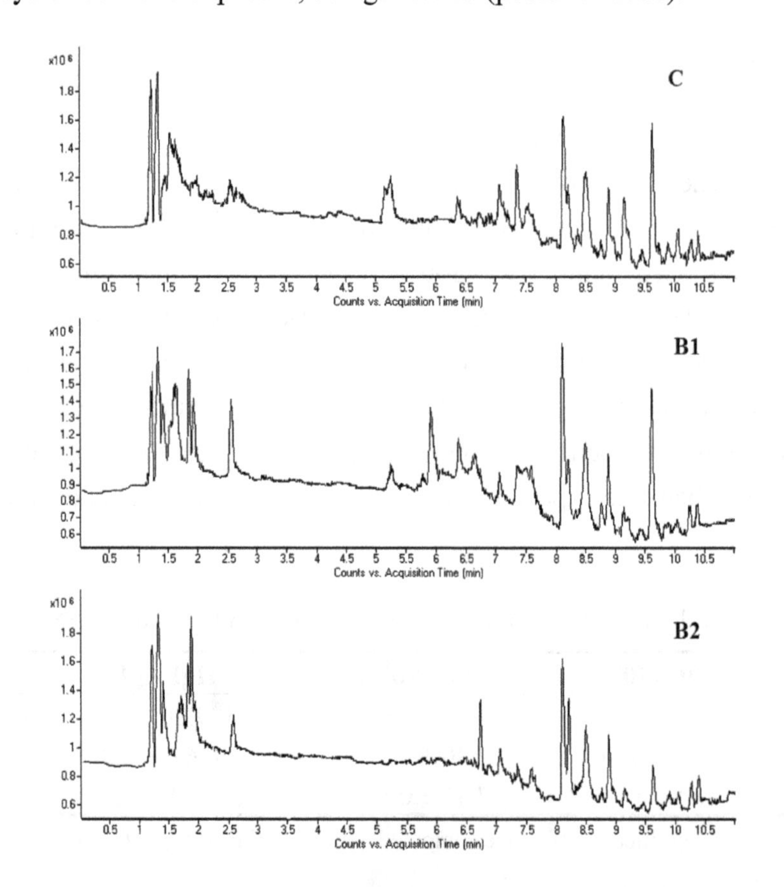

Figure 4. Total ion chromatograms (TICS) of plasma samples obtained from the control group (**C**), and benzene 1 (**B1**) and benzene 2 groups (**B2**) of male C3H/He mice following 7 days of benzene exposure, using LC/MS (positive mode).

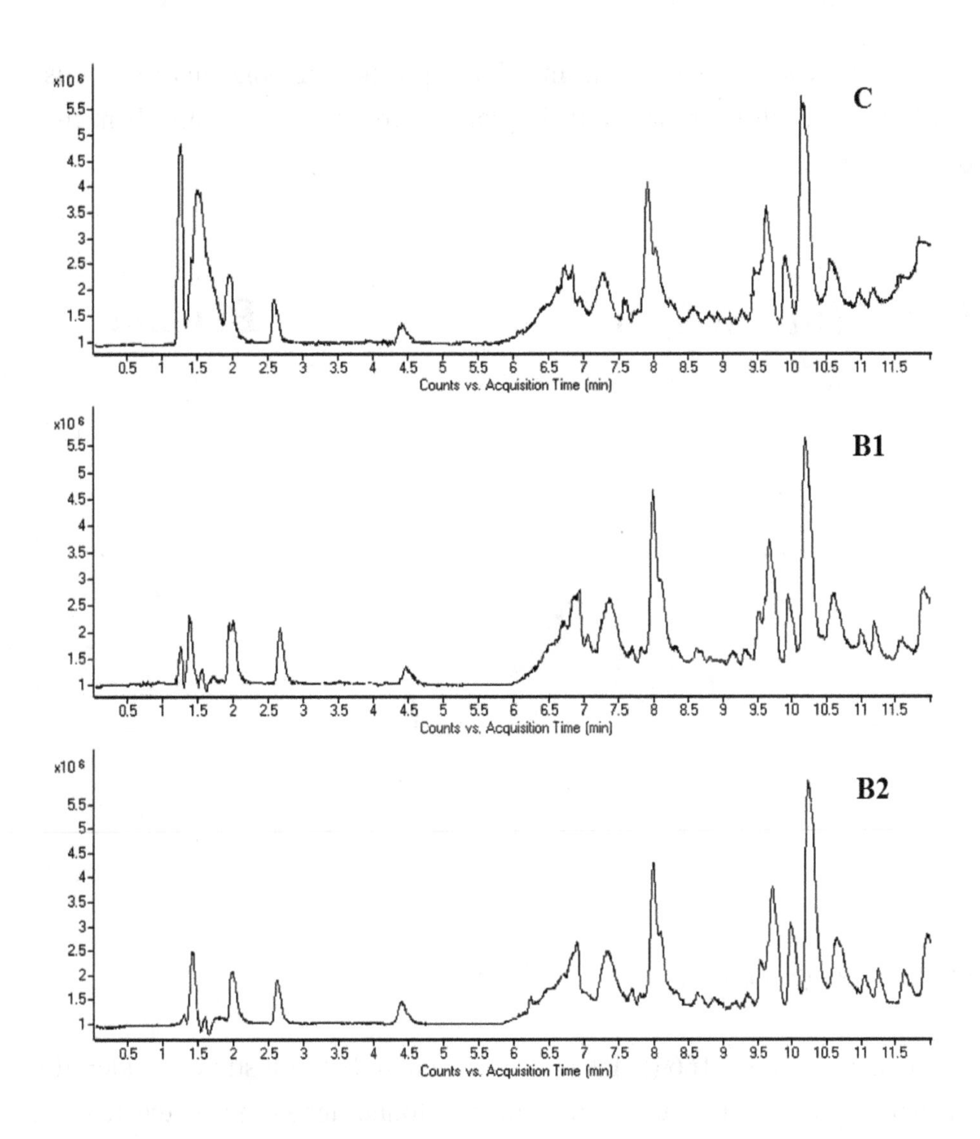

2.4. Principal Component Analysis and Discovery of Changed Endogenous Metabolites

We determined metabolites that were responsible for the changes illustrated above by using one-way ANOVA ($p < 0.05$, fold-change ≥ 2). Principal Component Analysis (PCA) was used to further select biomarkers that could discriminate between groups. PCA (an unsupervised method), is quite useful in distinguishing the several thousand biochemical endpoints retrieved from each sample. In the PCA score plots, each spot represents a metabonomic sample and each assembly of samples expresses a unique metabolic pattern at different time points. This analysis was successful in this experiment because more than 80% of the variability was explained using four components. PCA score plots derived from levels of 16 metabolites from bone marrow cells showed marked differences between the control and benzene-exposed mice on day seven. Similar results can be found in the PCA score map derived from plasma metabolites (Figure 5). The results showed that benzene exposure induced significant changes in 16 metabolites in bone marrow cells and 25 metabolites in plasma, with some metabolites changed in more than one compartment. These metabolites were considered to be potential biomarkers of benzene action.

Figure 5. 3D PCA score plot of the metabolic profiles of bone marrow cells (**A**) and plasma (**B**) in the control group (red), benzene 1 group (brown), and benzene 2 group (blue) on exposure day 7.

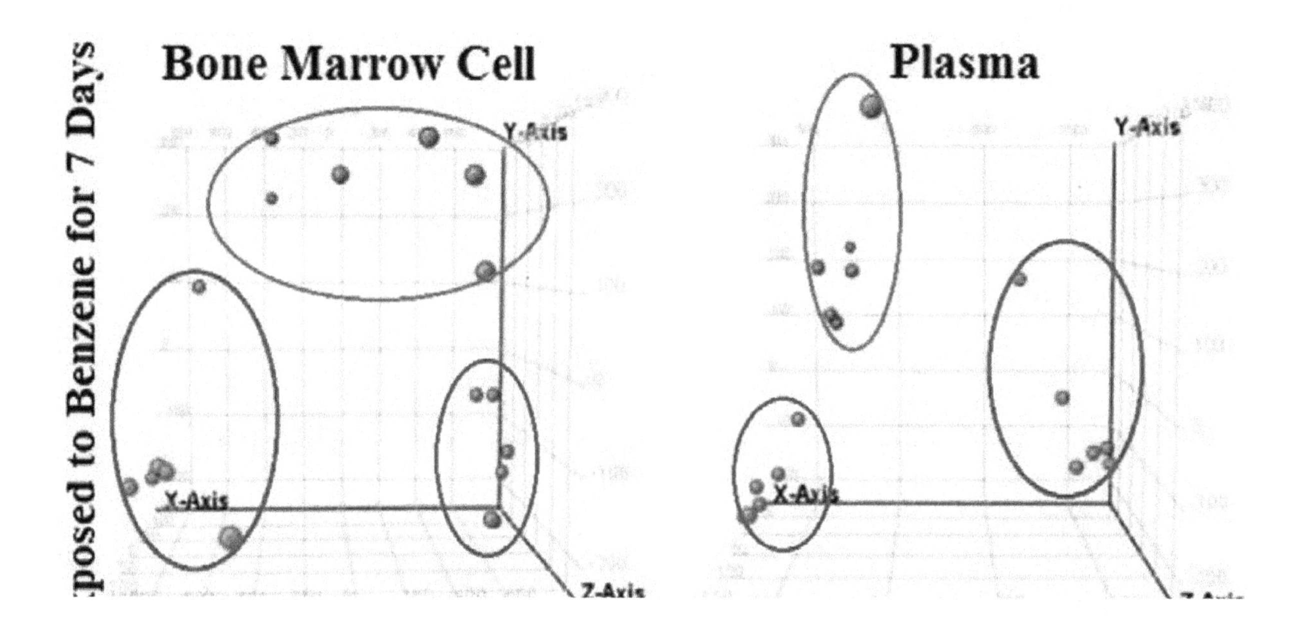

2.5. Identification of Changed Endogenous Metabolites

Herein, we took the ion at *m/z* 166 ($[M + H]^+$) as an example to illustrate the identification process. First, the corresponding quasi-molecular ion peak was found according to the retention time in the extracted ion chromatogram (EIC) of *m/z* 166 (Figure 6A). The accurate mass of the quasi-molecular ion was found as *m/z* 166.0863, and $C_9H_{11}NO_2$ was calculated as the most probable molecular formula using Agilent MassHunter software. Then, we conducted its fragmentation by tandem MS. Three major fragment ions were found at *m/z* 103.0545 and 120.0806, which represent the fragments of $[C_8H_7]^+$ and $[C_8H_{10}N]^+$, respectively. With the aforementioned information, we searched the freely accessible databases HMDB (http://www.hmdb.ca), METLIN (http://metlin.scripps.edu) and KEGG (http://www.kegg.jp). Finally, considering elemental composition, fragmentation patterns and chromatographic retention behavior, the *m/z* of 166 was identified as L-phenylalanine, which was then validated using a standard (Figure 6B,C). Likewise, other biomarkers have been identified and are listed in Table 3. However, the remaining biomarkers (data not shown) were unidentifiable due to insufficient intensity for the MS/MS experiments, or due to restrictions in the current metabolite databases.

Figure 6. Identification of a selected marker (*m/z* 166). (**A**) Extracted ion chromatogram (EIC) of *m/z* 166; (**B**) MS/MS spectrum of the ion; (**C**) MS/MS spectrum of a commercial standard L-phenylalanine. The collision energy was 20 V.

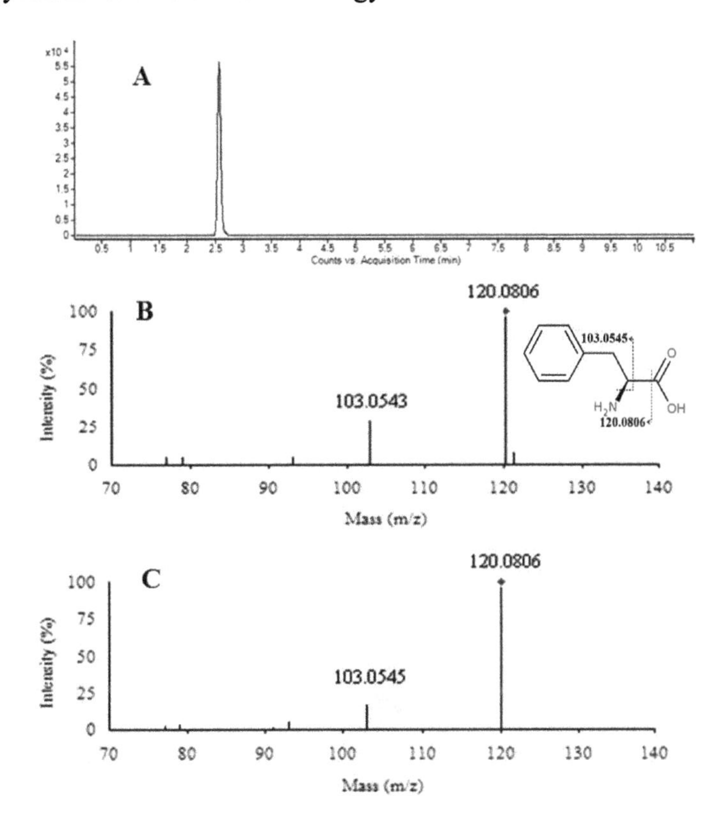

Table 3. Identified endogenous biomarkers in bone marrow cell and plasma on exposure day 7.

Compartment	*m/z*	RT (Retention time)	Trend [a]	*p*-value	Metabolites	Related pathway
Bone Marrow Cell	204.123	1.54	↓	9.30×10^{-4}	L-Acetylcarnitine	Oxidation of Fatty Acids
	165.0546	1.681	↑	5.14×10^{-3}	*p*-Coumaric acid	Unknown
	182.081	1.668	↑	3.97×10^{-4}	L-tyrosine	Tyrosine/Phenylalanine and Tyrosine Metabolism, Catecholamine Biosynthesis
	166.0863	2.548	↑	1.62×10^{-3}	L-Phenylalanine	Phenylalanine and Tyrosine Metabolism
	147.1168	9.205	↓	1.65×10^{-4}	Lysine	Lysine Degradation, Biotin Metabolism, Carnitine Synthesis
Plasma	192.0644	1.2775	↓	1.92×10^{-6}	5-Hydroxyindoleacetic acid	Tryptophan metabolism
	112.0869	1.3	↓	2.06×10^{-4}	Histamine	Histidine metabolism
	156.0765	1.321	↑	1.08×10^{-5}	L-Histidine	Histidine Metabolism, Ammonia Recycling Transcription/Translation
	126.1026	1.322	↓	1.32×10^{-4}	*N*-Methylhistamine	Histidine metabolism
	204.123	1.58	↓	6.01×10^{-5}	L-Acetylcarnitine	Oxidation of Fatty Acids
	130.0499	2.177	↑	2.88×10^{-6}	Pyrrolidonecarboxylic acid	Gamma-glutamyl cycle
	400.3421	9.195	↓	3.11×10^{-7}	Palmitoylcarnitine	Fatty acid Metabolism

[a] Change trend of benzene exposure mice *vs.* control mice. Variations compared to control samples: ↑, indicates relative increase in signal; ↓, relative decrease in signal ($p < 0.05$).

2.6. Biological Significance of Endogenous Metabolites Alternations in Bone Marrow Cells and Plasma

Metabolic profiling contributes diagnostic information and presents mechanistic insights into the biochemical effects of toxins [29]. The variation of endogenous metabolites in bone marrow may be indicative of benzene's toxic hematopoietic mechanisms, and the fluctuation may then be observed in plasma. Since blood is circulated around the body, the endogenous metabolites alternations in multi-organs induced by benzene can also be reflected in the blood. This may be helpful in identifying potential biomarkers of toxic effects that would relate to adverse health effects, and that could be monitored in blood. Therefore, in this study, we analyzed benzene-induced endogenous metabolites changes in bone marrow cells and plasma.

2.6.1. Significance of Changed Endogenous Metabolites in Bone Marrow Cells Induced by Benzene

Benzene is hematotoxic and leukemogenic in humans and induces bone marrow suppression in rodents. In recent years, research into the metabolic pathways involved in the renewal and differentiation of HSC and hematopoietic system diseases has intensified. In this study, we uncovered and identified five potential biomarkers in bone marrow cells, and expect to use these to further investigate the hematotoxic mechanisms of benzene. Our results showed in bone marrow cells increases in the levels of phenylalanine and tyrosine, and a decrease in L-lysine, three of the essential amino acids. In addition, a significant lowering of acetylcarnitine was found in bone marrow cells as compared to controls.

A recent study found an increased level of phenylalanine in stem cells from peripheral blood of t-MDS/AML patients, suggesting that there may exist an alteration in mitochondrial activity in these patients relative to controls [24]. In our study, the significant increase in phenylalanine levels in bone marrow cells might then be a manifestation of mitochondrial dysfunction. Phenylalanine is needed for the synthesis of protein, melanin and tyrosine. The increased bone marrow cell levels of phenylalanine may enhance the metabolic pathway from phenylalanine to tyrosine, resulting in a high concentration of bone marrow cell tyrosine. Higher serum levels of phenylalanine and tyrosine were observed in AML patients, which might be the result of enhanced degradation of proteins from the host experiencing a cancerous condition, and these two amino acids are needed for gluconeogenesis and for catabolism to provide intermediates for the tricarboxylic acid cycle (TCA) cycle [30]. It is possible that the toxicity to bone marrow with benzene exposure may be involved in the TCA cycle in mitochondria. The signaling pathway associated with increased phenylalanine and tyrosine may be one of the vital mechanisms intrinsic to benzene-induced hematotoxicity.

L-lysine is an essential amino acid and exerts antifibrinolytic activity by inhibition of fibrinolysis, and exerts a protective effect on platelets [31]. Lysine acetyltransferases were reported to play a key role in leukemogenesis and interact with Runx1 (or AML1), one of the most frequent targets of chromosomal translocations in leukemia [32]. Carnitine is synthesized from lysine residues in existing proteins, and then used to further synthesize acetylcarnitine via carnitine palmitoyl transferase I. Acetylcarnitine is an acetic acid ester of carnitine that facilitates movement of acetyl CoA into the matrices of mammalian mitochondria during the oxidation of fatty acids [33], and it has been observed that lower levels of acetylcarnitine are found in the blood of AML patients [30]. Ito *et al.* showed that fatty acid oxidation was associated with hematopoietic stem cell proliferation and differentiation, which determines whether they undergo symmetric or asymmetric cell division [34]. Therefore, decreases in bone marrow cell lysine and acetylcarnitine levels are likely associated with a down-regulation of carnitine synthesis, which then disturbs oxidation of fatty acids in hematopoietic stem cells that are exposed to benzene. In addition, significant changes in the levels of these compounds in bone marrow were traced in plasma, and we found specific metabolites that related to hematopoietic toxicity of benzene. It is worth noting that only acetylcarnitine levels in plasma decreased, consistent with the changes in bone marrow, indicating that lower levels of acetylcarnitine in plasma might be indicative of hematotoxic effects of benzene.

2.6.2. Significance of Changed Endogenous Metabolites in Plasma Induced by Benzene

The levels of five significantly changed metabolites in bone marrow were also traceable in plasma. Only a reduction in acetylcarnitine was found in plasma. In addition, benzene exposure caused an elevation of L-histidine and pyrrolidone carboxylic acid, concomitant with decreases in 5-hydroxyindoleacetic acid, histamine, *N*-methylhistamine, and palmitoyl-L-carnitine in plasma (Table 3). These significantly changed endogenous metabolites could be used to illustrate the multiple-organ toxicity induced by benzene.

Benzene was demonstrated to cause a disturbance in histidine-related metabolism, including a significant decline in histamine and *N*-methylhistamine, concomitant with an elevation in histidine. The high plasma histidine in the benzene-exposed groups may be caused by a diminished activity of histidine decarboxylase (HDC) [35], which results in inhibiting decarboxylation of histidine to histamine in benzene-exposed mice. Histamine is an amine derived by enzymatic decarboxylation of histidine, which plays a pivotal role in a number of processes, including inflammation, allergic reactions, gastric acid secretion and neurotransmission [36]. Histamine was reported to inhibit production of reactive oxygen species (ROS) in CML cells via the H_2-receptor (H_2R) [37]. A salt of histamine, histamine dihydrochloride (HDC), is used as a drug for the prevention of relapse in patients diagnosed with AML [38,39]. Recently, Aurelius *et al.* found that HDC acted on H_2R expressed by leukemia cells to reduce ROS formation, which might impact the effectiveness of histamine-based immunotherapy [40]. Phenolic metabolites of benzene accumulate in the bone marrow where myeloperoxidase and other peroxidases convert them to reactive semiquinones and quinines [41], which can further lead to the formation of ROS [42]. Histamine might thereby be consumed in order to respond to ROS generation induced by benzene.

Palmitoylcarnitine is a long-chain acyl fatty acid derivative ester of carnitine that facilitates the transfer of long-chain fatty acids from the cytoplasm into mitochondria during the oxidation of fatty acids. Palmitoylcarnitine was shown to stimulate the activity of caspases 3, 7 and 8, and the level of this long-chain acylcarnitine increased during apoptosis [43]. Ibuki *et al.* reported that benzene metabolites induced an anti-apoptotic effect, and that the effect was mainly due to the production of ROS by benzene metabolites (*p*-benzoquinone and hydroquinone) that inhibited caspase-3 activation. Inhibition of apoptosis, aberrantly prolonging cell survival, may contribute to cancer by facilitating the creation of mutations and by allowing a permissive environment for genetic instability [44]. Thus, the decreased plasma palmitoylcarnitine levels in the benzene-exposed groups may be related to disturbed fatty acid metabolism and the suppression of apoptosis by inhibiting caspase activation.

5-Hydroxyindoleacetic acid (5-HIAA) is a breakdown product of serotonin and levels of these substances may be measured in plasma to monitor progression of diseases such as carcinoid tumors [45] and pulmonary hypertension [46]. In the present study, plasma 5-HIAA was found to be notably reduced in benzene-exposed mice, and this may due to benzene toxicity.

Pyrrolidonecarboxylic acid can be irreversibly converted to glutamate, which is used to generate glutamine. Peng *et al.* [47] investigated amino acid concentrations during induction and preconsolidation therapy in cerebrospinal fluid (CSF) of children with lymphoblastic leukemia (ALL) with or without CNS involvement. However, they did not find any significantly changed amino acid levels except for higher baseline glutamine levels, indicative of a greater risk for CNS leukemia. Although increased levels of glutamine were not detected in the plasma of mice exposed to benzene, the increased pyrrolidonecarboxylic acid might indirectly reflect an early influence of benzene on glutamate metabolism.

3. Experimental Section

3.1. Chemicals and Reagents

Benzene was purchased from Sigma Co. (St. Louis, MO, USA); and corn oil from COFCO (Beijing, China). Ultrapure water (18.2 MO) was prepared with a Milli-Q water purification system (Millipore, Bedford, MA, USA). LC/MS grade methanol and acetonitrile were purchased from Spain Scharlau, Ltd. (Barcelona, Spain); and analytical grade formic acid was supplied by Dikma Corp. (Richmond, NY, USA).

3.2. Ethics Statement

This study was carried out in strict accordance with the recommendations of the Guide for the Care and Use of Laboratory Animals of the State Committee of Science and Technology of the People's Republic of China. The protocol of experiments was reviewed and approved by the Research Ethics Committee of the Southeast University (approval number: 20130027). Animals were maintained and experiments were conducted in accordance with the Institutional Animal Care and Use Committee of Southeast University.

3.3. Animals and Treatments

Eighteen male C3H/He mice (aged 4 weeks, and weighing 17.11 ± 1.03 g) obtained from Wei Tong Li Hua Laboratory Animal Co. Ltd. (Beijing, China) were acclimatized for one week in the Specific-Pathogen Free (SPF) animal facility prior to administration of substances. Animals were maintained under a 12-h light/12-h dark cycle at a temperature of 25 ± 2 °C with a relative humidity of 45%–65%. Animals had *ad libitum* access to a certified standard diet and to drinking water, and were divided randomly into a control group (vehicle, oil; n = 6), benzene group 1 (B1: 300 mg/kg b.w.; n = 6) and benzene group 2 (B2: 600 mg/kg b.w.; n = 6). Mice were injected subcutaneously with either corn oil or a benzene-corn oil mixture once daily for seven consecutive days. The aim of these doses is to examine the corresponding hematotoxicity of benzene. The route of benzene administration (injection, s.c.) was to allow us better control of benzene dosages [48,49].

The body weight of each mouse was recorded every other day during exposure periods. The mice were sacrificed on the seventh day of exposure. The mice were anesthetized with pelltobarbitalum natricum, blood was collected, and then were sacrificed by decapitation. Liver, spleen, lung, and kidneys were excised and weighed. Relative organ weight was calculated as the ratio between organ weight and body weight. Bone marrow cells were flushed from one tibia using a 26-gauge needle to make smears.

3.4. Collection of Plasma and Bone Marrow Cell Samples

The plasma and bone marrow cell samples of mice were collected after being exposed to benzene for seven days. Plasma was extracted from whole blood at 3000 rpm for 10 min at room temperature. After acquiring mouse femurs and tibias, the marrow cavities were washed with a 26-gauge needle (on ice). Mouse bone marrow cells were collected after centrifugation at 300× *g* for 10 min at 4 °C. To 5×10^6 bone marrow cells, 1 mL of quenching solution (iced, 0.9% [*w*/*v*] NaCl) was quickly added. All cell samples were centrifuged at 1000× *g* for 1 min. Cell pellets were resuspended in ice-cold 50% aqueous acetonitrile, vortexed, and incubated on ice for 10 min. The extracts were dried in a SpeedVac, and the dried cell extracts were resuspended in 500 μL water. All obtained samples were frozen immediately and stored at −80 °C until analysis.

3.5. Sample Preparation and HPLC/MS Analysis

Plasma samples were thawed at room temperature and then centrifuged at 13,000× *g* for 15 min at 4 °C. Each 100 μL aliquot of plasma was mixed with 300 μL of methanol and vortexed to allow for protein precipitation. After centrifugation at 13,000× *g* for 15 min at 4 °C, the combined supernatants were transferred to the auto-sample vials. The bone marrow cells were thawed and could be analyzed directly. For the metabonomic studies, aliquots of 1 μL of each sample were injected into a ZORBAX Eclipse Plus C18 column (3.00 mm × 100 mm × 1.8 μm, Agilent, Santa Clara, CA, USA) using an Agilent 6224 TOF LC-MS system (Agilent). The mobile phase was 0.1% formic acid in water (A) and 0.1% formic acid in acetonitrile (B). The optimized HPLC elution conditions were: (a) plasma: 0–1 min, 5% B; 1–3.5 min, 5%–80% B; 3.5–10 min, 80%–95% B; 10–12 min, 95% B; 12–12.5 min, 5% B; (b) bone marrow cell: 0–0.3 min, 5% B; 0.3–7 min, 5%–95% B; 7–9 min, 95% B; 9–10 min, 5% B. The flow rate was 0.4 mL/min. The column and autosample were maintained at 35 and 4 °C, respectively. The positive ion mode was used for the mass detection. The source parameters were set as follows: drying gas flow rate, 9 L/min; gas temperature, 350 °C; pressure of nebulizer gas, 40 psig; Vcap, 4000 V; fragmentor, 150 V; skimmer, 60 V; and scan range, *m*/*z* 50–1000. The tune mixture solution (Agilent) was employed as the lock mass (*m*/*z* = 121.050873, 922.009798) at a flow rate of 30 μL/min, via a lock spray interface for accurate mass measurement. To confirm the identity of the metabolites obtained after the non-targeted analysis (MS analysis and database search), a LC/6530 Q-TOF-MS (Agilent) was used. The MS/MS analysis was acquired in targeted MS/MS mode with collision energy from 10, 20 and 40 V; and a scan rate of 1 (MS/MS) scans/s.

3.6. Data Processing

The Masshunter Data Analysis Software (Ver B.02.01, Agilent Technologies, Barcelona, Spain) was used to analyze results; and the Masshunter Qualitative Analysis Software (Agilent Technologies) was used to obtain the molecular features of the samples, representing different, co-migrating ionic species of a given molecular entity using the Molecular Feature Extractor (MFE) algorithm. Finally, the Masshunter Mass Profiler Professional Software (Ver B.02.02, Agilent Technologies) was used to perform a non-targeted metabolomic analysis of the extracted features. Samples with a minimal absolute abundance of 2000 counts and with a minimum of 2 ions were selected. Multiple charge states were not considered. Compounds from different samples were aligned using a RT window of 0.2% ± 0.15 min and a mass window of 10 ppm ± 2.0 mDa. Only common features (found in at least 75% of the samples of the same condition) were analyzed, correcting for individual bias. Data for PCA analysis were obtained using this software. As a classic unsupervised method (no prior knowledge concerning groups or tendencies within the data sets was necessary) of pattern recognition, PCA was expected to discern through statistical protocols several distinct variables for use as potential biomarkers.

3.7. Statistics

Statistical analyses for non-targeted metabonomics analyses were performed using the Mass Profiler Professional Software (Agilent Technologies). The exact masses with significant differences in abundance were determined using a one-way analysis of variance (ANOVA); fold-change >2 was considered to be significant at $p < 0.05$, and was searched against various databases (METLIN, HMDB, LIPID MAPS and KEGG). Otherwise, statistical calculations were performed using the SPSS 15.0 software (SPSS, Chicago, IL, USA). Multiple comparisons were analyzed using one-way ANOVA. Statistical significance was established at a level of $p < 0.05$.

4. Conclusions

In conclusion, we applied an LC-MS-based metabonomics approach to investigate benzene-induced toxicity in male C3H/He mice. The combined experimental results of metabonomics, relative organ weights, blood parameters and bone marrow smears indicated benzene-induced hematotoxicity. The obvious metabolic alterations in mouse bone marrow cells and plasma indicated that benzene exposure disrupted metabolism of essential amino acids (lysine, phenylalanine and tyrosine) in bone marrow cells; resulting in benzene-induced hematotoxicity. Benzene also caused disturbance in the metabolism of fatty acids oxidation. The decreased acetylcarnitine in plasma was commensurate with that in bone marrow cells; suggesting that acetylcarnitine in plasma is very likely an appropriate biomarker of benzene hematotoxicity. Our work offers a new clue for further clarification of the mechanism(s) involved in benzene-induced toxicity via studying variation of endogenous metabolites.

Acknowledgments

We acknowledge the financial supports from the National Natural Science Foundation of China (Grant No: 81373034), the National Natural Science Foundation of Jiangsu province (Grant No: BK2011605), the Doctoral Fund of Ministry of Education of China (Grant No: 20100092120054) and Graduate Research and Innovation Program of Colleges and Universities of Jiangsu Province (Grant No: CXZZ13_0135).

References

1. Shen, M.; Lan, Q.; Zhang, L.; Chanock, S.; Li, G.; Vermeulen, R.; Rappaport, S.M.; Guo, W.; Hayes, R.B.; Linet, M.; *et al.* Polymorphisms in genes involved in DNA double-strand break repair pathway and susceptibility to benzene-induced hematotoxicity. *Carcinogenesis* **2006**, *27*, 2083–2089.
2. Glass, D.C.; Gray, C.N.; Jolley, D.J.; Gibbons, C.; Sim, M.R.; Fritschi, L.; Adams, G.G.; Bisby, J.A.; Manuell, R. Leukemia risk associated with low-level benzene exposure. *Epidemiology* **2003**, *14*, 569–577.
3. Hayes, R.B.; Songnian, Y.; Dosemeci, M.; Linet, M. Benzene and lymphohematopoietic malignancies in humans. *Am. J. Ind. Med.* **2001**, *40*, 117–126.
4. Smith, M.T.; Jones, R.M.; Smith, A.H. Benzene exposure and risk of non-Hodgkin lymphoma. *Cancer Epidemiol. Biomark. Prev.* **2007**, *16*, 385–391.
5. International Agency for Research on Cancer (IARC). Benzene. IARC monographs on the evaluation of the carcinogenic risk of chemicals to humans: Some industrial chemicals and dyestuffs. *Int. Agency Res. Cancer* **1982**, *29*, 93–148.
6. IARC (International Agency for Research on Cancer). Monographs on the Evaluation of Carcinogenic Risks to Humans Overall Evaluations of Carcinogenicity. Available online: http://monographs.iarc.fr/ENG/Monographs/vol71/mono71.pdf (accessed on 3 March 2014).
7. Qu, Q.; Shore, R.; Li, G.; Jin, X.; Chen, L.C.; Cohen, B.; Melikian, A.A.; Eastmond, D.; Rappaport, S.M.; Yin, S.; *et al.* Hematological changes among Chinese workers with a broad range of benzene exposure. *Am. J. Ind. Med.* **2002**, *42*, 275–285.
8. Rothman, N.; Li, G.L.; Dosemeci, M.; Bechtold, W.E.; Marti, G.E.; Wang, Y.Z.; Linet, M.; Xi, L.Q.; Lu, W.; Smith, M.T.; *et al.* Hematotoxicity among Chinese workers heavily exposed to benzene. *Am. J. Ind. Med.* **1996**, *29*, 236–246.
9. Snyder, R. Overview of the toxicology of benzene. *J. Toxicol. Environ. Health* **2000**, *61*, 339–346.
10. Wilbur, S.; Wohlers, D.; Paikoff, S.; Keith, L.S.; Faroon, O. ATSDR evaluation of potential for human exposure to benzene. *Toxicol. Ind. Health* **2008**, *24*, 399–442.
11. Subrahmanyam, V.V.; Doane-Setzer, P.; Steinmetz, K.L.; Ross, D.; Smith, M.T. Phenol-induced stimulation of hydroquinone bioactivation in mouse bone marrow *in vivo*: Possible implications in benzene myelotoxicity. *Toxicology* **1990**, *62*, 107–116.
12. Subrahmanyam, V.V.; Kolachana, P.; Smith, M.T. Hydroxylation of phenol to hydroquinone catalyzed by a human myeloperoxidase-superoxide complex: Possible implications in benzene-induced myelotoxicity. *Free Radic. Res. Commun.* **1991**, *15*, 285–296.
13. Zhang, L.; Eastmond, D.A.; Smith, M.T. The nature of chromosomal aberrations detected in humans exposed to benzene. *Crit. Rev. Toxicol.* **2002**, *32*, 1−42.
14. Longacre, S.L.; Kocsis, J.J.; Snyder, R. Influence of strain diffrences in mice on the metabolism and toxicity of benzene. *Toxicol. Appl. Pharmacol.* **1981**, *60*, 398−409.

15. Billet, S.; Paget, V.; Garcon, G.; Heutte, N.; Andre, V.; Shirali, P.; Sichel, F. Benzene-induced mutational pattern in the tumour suppressor gene Tp53 analysed by use of a functional assay, the functional analysis of separated alleles in yeast, in human lung cells. *Arch. Toxicol.* **2010**, *84*, 99–107.
16. McHale, C.M.; Zhang, L.; Lan, Q.; Vermeulen, R.; Li, G.; Hubbard, A.E.; Porter, K.E.; Thomas, R.; Portier, C.J.; Shen, M.; *et al.* Global gene expression profiling of a population exposed to a range of benzene levels. *Environ. Health Perspect.* **2011**, *119*, 628–634.
17. Kolachana, P.; Subrahmanyam, V.V.; Meyer, K.B.; Zhang, L.; Smith, M.T. Benzene and its phenolic metabolites produce oxidative DNA damage in HL60 cells *in vitro* and in the bone marrow *in vivo*. *Cancer Res.* **1993**, *53*, 1023–1026.
18. Zhang, L.; McHale, C.M.; Rothman, N.; Li, G.; Ji, Z.; Vermeulen, R.; Hubbard, A.E.; Ren, X.; Shen, M.; Rappaport, S.M.; *et al*. Systems biology of human benzene exposure. *Chem. Biol. Interact.* **2010**, *184*, 86–93.
19. Snyder, R. Benzene and leukemia. *Crit. Rev. Toxicol.* **2002**, *32*, 155–210.
20. Wang, I.M.; Stone, D.J.; Nickle, D.; Loboda, A.; Puig, O.; Roberts, C. Systems biology approach for new target and biomarker identification. *Curr. Top. Microbiol. Immunol.* **2013**, *363*, 169–199.
21. Gavaghan, C.L.; Wilson, I.D.; Nicholson, J.K. Physiological variation in metabolic phenotyping and functional genomic studies: Use of orthogonal signal correction and PLS-DA. *FEBS Lett.* **2002**, *530*, 191–196.
22. O'Connell, T.M.; Watkins, P.B. The application metabonomics to predict drug-induced liver injury. *Clin. Pharmacol. Ther.* **2010**, *88*, 394–399.
23. Robertson, D.G. Metabonomics in toxicology: A review. *Toxicol. Sci.* **2005**, *85*, 809–822.
24. Cano, K.E.; Li, L.; Bhatia, R.; Bhatia, R.; Forman, S.J.; Chen, Y. NMR-based metabolomic analysis of the molecular pathogenesis of therapy-related myelodysplasia/acute myeloid leukemia. *J. Proteome Res.* **2011**, *10*, 2873–2881.
25. Wang, Y.; Gao, D.; Chen, Z.; Li, S.; Gao, C.; Cao, D.; Liu, F.; Liu, H.; Jiang, Y. Acridone derivative 8a induces oxidative stress-mediated apoptosis in CCRF-CEM leukemia cells: Application of metabolomics in mechanistic studies of antitumor agents. *PLoS One* **2013**, *8*, e63572.
26. Huang, Z.; Lin, L.; Cao, Y.; Chen, Y.; Yan, X.; Xing, J.; Hang, W. Bladder cancer determination via two metabolites: A biomarker pattern approach. *Mol. Cell. Proteomics* **2011**, *10*, doi:10.1074/mcp.M111.007922.
27. Theodoridis, G.A.; Gika, H.G.; Want, E.J.; Wilson, I.D. Liquid chromatography–mass spectrometry based global metabolite profiling: A review. *Anal. Chim. Acta* **2012**, *711*, 7–16.
28. Sun, R.; Zhang, J.; Xiong, M.; Chen, Y.; Yin, L.; Pu, Y. Metabonomics biomarkers for subacute toxicity screening for benzene exposure in mice. *J. Toxicol. Environ. Health A* **2012**, *75*, 1163–1173.
29. Coen, M.; Holmes, E.; Lindon, J.C.; Nicholson, J.K. NMR-based metabolic profiling and metabonomic approaches to problems in molecular toxicology. *Chem. Res. Toxicol.* **2008**, *21*, 9–27.
30. Wang, Y.; Zhang, L.; Chen, W.L.; Wang, J.H.; Li, N.; Li, J.M.; Mi, J.Q.; Zhang, W.N.; Li, Y.; Wu, S.F.; *et al.* Rapid diagnosis and prognosis of *de novo* acute myeloid leukemia by serum metabonomic analysis. *J. Proteome Res.* **2013**, *12*, 4393–4401.

31. Levi, M.M.; Vink, R.; de Jonge, E. Management of bleeding disorders by prohemostatic therapy. *Int. J. Hematol.* **2002**, *76*, 139–144.
32. Yang, X.J. The diverse superfamily of lysine acetyltransferases and their roles in leukemia and other diseases. *Nucleic Acids Res.* **2004**, *32*, 959–976.
33. Koves, T.R.; Ussher, J.R.; Noland, R.C.; Slentz, D.; Mosedale, M.; Ilkayeva, O.; Bain, J.; Stevens, R.; Dyck, J.R.; Newgard, C.B.; *et al.* Mitochondrial overload and incomplete fatty acid oxidation contribute to skeletal muscle insulin resistance. *Cell Metab.* **2008**, *7*, 45–56.
34. Ito, K.; Carracedo, A.; Weiss, D.; Arai, F.; Ala, U.; Avigan; D.E.; Schafer, Z.T.; Evans, R.M.; Suda, T.; Lee, C.H.; *et al.* A PML–PPAR-δ pathway for fatty acid oxidation regulates hematopoietic stem cell maintenance. *Nat. Med.* **2012**, *18*, 1350–1358.
35. Hough, L.B. Genomics meets histamine receptors: New subtypes, new receptors. *Mol. Pharmacol.* **2001**, *59*, 415–419.
36. Kletke, O.; Sergeeva, O.A.; Lorenz, P.; Oberland, S.; Meier, J.C.; Hatt, H.; Gisselmann, G. New insights in endogenous modulation of ligand-gated ion channels: Histamine is an inverse agonist at strychnine sensitive glycine receptors. *Eur. J. Pharmacol.* **2013**, *710*, 59–66.
37. Mellqvist, U.H.; Hansson, M.; Brune, M.; Dahlgren, C.; Hermodsson, S.; Hellstrand, K. Natural killer cell dysfunction and apoptosis induced by chronic myelogenous leukemia cells: Role of reactive oxygen species and regulation by histamine. *Blood* **2000**, *96*, 1961–1968.
38. Brune, M.; Castaigne, S.; Catalano, J.; Gehlsen, K.; Ho, A.D.; Hofmann, W.K.; Hogge, D.E.; Nilsson, B.; Or, R.; Romero, A.I.; *et al.* Improved leukemia-free survival after postconsolidation immunotherapy with histamine dihydrochloride and interleukin-2 in acute myeloid leukemia: Results of a randomized phase 3 trial. *Blood* **2006**, *108*, 88–96.
39. Smits, E.L.; Berneman, Z.N.; van Tendeloo, V.F. Immunotherapy of acute myeloid leukemia: Current approaches. *Oncologist* **2009**, *14*, 240–252.
40. Aurelius, J.; Martner, A.; Brune, M.; Palmqvist, L.; Hansson, M.; Hellstrand, K.; Thoren, F.B. Remission maintenance in acute myeloid leukemia: Impact of functional histamine H2 receptors expressed by leukemic cells. *Haematologica* **2012**, *97*, 1904–1908.
41. Subrahmanyam, V.V.; Ross, D.; Eastmond, D.A.; Smith, M.T. Potential role of free radicals in benzene-induced myelotoxicity and leukemia. *Free Radic. Biol. Med.* **1991**, *11*, 495–515.
42. Hiraku, Y.; Kawanishi, S. Oxidative DNA damage and apoptosis induced by benzene metabolites. *Cancer Res.* **1996**, *56*, 5172–5178.
43. Mutomba, M.C.; Yuan, H.; Konyavko, M.; Adachi, S.; Yokoyama, C.B.; Esser, V.; McGarry, J.D.; Babior, B.M.; Gottlieb, R.A. Regulation of the activity of caspases by L-carnitine and palmitoylcarnitine. *FEBS Lett.* **2000**, *478*, 19–25.
44. Ibuki, Y.; Goto, R. Dysregulation of apoptosis by benzene metabolites and their relationships with carcinogenesis. *Biochim. Biophys. Acta* **2004**, *1690*, 11–21.
45. Allen, K.R.; Degg, T.J.; Anthoney, D.A.; Fitzroy-Smith, D. Monitoring the treatment of carcinoid disease using blood serotonin and plasma 5–hydroxyindoleacetic acid: Three case examples. *Ann. Clin. Biochem.* **2007**, *44*, 300–307.
46. Kirillova, V.V.; Nigmatullina, R.R.; Dzhordzhikiya, R.K.; Kudrin, V.S.; Klodt, P.M. Increased concentrations of serotonin and 5-hydroxyindoleacetic acid in blood plasma from patients with pulmonary hypertension due to mitral valve disease. *Bull. Exp. Biol. Med.* **2009**, *147*, 408–410.

47. Peng, C.T.; Wu, K.H.; Lan, S.J.; Tsai, J.J.; Tsai, F.J.; Tsai, C.H. Amino acid concentrations in cerebrospinal fluid in children with acute lymphoblastic leukemia undergoing chemotherapy. *Eur. J. Cancer* **2005**, *41*, 1158–1163.
48. Tunek, A.; Olofsson, T.; Berlin, M. Toxic effects of benzene and benzene metabolites on granulopoietic stem cells and bone marrow cellularity in mice. *Toxicol. Appl. Pharmacol.* **1981**, *59*, 149–156.
49. Velasco Lezama, R.; Barrera Escorcia, E.; Muñoz Torres, A.; Tapia Aguilar, R.; González Ramírez, C.; García Lorenzana, M.; Ortiz Monroy, V.; Betancourt Rule, M. A model for the induction of aplastic anemia by subcutaneous administration of benzene in mice. *Toxiology* **2001**, *162*, 179–191.

11

Continuous Flow Atmospheric Pressure Laser Desorption/Ionization Using a 6–7-µm-Band Mid-Infrared Tunable Laser for Biomolecular Mass Spectrometry

Ryuji Hiraguchi [1], Hisanao Hazama [1,*], Kenichirou Senoo [2], Yukinori Yahata [2], Katsuyoshi Masuda [3] and Kunio Awazu [1,4,5]

[1] Graduate School of Engineering, Osaka University, 2-1 Yamadaoka, Suita, Osaka 565-0871, Japan; E-Mails: hiraguchi-r@mb.see.eng.osaka-u.ac.jp (R.H.); awazu@see.eng.osaka-u.ac.jp (K.A.)

[2] JEOL Ltd., 1156 Nakagamicho, Akishima, Tokyo 196-0022, Japan; E-Mails: ksenoo@jeol.co.jp (K.S.); yyahata@jeol.co.jp (Y.Y.)

[3] Suntory Institute for Bioorganic Research, Suntory Foundation for Life Sciences, 1-1-1 Wakayamadai, Shimamotocho, Mishimagun, Osaka 618-0024, Japan; E-Mail: katsuyoshimasuda@hotmail.co.jp

[4] Graduate School of Frontier Biosciences, Osaka University, 1-1 Yamadaoka, Suita, Osaka 565-0871, Japan

[5] The Center of Advanced Medical Engineering and Informatics, Osaka University, 2-2 Yamadaoka, Suita, Osaka 565-0871, Japan

* Author to whom correspondence should be addressed; E-Mail: hazama-h@see.eng.osaka-u.ac.jp

---

Abstract: A continuous flow atmospheric pressure laser desorption/ionization technique using a porous stainless steel probe and a 6–7-µm-band mid-infrared tunable laser was developed. This ion source is capable of direct ionization from a continuous flow with a high temporal stability. The 6–7-µm wavelength region corresponds to the characteristic absorption bands of various molecular vibration modes, including O–H, C=O, CH_3 and C–N bonds. Consequently, many organic compounds and solvents, including water, have characteristic absorption peaks in this region. This ion source requires no additional matrix, and utilizes water or acetonitrile as the solvent matrix at several absorption peak wavelengths (6.05 and 7.27 µm, respectively). The distribution of multiply-charged peptide ions is extremely sensitive to the temperature of the heated capillary, which is the inlet of the mass spectrometer. This ionization technique has potential for the interface of liquid chromatography/mass spectrometry (LC/MS).

Keywords: mass spectrometry; atmospheric pressure laser ionization; mid-infrared tunable laser; peptides

---

1. Introduction

Currently, electrospray ionization (ESI) [1,2] and matrix-assisted laser desorption/ionization (MALDI) [3,4] are the two major ionization methods for large biomolecules. Although these methods are used in combination with other separation techniques (e.g., liquid chromatography) for proteome analysis, they have some limitations. ESI can continuously ionize flowing liquid samples, but ionization is unstable for low volatility solvents, such as buffer solutions. On the other hand, conventional MALDI is limited to dry samples, because the MALDI process is conducted under a high vacuum, and co-crystallization of the analyte and aromatic matrix leads to spatial inhomogeneity and a low reproducibility. In fact, conventional soft ionization methods for biological polymers cannot produce a highly stable ionization with low volatility solvents. Consequently, novel ionization techniques are necessary for high-throughput and reproducible proteome analysis.

To overcome the above issues, the solvent, the state in conventional ionization methods (ESI and MALDI) and various original ambient ionization methods that can simplify or omit the sample preparation process have been intensively investigated. A few methods have been reported at atmospheric pressure. Ionization under atmospheric pressure can tolerate various sample states, because the samples do not have to be dry. Additionally, electrospray laser desorption ionization (ELDI) [5] generates gas phase ions when an untreated tissue sample is irradiated with an ultraviolet (UV) laser under atmospheric pressure (AP), and the charged liquid droplets by the electrospray act to desorbed neutral molecules and ions. Unlike ELDI, matrix-assisted laser desorption electrospray ionization (MALDESI) [6] increases the ionization efficiency by adding an aromatic matrix. Laser ablation electrospray ionization (LAESI) [7] enables lipids from untreated biological samples to be directly ionized using the intensive absorption of an infrared (IR) laser by a tissue and electrospray, while modified ESI techniques without laser ablation/desorption have been developed for surface analysis. Desorption electrospray ionization (DESI), which was initially reported by Cooks *et al.*, enables proteins, peptides, *etc.*, to be directly analyzed from the surfaces of tissues and plants under ambient conditions [8]. DESI can also be used with thin-layer chromatography [9]. Probe electrospray ionization (PESI), which was developed by Hiraoka *et al.*, has a sampling system where a solid needle moves along a vertical axis by a motor-driven system [10]. PESI has the potential to analyze proteins directly from salt/urea-contaminated solutions [11].

However, these ambient ionization methods using laser desorption cannot ionize a liquid state sample directly. To realize high-throughput proteomics with a high reproducibility, an ambient laser ionization method for liquid state samples with low volatility solvents is desired.

Although conventional MALDI has been performed using UV lasers, IR-MALDI [12–14] has been investigated, because it may utilize more diverse matrices; many organic compounds with strong characteristic absorption peaks in the mid-IR wavelength range have been proposed as new matrices.

On the other hand, attempts to conduct the MALDI process under atmospheric pressure have been executed in order to expand the range of the analytical protocol. Laiko *et al.* presented an AP-MALDI technique coupled with an orthogonal acceleration time-of-flight (oaTOF) mass spectrometer and an ion trap mass spectrometer [15–17]. AP-MALDI has a better softness than conventional vacuum MALDI due to the collisional cooling of the expanding plume. In addition, alternating the surrounding condition to atmospheric pressure may expand AP-MALDI to include samples containing volatile compounds. In fact, online analysis of a solution sample by AP-MALDI has been demonstrated [18,19].

One of the most promising IR matrices is water, because it is abundant, environmentally benign and has a strong absorption in the mid-IR region. In 2002, Laiko *et al.* demonstrated the atmospheric pressure ionization of peptides in an aqueous solution using a Yb:YAG laser-pumped optical parametric oscillator (OPO) with a fixed wavelength of 3 μm [20]. Their work strongly suggests that utilizing a solvent as a light-absorbing matrix can expand the target samples in atmospheric pressure infrared matrix-assisted laser desorption ionization (AP-IR-MALDI) to include natural (aqueous) solutions. AP-IR-MALDI using a mid-IR OPO with a 2.9-μm wavelength has been applied to online liquid chromatography/mass spectrometry (LC/MS) analysis of the tryptic digest of bovine serum albumin [21]. Several works have employed 3-μm-band lasers, including OPO pumped by a Nd:YAG laser and an Er:YAG laser with a 2.94-μm wavelength for AP-IR-MALDI. This wavelength corresponds to the absorption peak of the O–H stretching vibration of liquid water.

The mid-IR wavelength region corresponds to the characteristic absorption bands of various molecular vibration modes, including O–H, C=O, CH_3 and C–N bonds. Consequently, many organic compounds and solvents, including water, have characteristic absorption peaks in this region (e.g., acetonitrile, methanol, acetone and several buffer solutions, such as a phosphate buffer, acetic acid buffer and Tris buffer). For example, the O–H bending vibration for liquid water has an absorption peak at a wavelength of 6.07 μm. Therefore, 6–7-μm-band mid-IR lasers should be able to utilize diverse solvents as a matrix compared to 3-μm-band lasers.

In fact, a 6-μm-band mid-infrared tunable laser using difference-frequency generation (DFG), which utilizes energy absorption in the C=O stretching region, has been applied to vacuum MALDI [22], and the softness for labile molecules (e.g., polysulfated oligosaccharides or polysialylated gangliosides) appears to exceed those of all other UV MALDI methods [23]. Simultaneous irradiations of a UV laser and 6-μm-band mid-IR free electron laser (FEL) enable protein samples containing a denaturant at a high concentration to be analyzed [24]. Awazu *et al.* have demonstrated a promising technique of IR-MALDI using a DFG laser utilizing various compounds (e.g., urea) as a matrix [25]. Although a 6–7-μm-band mid-infrared laser may utilize organic compounds, as well as various other solvents as a matrix when the MALDI process is conducted at atmospheric pressure, an AP-IR-MALDI method using a 6–7-μm-band infrared laser has yet to be reported.

In this study, we developed a new atmospheric pressure laser ionization method using a novel 6–7-μm-band mid-IR tunable laser. Our method allows the mass spectra of peptides to be directly measured from the continuous flow of several solutions.

2. Results and Discussion

2.1. Temporal Stability of Continuous Flow (CF) Ionization of Peptides

We initially examined the adequacy of the temporal stability of the ion signal intensity. Figure 1 shows a typical result of continuous ionization with a 5-μL/min sample flow and laser irradiation at 6.05 or 7.27 μm. The singly protonated ion $[M + H]^+$ of angiotensin II is predominantly observed. In addition, the extracted ion chromatogram of $[M + H]^+$ obtained from 50 mass spectra shows a superior stability of the ionization process compared with conventional MALDI using a solid matrix (Figure 2). The variability of the ion signal intensity has a standard deviation of 12.7%. On the other hand, the DFG laser has a variation of up to 10% in the output energy. In addition, the temporal relationship between laser pulses and injection time of the ion trap may be responsible for these variations, because the lasers were operated without synchronization. Thus, the sequential ion signal from the solution at a surface near the frit is responsible for the continuous flow ionization. This result suggests that this technique may be a novel interface in LC and MS. On the other hand, the efficiency of sample desorption depends strongly on the parameters of the laser and solvent. Thus, we then investigated the relationships between the laser wavelength, pulse energy, absorption coefficient of the solvent and ionization efficiency.

Figure 1. Typical mass spectra from angiotensin II dissolved in an 80% acetonitrile aqueous solution upon irradiation with a mid-infrared laser at wavelengths of (**a**) 6.05 μm and (**b**) 7.27 μm. The electric potential of the frit and the capillary temperature were set to 2.5 kV and 270 °C, respectively.

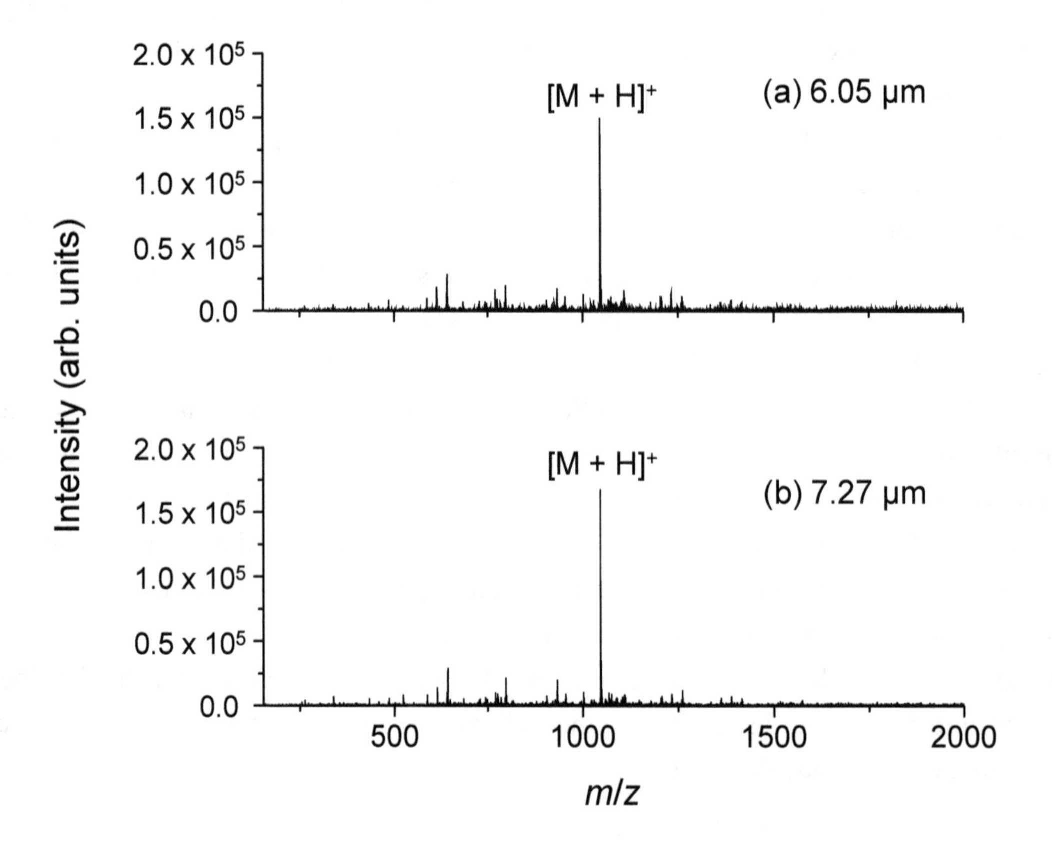

Figure 2. Extracted ion chromatogram of the protonated ion $[M + H]^+$ of angiotensin II. The wavelength of the DFG laser is set at 6.05 μm, and the pulse energy is 400 μJ. Each data point was extracted from the raw mass spectra, which were averaged from three micro-scans. The electric potential of the frit and the capillary temperature were set to 2.5 kV and 270 °C, respectively.

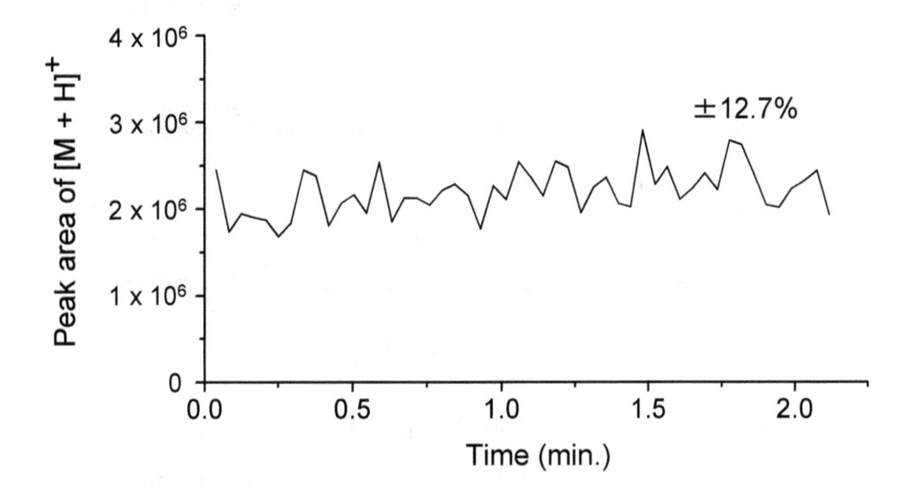

2.2. Wavelength Dependence of the Ion Signal Intensity

The laser wavelength must be properly selected for efficient and subsequent ion yield due to the characteristic absorption of the solvent in the mid-infrared region. Figure 3 shows the absorption spectra of a typical mobile-phase in reversed phase liquid chromatography, which includes water, acetonitrile and a small amount of formic acid. The spectral shape of the mixed solvent depends on the mixing ratio. The O–H bending vibration around 6 μm is due to water, whereas the CH_3 symmetric and degenerate bending vibration modes around 7.3 and 6.9 μm, respectively, are due to acetonitrile. Previous reports about laser ionization without additional matrix elements depended on the strong absorption of water [20,21]. Herein, we tried to utilize a matrix containing water and acetonitrile using the mid-infrared tunable laser, which covers a wide wavelength range of 6–7 μm.

Figure 4 shows the relationship between the ion signal intensity of $[M + H]^+$ and the IR absorption spectrum. Mainly two local maxima are observed in the plot of the ion signal intensity at the respective absorption peak wavelengths of water and acetonitrile. In addition, these peak wavelengths are very similar to the case of an aqueous solution on the sample plate. These results indicate that acetonitrile can act as a matrix by using a laser with the wavelength that corresponds to the matrix absorption peak.

In general, ESI is taken as one of the most useful techniques for the production of molecular ions also in LC/MS. In the gradient analysis in which the mixing ratio of the organic solvent in an aqueous solution is elevated as the analysis proceeds, an excessive concentration of acetonitrile decreases the ionization efficiency of ESI. On the other hand, our novel method using a mid-infrared laser can utilize both water and acetonitrile as the "ionization support agent". Therefore, the appropriate wavelength can be selected based on the eluent condition of HPLC in LC/MS with this ionization technique. Even in the case of gradient LC/MS, the center wavelength of the peaks is almost constant; even the mixing ratio is changed, as shown in Figure 3. Thus, gradient LC/MS should be possible using a fixed laser wavelength.

Figure 3. IR absorption spectra of the acetonitrile mixture used in reversed-phase liquid chromatography.

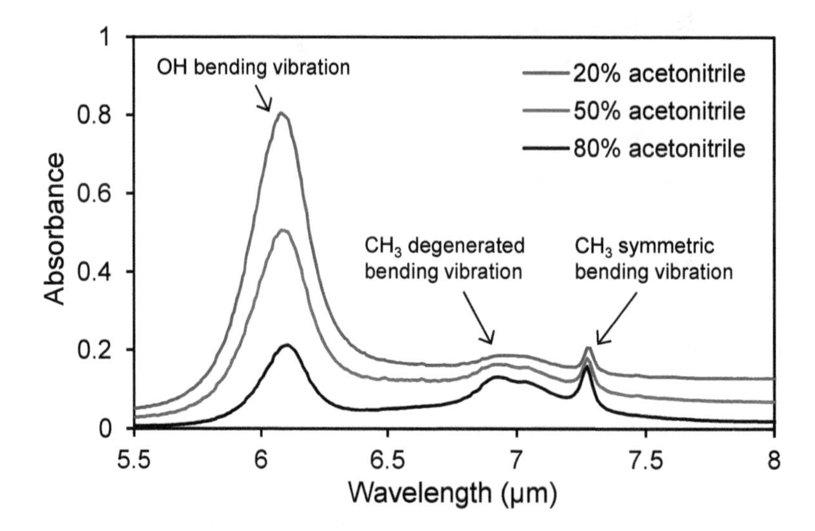

Figure 4. Wavelength dependence of the peak intensity of $[M + H]^+$ of angiotensin II from an 80% acetonitrile aqueous solution and the IR absorption spectrum of the 80% acetonitrile aqueous solution. The electric potential of the frit and the capillary temperature were set to 2.5 kV and 270 °C, respectively.

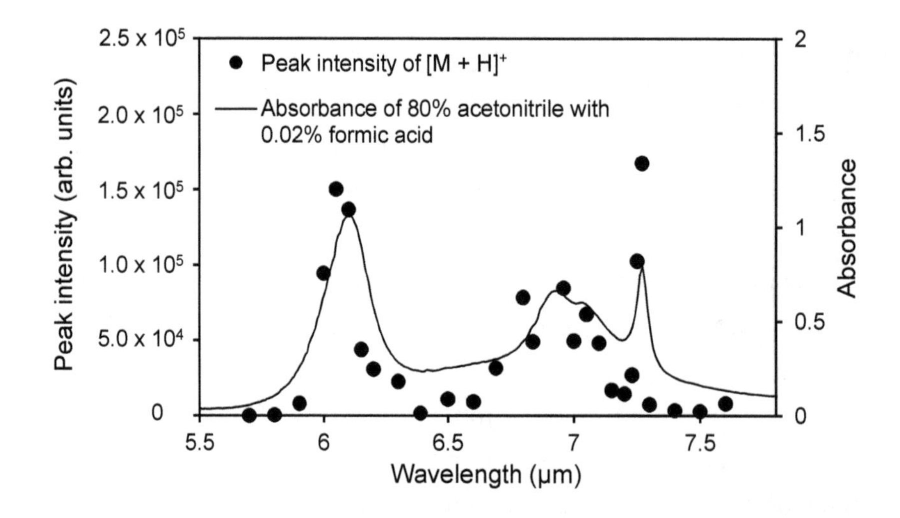

2.3. Dependence of the Ionization Efficiency on the Mixing Ratio of Water and Organic Solvent

It has been previously suggested that laser absorption at 7.27 μm by acetonitrile contributes to the ionization of a dissolved peptide sample. On the other hand, the residual water solvent contained in the dried matrix crystal has been reported to influence the ionization processes in UV-MALDI [26]. Thus, we investigated whether water is necessary for the laser wavelength corresponding to the specific absorption peak of acetonitrile. Figure 5 shows typical mass spectra obtained from angiotensin II dissolved in 100% acetonitrile and a 90% acetonitrile aqueous solution. $[M + H]^+$ of angiotensin II is clearly observed for the 90% acetonitrile containing 10% water, but not for 100% acetonitrile, suggesting that the presence of water is essential, regardless of whether the laser wavelength is fixed at the absorption peak of another solvent (acetonitrile in this case).

Figure 5. Typical mass spectra of angiotensin II dissolved in (**a**) 100% acetonitrile and (**b**) a 90% acetonitrile aqueous solution upon irradiation with a mid-IR laser at a wavelength of 7.27 μm, which corresponds to the absorption peak of the CH_3 symmetric bending vibration mode in acetonitrile. The electric potential of the frit and the capillary temperature were set to 2.5 kV and 270 °C, respectively.

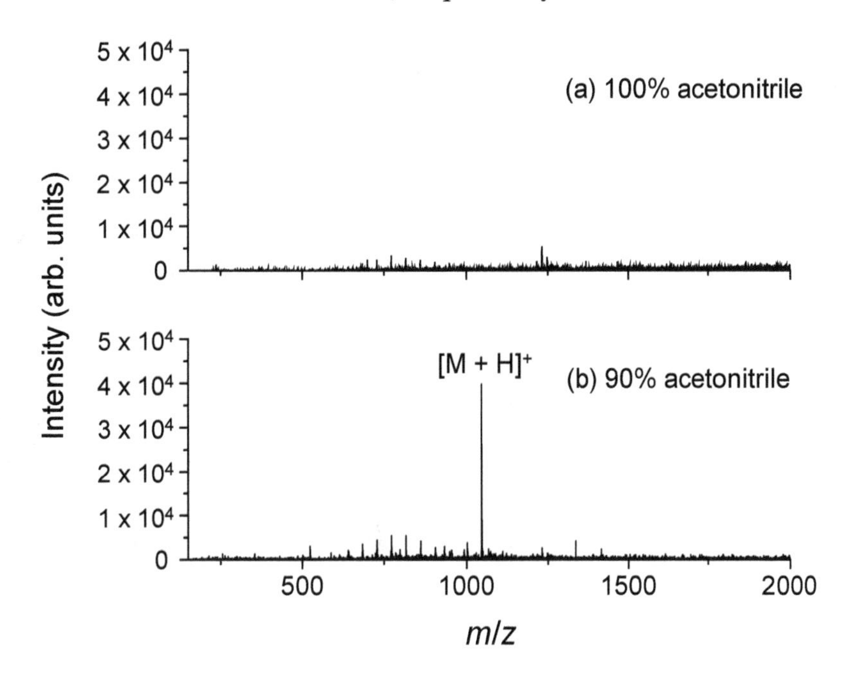

The absorption coefficients of the solvent at several wavelengths depend on the mixing ratios of the solvents. A larger mixing ratio of water leads to a stronger absorption peak in the 6-μm range, which corresponds to the O–H bending vibration mode. Water also has a broadband absorption over the 6-μm peak. The decrease in the absorption intensity at 7.27 μm, which corresponds to the CH_3 symmetric bending vibration mode in acetonitrile, is almost countered by increasing the broadband absorption of water. Therefore, laser absorption by the solvent matrix may become constant even if the mixing ratio of the solvent (the eluent of HPLC) changes temporally according to the gradient program.

In conventional LC/MS using an ESI source, the fluctuating ionization efficiency due to the solvent mixing ratio makes quantitative analysis difficult. Hence, we investigated the dependence of the solvent mixing ratio on the ionization efficiency when using a 7.27-μm laser wavelength. Figure 6 shows the relationship between the ion signal intensity of $[M + H]^+$ and three mixing ratios of water and acetonitrile. The ion signal intensities depend greatly on both the mixing ratio and the laser pulse energy.

We also investigated the influence of the solvent mixing ratio on the ion efficiency for the wavelength corresponding to water absorption (6.05 μm). Similarly, the mixing ratio affects the ion signal intensity of $[M + H]^+$ (Figure 6). The 80% acetonitrile aqueous solution gives the strongest ion signal of $[M + H]^+$ despite having the weakest absorption at 6.05 μm (Figure 3), indicating that the absorption intensity is not the only factor contributing to the ionization process. It is possible that the volatility and surface tension of the solvent may influence the desorption efficiency. Water has a relatively high surface tension and a lower volatility compared to other organic solvent. If the surface tension and volatility influence the ionization process, desolvation would produce ions. All mass spectrometers with an atmospheric pressure ion source have a heated capillary or a skimmer to promote desolvation from charged droplets containing sample molecules. Thus, the capillary temperature may have a considerable effect on ion production.

Figure 6. Relationships between the ion signal intensity of $[M + H]^+$ of angiotensin II and the mixing ratio of acetonitrile in an aqueous solvent with laser wavelengths of (**a**) 7.27 μm and (**b**) 6.05 μm. The electric potential of the frit and the capillary temperature were set to 2.5 kV and 270 °C, respectively.

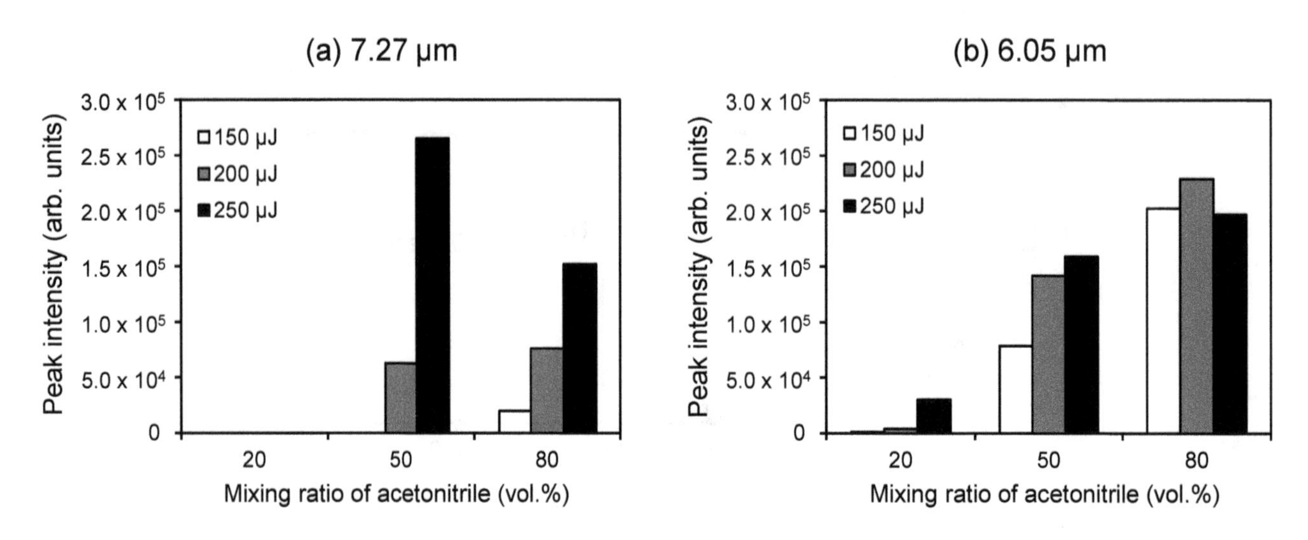

2.4. Relationship between the Generation of Multiply Charged Ions and the Desolvation Temperature

The above results suggest that the desolvation temperature and ion production are related. Thus, we investigated the relationship between the temperature of the heated capillary in an ion trap mass spectrometer (LCQ Classic, Thermo Finnigan, CA, USA) and the production of multiply-charged ions of peptides. Mass spectra were obtained from a mixture of three peptides (angiotensin II, $P_{14}R$ and ACTH (adrenocorticotropic hormone) Fragment 18–39) dissolved in an 80% acetonitrile aqueous solution with 0.01% formic acid.

In previous research using the AP-IR-MALDI of peptides and proteins [27], multiply-charged ions were produced instead of singly-charged ions, which are chiefly observed in UV-MALDI. In addition, it was recently reported that liquid AP-UV-MALDI enables stable ion yields of multiply-charged ions of peptides and proteins [28].

This ionization method produces multiply-charged ions, whose distribution drastically depends on the capillary temperature (Figure 7). For a capillary temperature of 270 °C, the mass spectrum contains mainly singly-protonated ions $[M + H]^+$, but the intensities of doubly-charged ion $[M + 2H]^{2+}$ and triply-charged ion $[M + 3H]^{3+}$ increase as the temperature decreases. These results indicate that the ion production process in 6–7-μm atmospheric pressure laser desorption/ionization using a solvent matrix occurs inside the heated capillary and not at the laser irradiation point. Usually, in UV-MALDI, ions are produced in or form a dense plume containing the matrix and analytes in the gas phase. Thus, the mechanisms of the ionization process in conventional MALDI and 6–7-μm atmospheric pressure laser desorption/ionization using a solvent matrix should differ.

On the other hand, the distribution of the ion valences produced by an electrospray shifts to the high-valence side, according to the increment of the vaporization rate from charged droplets [29,30]. On the other hand, a higher vaporization rate from charged droplets causes the formation of more highly charged ions in electrospray ionization [29,30]. In ESI, multiply-charged ions are produced due to the rapid desolvation from the charged droplets and the condensation of charge. Conversely, slow desolvation may result in poorly charged ions. ESI and continuous flow AP-IR-MALDI atmospheric pressure laser desorption/ionization are completely opposite in this regard. However, both have a commonality: they need the desolvation from charged droplets, which is not required for conventional MALDI. However, the mechanism is unclear at this time, and a more detailed study is necessary.

These investigations suggested that the laser might induce not only the desorption of a charged droplet, but ionization. On the other hand, for example, simply heating is one of the choices for just desorption. However, probe heating preferentially would desorb highly volatile compounds. As a result, non-volatile compounds remain on the probe. In this regard, the use of laser desorption could provide an important benefit that every compounds in the solution would be desorbed regardless of the volatility.

Figure 7. Relationship between the signal intensity of the protonated peptide ions and the temperature of the heated capillary in the mass spectrometer. The electric potential of the frit was set to 2.5 kV.

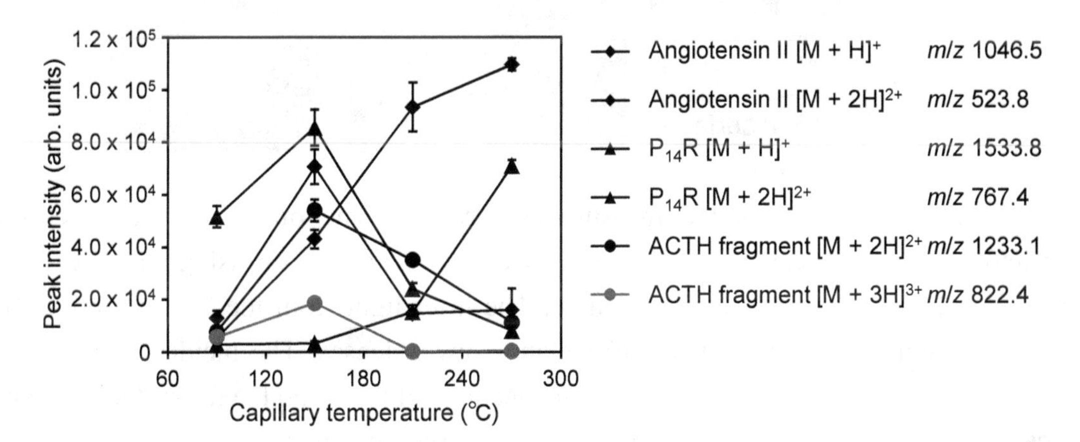

3. Experimental Section

3.1. Mid-Infrared Tunable Laser Using Difference-Frequency Generation (DFG)

A mid-IR tunable laser using difference-frequency generation (DFG) was used for ionization [31]. The DFG laser was developed by Kawasaki Heavy Industries, Ltd. (Kobe, Hyogo, Japan), and RIKEN (Wako, Saitama, Japan). Laser pulses from a Nd:YAG laser with a wavelength λ_1 of 1.064 μm and a tunable Cr:forsterite laser with a wavelength λ_2 of 1.19–1.32 μm were synchronized and mixed in two nonlinear optical crystals ($AgGaS_2$). Consequently, the mid-IR output of DFG with a wavelength $\lambda_{DFG} = (1/\lambda_1 - 1/\lambda_2)^{-1}$ was obtained. This laser system had a tunable wavelength range of 5.50–10.00 μm, and wavelength tuning with a minimum interval of 0.01 μm was automatically controlled by a computer. The laser pulse width was about 5 ns, and the pulse repetition rate was 10 Hz. In the experiments, the laser pulse energy ranged between 250 and 400 μJ at the sample surface.

3.2. Ion Trap Mass Spectrometer with an Atmospheric Pressure Laser Ion Source

All experiments were conducted with an ion trap mass spectrometer (LCQ Classic, Thermo Finnigan, CA, USA) integrated with a self-assembled continuous flow atmospheric pressure laser ion source.

The sample probe was a porous stainless steel substance called a "frit" (Figure 8). A frit has been used with a conventional ionization technique using a neutral beam and some viscous matrices, called fast atom bombardment (FAB) [32,33]. These conditions realized continuous flow ionization, which was used as the interface of LC/MS [34–36]. In addition, a frit has applied to a modified MALDI technique using an infrared laser, named continuous flow IR-MALDI [37]. Although ionization by frit-FAB or continuous flow IR-MALDI was conducted under a high vacuum, in this work, the frit was subjected to atmospheric pressure in continuous flow laser ionization.

Figure 8. (**a**) Photograph of the frit probe and (**b**) scanning electron microscope image of the frit surface.

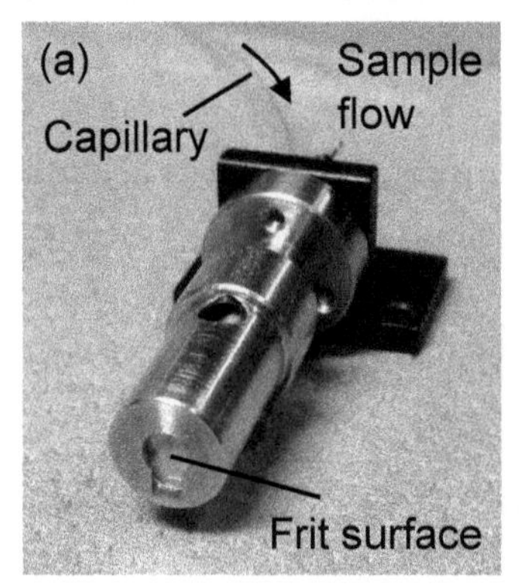

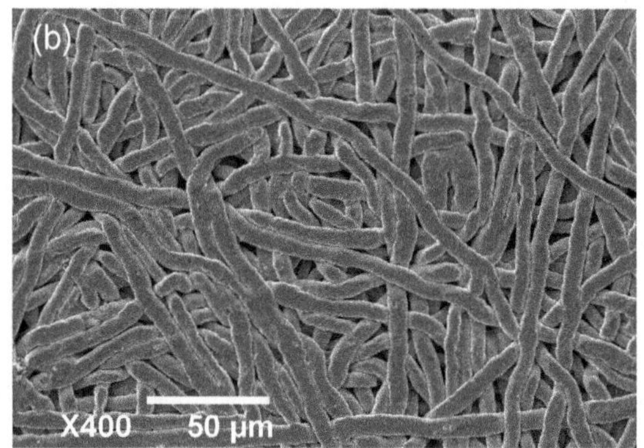

Figure 9 schematically depicts the measurement system. The output of the DFG laser was introduced into a hollow optical fiber with an inner diameter of 700 µm using an off-axis parabolic mirror to deliver the DFG laser to the ion source. The laser light from the hollow optical fiber was focused onto the sample plate using two ZnSe plano-convex lenses. The incident angle of the laser against the surface of the frit was about 45°. To calculate laser fluences on the frit surface, the knife-edge method measured the focused laser spot size, which was approximately 0.16 mm^2.

An extension capillary was connected to the heated capillary of the mass spectrometer to bring the ion inlet close to the frit surface. The length and inner diameter of the extension capillary were 80 and 0.6 mm, respectively, and the temperature of the heated capillary was set to 270 °C. The frit probe mounted on a manual *XY* translation stage was located 2 mm from the tip of the extension capillary. It is supposed that the potential of the frit probe should have an influence on the ion production. However, the effect of the sample plate voltage on the atmospheric ionization using a 3-µm-band mid-infrared laser has already been reported elsewhere [38]. Thus, a static high voltage of 2.5 kV was applied to the frit probe by connecting the voltage source originally used for the ESI nozzle [39].

The ion injection time of the ion trap was set to 300 ms, and ions were produced by the irradiation of the DFG laser in the meantime. A raw mass spectrum was acquired by three micro-scans of the ion trap, and the mass spectra shown in Figures 1 and 5 were obtained from the averaging of 50 raw mass spectra.

Figure 9. Schematic of the ion source for continuous flow atmospheric pressure laser/desorption ionization using a frit probe integrated with the ion trap mass spectrometer.

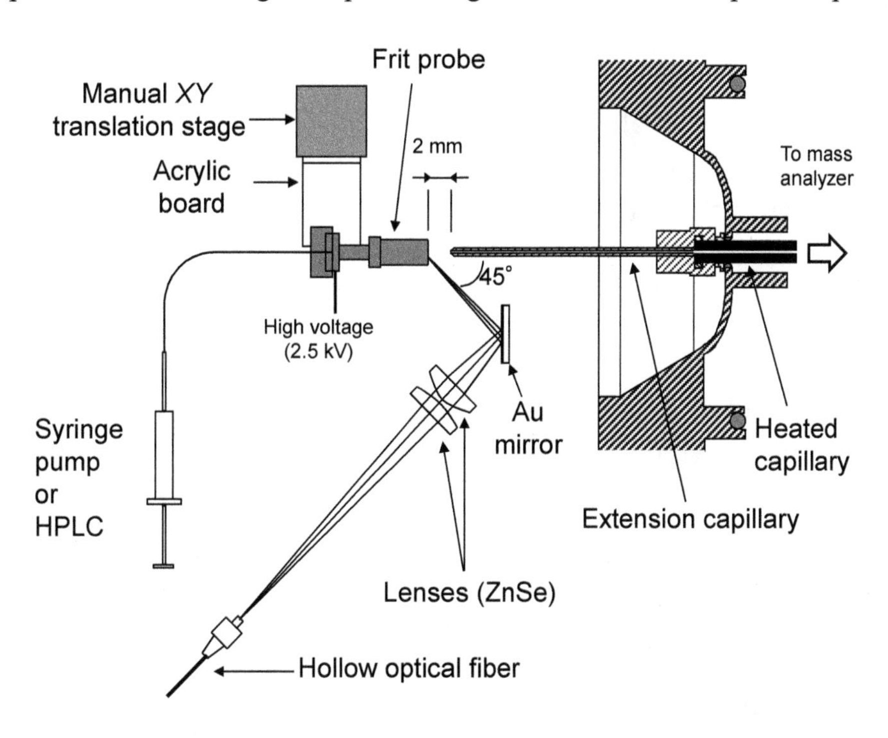

3.3. Materials and Methods

Human angiotensin II (A8846), $P_{14}R$ (synthetic) (P2613) and human ACTH Fragment 18–39 (A8346) were purchased from Sigma-Aldrich (St. Louis, MO, USA). Distilled water (049-16787) was purchased from Wako Pure Chemical Industries Ltd. (Osaka, Japan). All were used as received. These peptides were dissolved in 20%–90% acetonitrile aqueous solutions with 0.005%–0.04% formic acid (11-0780-5, Katayama Chemical Industries Co., Ltd., Osaka, Japan) at a concentration of 10 pmol/μL. Otherwise, a peptide was dissolved in 100% acetonitrile.

Sample solutions were delivered to the frit probe via a syringe with a 250-μL capacity (Unimetrics, IL, USA) and an LCQ classic syringe pump. The sample flow rate was controlled in the range of 3–20 μL/min. A sample solution was pumped at a constant flow rate during laser irradiation.

The mass spectra shown in Figures 1 and 5 were averaged 50 times for only the reduction of noise. The temporal changes in the ion signal intensity were investigated by plotting the intensities in the respective spectra. The heated capillary was set at 90, 150, 210 or 270 °C.

4. Conclusions

The continuous flow atmospheric pressure laser ion source using a frit and a 6–7-μm-band mid-infrared tunable laser is capable of direct ionization from a continuous flow with a high temporal stability. This modified ion source requires no additional matrix and utilizes water or acetonitrile as the solvent matrix at several absorption peak wavelengths (6.05 and 7.27 μm).

In addition, the effects of the solvent mixing ratio on the ionization efficiency were investigated. The ion signal intensity depends on the mixing ratio at wavelengths of 6.05 and 7.27 μm. In the case of 6.05 μm, the ion signal intensity obtained from the 20% acetonitrile aqueous solution is lower than that from the 50% and 80% acetonitrile aqueous solutions despite their relatively high absorbance at 6.05 μm, indicating that not only the intensity of laser absorption, but also the volatility and surface tension may affect the desorption efficiency.

The distribution of multiply-charged peptide ions is extremely sensitive to the temperature of the heated capillary, which is the inlet of the mass spectrometer. This suggests that ions are produced from the charged droplet after desorption of the solvent matrices (water or acetonitrile). The solvent matrices used in this work were broadly used as mobile-phase in reversed-phase liquid chromatography. Thus, this ionization method has the potential for the interface of LC/MS as an alternative for ESI.

Acknowledgments

This work was supported by a Grant-in-Aid for Scientific Research (C) of the Japan Society for the Promotion of Science (JSPS) (Grant Number 25513004).

Author Contributions

The authors' contributions were as follows: Ryuji Hiraguchi and Hisanao Hazama designed and conducted the experiments; Kenichirou Senoo, Yukinori Yahata, Hisanao Hazama and Ryuji Hiraguchi constructed the experimental systems including the ion source; Katsuyoshi Masuda and Kunio Awazu supervised this work and suggested the experiments 2.2 and 2.3; Ryuji Hiraguchi and Hisanao Hazama prepared and edited the manuscript. All authors read and approved the final manuscript.

References

1. Fenn, J.B.; Mann, M.; Meng, C.K.; Wong, S.F.; Whitehouse, C.M. Electrospray ionization for mass spectrometry of large biomolecules. *Science* **1989**, *246*, 64–71.
2. Fenn, J.B.; Mann, M.; Meng, C.K.; Wong, S.F. Electrospray ionization–principles and practice. *Mass Spectrom. Rev.* **1990**, *9*, 37–70.
3. Karas, M.K.; Bachmann, D.; Bahr, U.; Hillenkamp, F. Matrix-assisted ultraviolet laser desorption of non-volatile compounds. *Int. J. Mass Spectrom. Ion Process.* **1987**, *78*, 53–68.
4. Karas, M.; Hillenkamp, F. Laser desorption ionization of proteins with molecular masses exceeding 10,000 daltons. *Anal. Chem.* **1988**, *60*, 2299–2301.
5. Huang, M.Z.; Hsu, H.J.; Lee, J.Y.; Jeng, J.J.; Shiea, J. Direct protein detection from biological media through electrospray-assisted laser desorption ionization/mass spectrometry. *J. Proteome Res.* **2006**, *5*, 1107–1116.
6. Sampson, J.S.; Hawkridge, A.M.; Muddiman, D.C. Generation and detection of multiply-charged peptides and proteins by matrix-assisted laser desorption electrospray ionization (MALDESI) fourier transform ion cyclotron resonance mass spectrometry. *J. Am. Soc. Mass Spectrom.* **2006**, *17*, 1712–1716.
7. Nemes, P.; Vertes, A. Laser ablation electrospray ionization for atmospheric pressure, *in vivo*, and imaging mass spectrometry. *Anal. Chem.* **2007**, *79*, 8098–8106.
8. Takats, Z.; Wiseman, J.M.; Gologan, B.; Cooks, R.G. Mass spectrometry sampling under ambient conditions with desorption electrospray ionization. *Science* **2004**, *306*, 471–473.

9. Kertesz, V.; Ford, M.J.; Berkel, G.V. Automation of a surface sampling probe/electrospray mass spectrometry system. *Anal. Chem.* **2005**, *77*, 7183–7189.
10. Hiraoka, K.; Nishidate, K.; Mori, K.; Asakawa, D.; Suzuki, S. Development of probe electrospray using a solid needle. *Rapid Commun. Mass Spectrom.* **2007**, *21*, 3139–3144.
11. Mandal, M.K.; Chen, L.C.; Hashimoto, Y.; Yu, Z.; Hiraoka, K. Detection of biomolecules from solutions with high concentration of salts using probe electrospray and nano-electrospray ionization mass spectrometry. *Anal. Methods* **2010**, *2*, 1905–1912.
12. Overberg, A.; Karas, M.; Bahr, U.; Kaufmann, R.; Hillenkamp, F. Matrixassisted infrared-laser (2.94 µm) desorption/ionization mass spectrometry of large biomolecules. *Rapid Commun. Mass Spectrom.* **1990**, *4*, 293–296.
13. Overberg, A.; Karas, M.; Hillenkamp, F. Matrix-assisted laser desorption of large biomolecules with a TEA-CO_2-laser. *Rapid Commun. Mass Spectrom.* **1991**, *5*, 128–131.
14. Sheffer, J.D.; Murray, K.K. Infrared matrix-assisted laser desorption/ionization using OH, NH and CH vibrational absorption. *Rapid Commun. Mass Spectrom.* **1998**, *12*, 1685–1690.
15. Laiko, V.V.; Baldwin, M.A.; Burlingame, A.L. Atmospheric pressure matrix-assisted laser desorption/ionization mass spectrometry. *Anal. Chem.* **2000**, *72*, 652–657.
16. Laiko, V.V.; Moyer, S.C.; Cotter, R.J. Atmospheric pressure MALDI/ion trap mass spectrometry. *Anal. Chem.* **2000**, *72*, 5239–5243.
17. Moyer, S.C.; Marzilli, L.A.; Woods, A.S.; Laiko, V.V.; Doroshenko, V.M.; Cotter, R.J. Atmospheric pressure matrix-assisted laser desorption/ionization (AP MALDI) on a quadrupole ion trap mass spectrometer. *Int. J. Mass Spectrom.* **2003**, *226*, 133–150.
18. Daniel, J.M.; Ehala, S.; Friess, S.D.; Zenobi, R. On-line atmospheric pressure matrix-assisted laser desorption/ionization mass spectrometry. *Analyst* **2004**, *129*, 574–578.
19. Daniel, J.M.; Laiko, V.V.; Droshenko, V.M.; Zenobi, R. Interfacing liquid chromatography with atmospheric pressure MALDI-MS. *Anal. Bioanal. Chem.* **2005**, *383*, 895–902.
20. Laiko, V.V.; Taranenko, N.I.; Berkout, V.D.; Yakshin, M.A.; Prasad, C.R.; Lee, H.S.; Doroshenko, V.M. Desorption/ionization of biomolecules from aqueous solution at atmospheric pressure using an infrared laser at 3 µm. *J. Am. Soc. Mass Spectrom.* **2002**, *13*, 354–361.
21. Rapp, E.; Charvat, A.; Beinsen, A.; Reichl, U.; Morgenstern, A.S.; Urlaub, H.; Abel, B. Atmospheric pressure free liquid infrared MALDI mass spectrometry: Toward a combined ESI/MALDI-liquid chromatography interface. *Anal. Chem.* **2009**, *81*, 443–452.
22. Tajiri, M.; Takeuchi, T.; Wada, Y. Distinct features of matrix-assisted 6 µm infrared laser desorption/ionization mass spectrometry in biomolecular analysis. *Anal. Chem.* **2009**, *81*, 6750–6755.
23. Tajiri, M.; Wada, Y. Infrared matrix-assisted laser desorption/ionization mass spectrometry for quantification of glycosaminoglycans and gangliosides. *Int. J. Mass Spectrom.* **2011**, *305*, 164–169.
24. Naito, Y.; Yoshihashi-Suzuki, S.; Ishii, K.; Kanai, T.; Awazu, K. Matrix-assisted laser desorption/ionization of protein samples containing a denaturant at high concentration using a mid-infrared free-electron laser (MIR-FEL). *Int. J. Mass Spectrom.* **2005**, *241*, 49–56.
25. Yoshihashi-Suzuki, S.; Sato, I.; Awazu, K. Wavelength dependence of matrix-assisted laser desorption and ionization using a tunable mid-infrared laser. *Int. J. Mass Spectrom.* **2008**, *270*, 134–138.

26. Hazama, H.; Furukawa, S.; Awazu, K. Effect of solvent on ionization efficiency in matrix-assisted laser desorption/ionization mass spectrometry of peptides. *Chem. Phys.* **2013**, *419*, 196–199.
27. Konig, S.; Kollas, O.; Dreisewerd, K. Generation of highly charged peptide and protein ions by atmospheric pressure matrix-assisted infrared laser desorption/ionization ion trap mass Spectrometry. *Anal. Chem.* **2007**, *79*, 5484–5488.
28. Cramer, R.; Pirkl, A.; Hillenkamp, F.; Dreisewerd, K. Liquid AP-UV-MALDI enables stable ion yields of multiply charged peptide and protein ions for sensitive analysis by mass spectrometry. *Angew. Chem. Int. Ed.* **2013**, *52*, 2364–2367.
29. Winger, B.E.; Light-Wahl, K.J.; Loo, R.R.O.; Udseth, H.R.; Smith, R.D. Observation and implications of high mass-to-charge ratio ions from electrospray ionization mass spectrometry. *J. Am. Soc. Mass Spectrom.* **1993**, *4*, 536–545.
30. Fenn, J.B. Ion formation from charged droplets: Roles of geometry, energy, and time. *J. Am. Soc. Mass Spectrom.* **1993**, *4*, 524–535.
31. Hazama, H.; Takatani, Y.; Awazu, K. Integrated ultraviolet and tunable mid-infrared laser source for analyses of proteins. *Proc. SPIE* **2007**, *6455*, 645507.
32. Caprioli, R.M.; Fan, T.; Fan, T. Continuous-flow sample probe for fast atom bombardment mass spectrometry. *Anal. Chem.* **1986**, *58*, 2949–2954.
33. Hau, J.; Schrader, W.; Linscheid, M. Continuous-flow fast atom bombardment mass spectrometry: A concept to improve the sensitivity. *Org. Mass Spectrom.* **1993**, *28*, 216–222.
34. Ito, Y.; Takeuchi, T.; Ishii, D.; Goto, M. Direct coupling of micro high-performance liquid chromatography with FAB-MS. *J. Chromatogr.* **1985**, *346*, 161–166.
35. Takeuchi, T.; Watanabe, S.; Kondo, N.; Ishii, D.; Goto, M. Improvement of the interface for coupling of FAB-MS and micro high-performance liquid chromatography. *J. Chromatogr.* **1988**, *435*, 482–488.
36. Siethoff, C.; Nigge, W.; Linscheid, M.W. The determination of ifosfamide in human blood serum using LC/MS. *Fresenius J. Anal. Chem.* **1995**, *352*, 801–805.
37. Lawson, S.J.; Murray, K.K. Continuous flow infrared matrix-assisted laser desorption/ionization with a solvent matrix. *Rapid Commun. Mass Spectrom.* **2000**, *14*, 129–134.
38. Laiko, V.V.; Taranenko, N.I.; Droshenko, V.M. On the mechanism of ion formation from the aqueous solutions irradiated with 3 μm IR laser pulses under atmospheric pressure. *J. Mass Spectrom.* **2006**, *41*, 1315–1321.
39. Droshenko, V.M.; Laiko, V.V.; Taranenko, N.I.; Berkout, V.D.; Lee, H.S. Recent developments in atmospheric pressure MALDI mass spectrometry. *Int. J. Mass Spectrom.* **2002**, *221*, 39–58.

12

MALDI Q-TOF CID MS for Diagnostic Ion Screening of Human Milk Oligosaccharide Samples

Marko Jovanović [1,*], Richard Tyldesley-Worster [2], Gottfried Pohlentz [3] and Jasna Peter-Katalinić [1,4]

[1] Department of Biotechnology, University of Rijeka, Radmile Matejčić 2, Rijeka 51000, Croatia
[2] Waters Corporation, Stamford Avenue, Altrincham Road, Wilmslow SK9 4AX, UK; E-Mail: richard_tyldesley-worster@waters.com
[3] Institute for Hygiene, University of Muenster, Robert-Koch-Strasse 41, Muenster D-48149, Germany; E-Mail: pohlentz@uni-muenster.de
[4] Institute for medical Physics and Biophysics, University of Muenster, Robert-Koch-Strasse 31, Muenster D-48149, Germany; E-Mail: jkp@uni-muenster.de or jasnapk@biotech.uniri.hr

* Author to whom correspondence should be addressed; E-Mail: mjovanovic@biotech.uniri.hr

---

Abstract: Human milk oligosaccharides (HMO) represent the bioactive components of human milk, influencing the infant's gastrointestinal microflora and immune system. Structurally, they represent a highly complex class of analyte, where the main core oligosaccharide structures are built from galactose and *N*-acetylglucosamine, linked by 1-3 or 1-4 glycosidic linkages and potentially modified with fucose and sialic acid residues. The core structures can be linear or branched. Additional structural complexity in samples can be induced by endogenous exoglycosidase activity or chemical procedures during the sample preparation. Here, we show that using matrix-assisted laser desorption/ionization (MALDI) quadrupole-time-of-flight (Q-TOF) collision-induced dissociation (CID) as a fast screening method, diagnostic structural information about single oligosaccharide components present in a complex mixture can be obtained. According to sequencing data on 14 out of 22 parent ions detected in a single high molecular weight oligosaccharide chromatographic fraction, 20 different oligosaccharide structure types, corresponding to over 30 isomeric oligosaccharide structures and over 100 possible HMO isomers when biosynthetic linkage variations were taken into account, were postulated. For MS/MS data analysis, we used the *de novo* sequencing approach using diagnostic ion analysis on reduced oligosaccharides by following known biosynthetic rules. Using this approach, *de novo* characterization has been achieved also for the structures, which could not have been predicted.

Keywords: human milk oligosaccharides; MALDI Q-TOF (matrix-assisted laser desorption/ionization quadrupole-time-of-flight) MS (mass spectrometry); CID (collision-induced dissociation); diagnostic ion MS; *de novo* sequencing

1. Introduction

Human milk oligosaccharides (HMO) are the third most abundant type of component in human milk [1]. Their core structure is biosynthesized by adding *N*-acetyllactosamine units to the single lactose core at the reducing end. Apart from variable branching patterns, even a more significant layer of structural diversity is provided by extensive fucosylation. Recent studies have shown that HMO consist of around 200 oligosaccharide structures, including both neutral and negatively charged structures [2–4]. Their most understood function is the interaction with infant's gut microflora, stimulating the growth of probiotic bacteria [5,6] and providing immunological benefits to the infant [7]. Based on animal studies, a role of sialylated oligosaccharides in brain development has been postulated [8,9]. In recent studies, milk oligosaccharide patterns from women delivering preterm and at term were compared [10], focusing on differences in fucosylation patterns and lacto-*N*-tetraose (LNT) abundance. In spite of significant advances in the analytical techniques used for oligosaccharide structure elucidation over the past two decades [2–4,11–19], all individual components of HMO cannot be fully revealed.

In addition to the complex nature of HMO biosynthesis, additional variety in HMO structures can arise from chemical procedures carried out during sample preparation steps or due to biological enzymatic activity. To these belongs also the chemical procedure of mild acid hydrolysis in order to remove fucose and sialic acid residues [20]. This may be necessary to simplify a detailed structural analysis of the oligosaccharide core structure. It has also been found that sugars on the nonreducing terminus of the oligosaccharide are susceptible to cleavage by glycosidases in breast and during the storage of milk, although this degradation was found to be modest [21]. Other experimental parameters and biological specificities of the milk donor may influence HMO composition. For targeted (data-dependent acquisition or DDA) MS analysis methods, preliminary characterization of the sample is necessary. Matrix-assisted laser desorption/ionization (MALDI) MS is the method of choice for oligosaccharide screening, in particular in the context of high-throughput options. In most cases, MALDI-MS fingerprints of glycan mixtures display a singly charged ionic signal per component, which tends to simplify spectral interpretation in comparison to electrospray ionization (ESI). For most experiments conducted on glycans, ubiquitous sodium ions from glass and other sources attach to glycan molecules to form $[M + Na]^+$ ions. For direct measurements on neutral glycans, $[M + Na]^+$ abundances (peak area) may be compared in a semi-quantitative way within the same sample [22]. Combined with collision-induced dissociation (CID) available on quadrupole-time-of-flight (Q-TOF) mass spectrometers, it can quickly provide sufficient information for subsequent, more detailed MS analyses for a large number of precursor ions.

For large molecular weight HMO, which can possess more complex branching sites, as well as multiple fucosylation sites, the structures are predominantly difficult to define, also due to isobaric mixtures. This is frequently solved by the application of liquid chromatography (LC) protocols, which allow isomer separation prior to MS analysis [2,3]. More direct is the MALDI Q-TOF CID *de novo* approach. We show that the already known oligosaccharide structures can be assigned according to specific diagnostic ions in combination with the HMO biosynthetic rules, but also, novel structure types can be proposed *de novo*. The additional advantages of the MALDI Q-TOF MS analysis are the sensitivity, speed, high dynamic range and ability to analyse data in more detail on the same sample spot once the preliminary data analysis is completed.

2. Results and Discussion

2.1. MALDI TOF (Matrix-Assisted Laser Desorption/Ionization Time-of-Flight) Mapping of Complex Oligosaccharide Mixtures

The MALDI TOF map of the sample containing human milk oligosaccharides is shown in Figure 1. Twenty two different ions were detected that could be assigned according to monosaccharide compositions (in terms of the building blocks, Hex (H, hexose), HexNAc (HN, *N*-acetylhexosamine) and dHex (F, deoxyhexose)). The oligosaccharides ranged from tetra- to dodeca-saccharides, including eight fucosylated structures. The MALDI spectrum was acquired from the sample without further purification procedures as an initial oligosaccharide mapping experiment. The most intense ions represent octa-, nona- and deca-saccharides. The undeca- and dodeca-saccharide-related ions were at a lower intensity. Tetra- to hepta-saccharide series were detected, as well, with similar relative abundances.

Figure 1. The (+) matrix-assisted laser desorption/ionization (MALDI) quadrupole-time-of-flight (Q-TOF) spectrum of the sample containing human milk oligosaccharide alditols. The *m/z* values represent $[M + Na]^+$ ions.

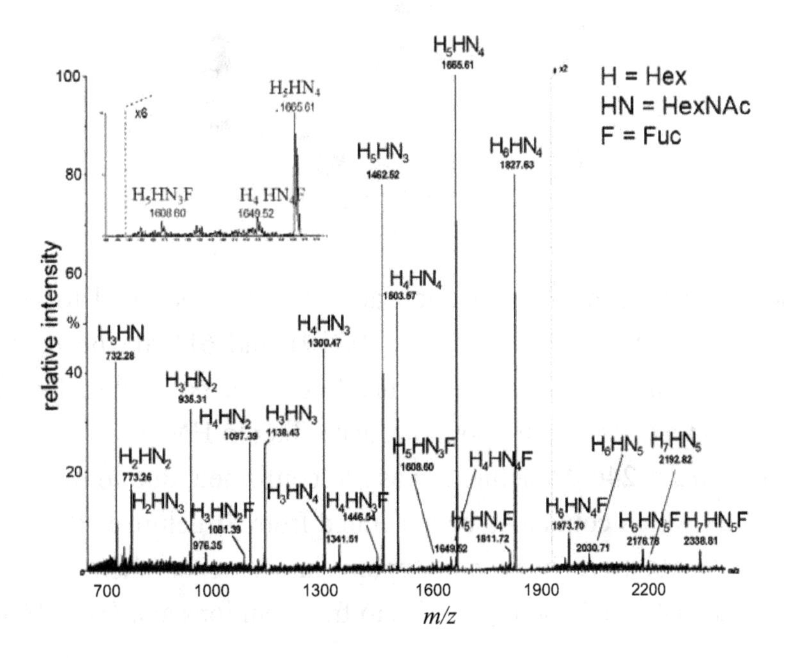

2.2. *De Novo* MALDI Q-TOF CID Data Analysis Using Diagnostic Ions

The fragmentation spectra (MS/MS) were assigned using the theoretical cleavage ions according to Domon and Costello [23]. In Figure 2A, the CID spectrum of the basic tetrasaccharide precursor ion at *m/z* = 732 is shown. Due to the reduced reducing end, it was possible to unambiguously assign $Y_{1\text{-}3}$ ions for the full sequence information, along with $B_{1\text{-}3}$ ions (*m/z* = 185, 388 and 550), in agreement with the known biosynthetic core structure of human milk oligosaccharides (HMO) belonging to LNT and LNnT tetrasaccharides. No intra-ring fragmentation was observed under these conditions.

Figure 2. The (+) MALDI Q-TOF collision-induced dissociation (CID) spectra of two low mass oligosaccharide alditol ions from the mixture (Figure 1). (**A**) Tetrasaccharide alditol lacto-*N*-tetraose (732.23); and (**B**) isobaric mixture of tetrasaccharide alditols (773.27) of the sum composition $Hex_2HexNAc_2$.

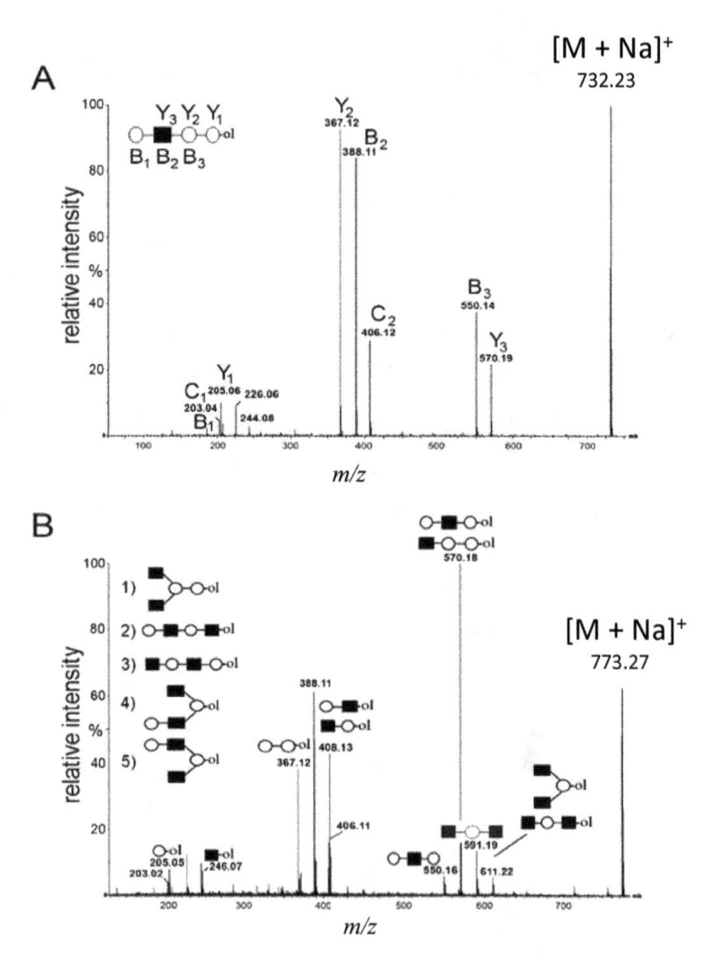

In Figure 2B, the CID spectrum of parent ion at *m/z* = 773 assigned to H_2HN_2 is shown. The cleavage ions at *m/z* = 205, 246, 367, 408, 550, 570, 591 and 611 are postulated to be the partial structures shown. From the ion at *m/z* = 367, Structure 1 could be postulated. For the ion at *m/z* = 408, two tetrasaccharide structures (2 and 3) are possible, according to HMO biosynthetic rules. Structure 2 corresponds to the ion at *m/z* = 246. Structure 3 was not confirmed, due to the ambiguity of fragment ions at *m/z* = 205 and 570, which could also be generated from Structure 1. It is important to note that both Structure 2 and 3 could have been built upon enzymatic or chemical hydrolysis of HMO. Furthermore, both Structures 4 and 5 could give rise to fragment ions at *m/z* = 408 and 611.

In Figure 2, MS analysis of oligosaccharides when isomeric mixtures are present is presented. For manual *de novo* tandem MS data analysis, the pragmatic first step is the assignment of diagnostic ions (Table 1). Accordingly, the fragment ion at *m*/*z* = 367 as a diagnostic one is relevant for Structure 1, indicating the presence of the lactose core on the reducing end. The fragment ion at *m*/*z* = 246 is diagnostic for the presence of HexNAc on the reducing end, typical for the truncated structure. The diagnostic ion at *m*/*z* = 408 in combination with other diagnostic ions, such as at *m*/*z* = 205 (indicating the presence of a hexose on the reducing end) and at *m*/*z* = 246 (indicating the presence of an *N*-acetylhexosamine on the reducing end), represents the "conditional" diagnostic ion, because additional information was necessary to fully assign its structure. In the case of isomeric mixtures, this information can be probed with MS^3 experiments on appropriate instrument types for full structural elucidation. A list of diagnostic and conditional diagnostic ions used in this study is shown in Table 1. Ions at *m*/*z* = 246, 367 and 794 in Table 1 are diagnostic ones. Additionally, two examples of non-diagnostic ions are given, illustrating that no significant information can be deduced from them, alone or in combination with other fragment ions.

Table 1. Diagnostic and conditional diagnostic ions found and monosaccharide symbols used in this study. ■ denotes *N*-acetylglucosamine, ○ galactose and ● glucose.

Type of ions	*m*/*z*	Type	Structure	
non-diagnostic ions	753	B_4 or internal B/Y fragment ion		○-■-○-■ ■-○-■-○
	771	C_4 or internal C/Z fragment ion		○-■-○-■- ■-○-■-○-
diagnostic ions	205	hexose at reducing end		●-ol ○-ol
	246	HexNAc at reducing end (non-biosynthetic)		■-ol
	367	contains biosynthetic core lactose unit		○-●-ol
	794	branching at $Hex_{n≥4}$		■,■>○-■
conditional diagnostic ions	408	non-biosynthetic reducing end structure	no 246	■-○-ol
			no 205	○-■-ol
	611	non-biosynthetic reducing end structure	no 246	■,■>○-ol
			no 205	■-○-■-ol
	732	Y4 with core lactose—linear oligosaccharide	no 773 and no 794	○-■-○-●-ol
	773	branching at Hex_2 (biosynthetic structure)	yes 367 and no 408	■,■>○-●-ol

Efficient fragmentation has been obtained by MALDI Q-TOF CID (Micromass, Manchester, UK) from major parent ions at *m*/*z* = 732, 773, 935, 1097, 1138, 1300, 1462, 1503, 1665 and 1827. The parent ion at *m*/*z* = 1341 fragmented poorly. The parent ion at *m*/*z* = 1503, was further tested along low-intensity parent ions at m/z = 1608, 1973 and 2338 on an AB SCIEX QSTAR® Pulsar i instrument (AB SCIEX, Toronto, ON, Canada).

In Figure 3A, the fragmentation spectra of parent ions at m/z = 773, 935 and 1138 are depicted. A structural assignment has been carried out using the diagnostic ion at m/z = 367 (blue asterisk) and conditional diagnostic ions at m/z = 408 (red asterisk) and 773 (green asterisk) (see Table 2). For the parent ion at m/z = 935, the assignment to two possible isomeric pentasaccharide structures is ambiguous. Based on the diagnostic ion at m/z = 367 and the conditional diagnostic ion at m/z = 773 (which in the absence of the conditional diagnostic ion at m/z = 408 and the presence of the diagnostic ion at m/z = 367 indicates the presence of a branching point at Hex_2 (see Table 2)), the presence of a single tetrasaccharide pattern (structure type) could be postulated. However, the location of the final galactose residue cannot be deduced from our data. Furthermore, the presence of a linear pentasaccharide structure cannot be proven in the presence of branched structure(s). Finally, in the case of the CID of parent ion at m/z = 1138, we detected all three ion types at m/z = 367, 408 and 773. According to the diagnostic ion at m/z = 794, a branching point is at Hex_4. A branching point at Hex_2 (supported by the presence of the conditional diagnostic ion at m/z = 773) could not be deduced due to the presence of another conditional diagnostic ion at m/z = 408.

Figure 3. The assignment of CID MS data using diagnostic (367 and 794) and conditional diagnostic (408 and 773) fragment ions. (**A**) Ions at m/z = 773, 935 and 1138; and (**B**) zoom-in of the mass range of 385–410 and 770–800 Da. Possible structures are all in accord with the fragmentation pattern. ■ denotes *N*-acetylglucosamine, ○ galactose and ● glucose. *, * and * denote ions relevant to the explanation in the text.

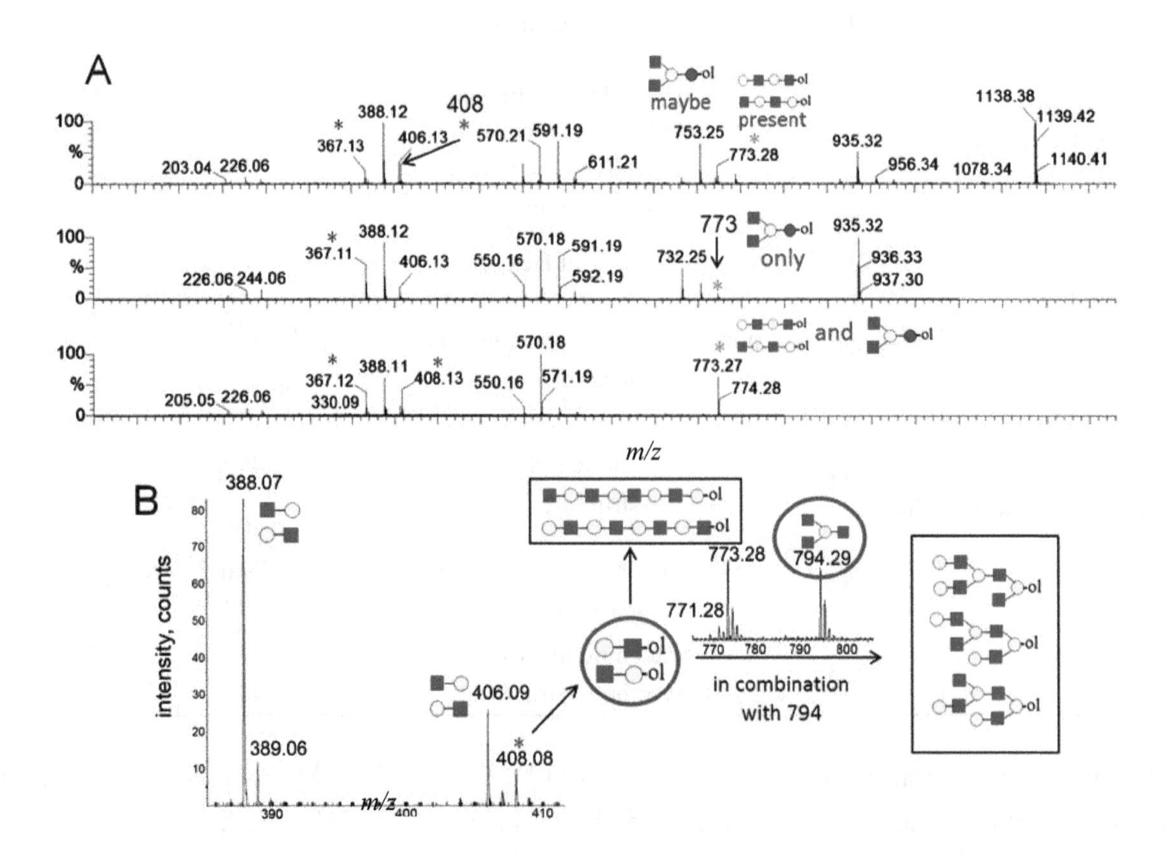

Table 2. Confirmed and proposed oligosaccharide alditol structures along with respective diagnostic ions. Additional branched isomers are possible, due to galactose 1-3 and 1-4 linkages, and most of them are further fucosylated *in vivo* with fucose 1-2, 1-3 or 1-4. ■ denotes *N*-acetylglucosamine, ○ galactose, ● glucose and ▲ fucose. * structures that could be neither confirmed nor excluded. * present in very small amounts.

Monosaccharide composition	Biosynthetic	Truncated with lactose core	Truncated without lactose core	Diagnostic ions
H_3HN (732)	 ol	-	-	367
H_2HN_2 (773)	-	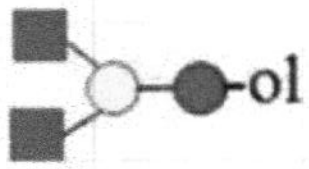 ol	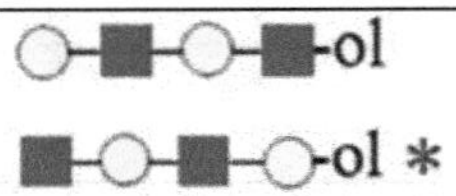 ol ol *	246, 367, 408
H_3HN_2 (935)	-	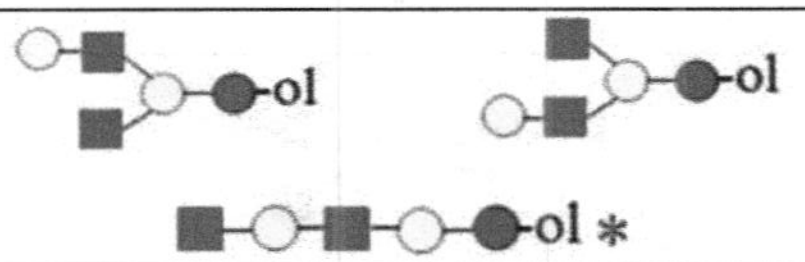 ol ol ol *	-	367, 773
H_2HN_3 (976)	-	-	 ol	-
H_4HN_2 (1097)	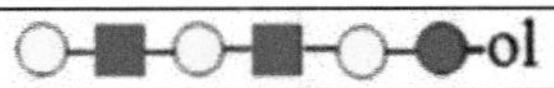 ol	-	-	367
H_3HN_3 (1138)	-	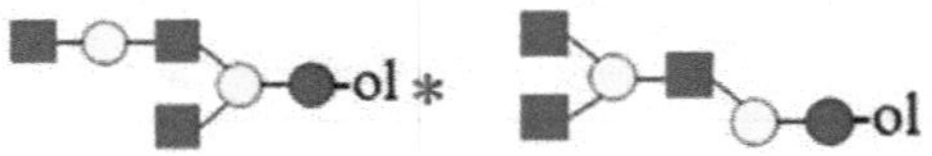 ol * ol	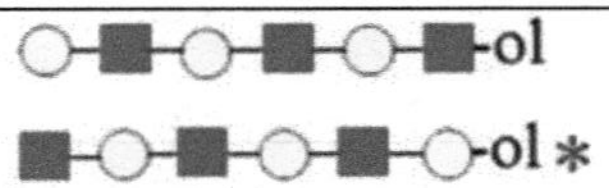 ol ol *	246, 367, 408, 773, 794
H_4HN_3 (1300)	-	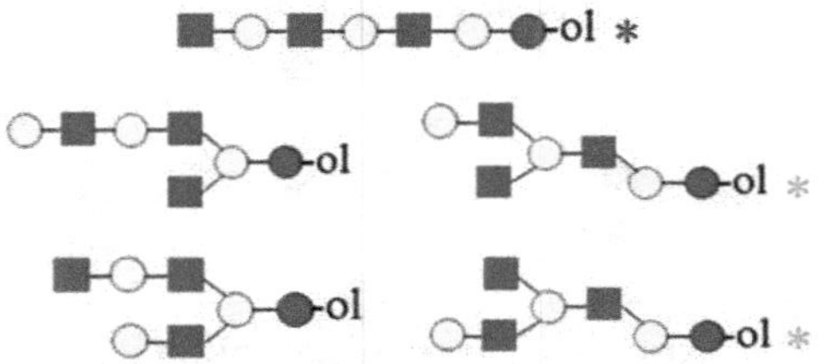 ol * ol ol * ol ol *	-	367, 773, 794
H_3HN_4 (1341)	-	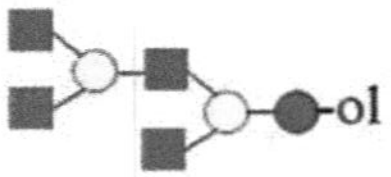 ol	-	773, 794
H_5HN_3 (1462)	 ol	-	-	367

Table 2. *Cont.*

Monosaccharide composition	Biosynthetic	Truncated with lactose core	Truncated without lactose core	Diagnostic ions
H_4HN_4 (1503)	-	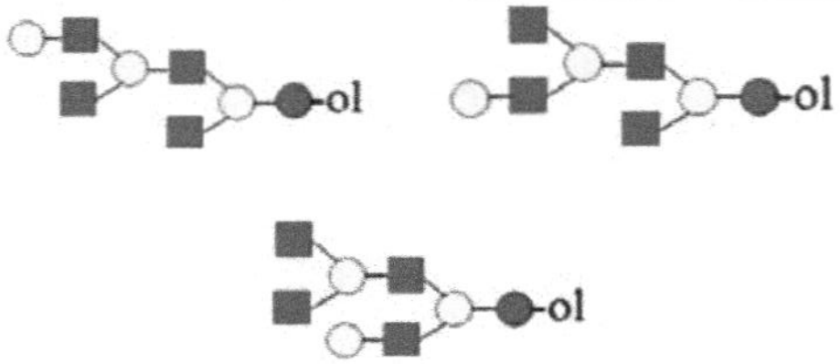	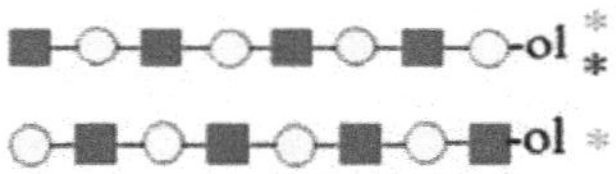	367, 408, 773, 794
H_5HN_4 (1665)	-	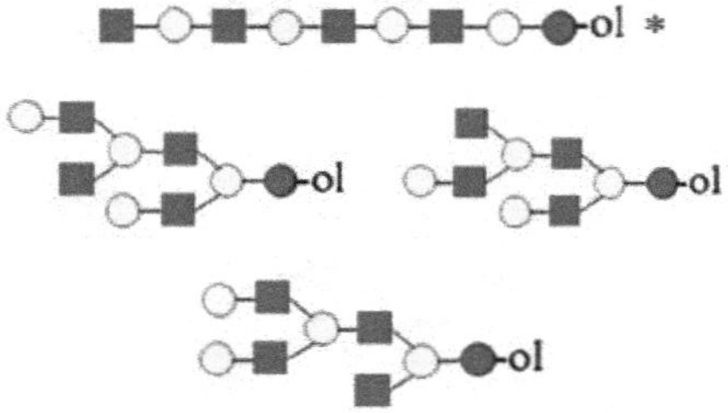	-	367, 773, 794
H_6HN_4 (1827)		-	-	367
H_5HN_3F (1609)	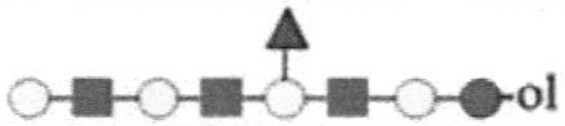	-	-	878
H_6HN_4F (1973)	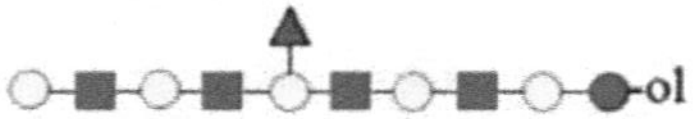	-	-	1243
H_7HN_5F (2338)	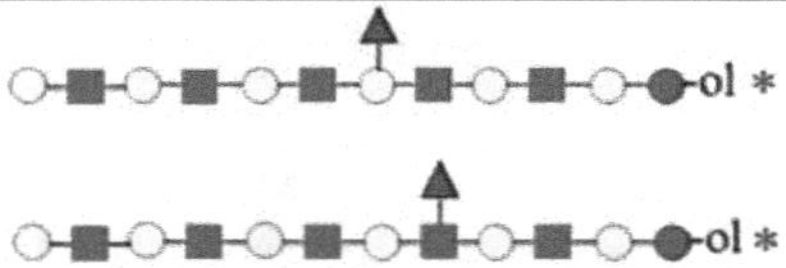	-	-	1243 ?

In Figure 3B, the inserts of the MALDI Q-TOF CID spectrum of the parent ion at m/z = 1503 are shown, obtained by using the "enhance all" function on the AB SCIEX QSTAR® Pulsar i (see the Experimental Section). In this experiment, the conditional diagnostic ion at m/z = 408 was detected (left insert) and allowed one to deduce the presence of two further oligosaccharide structures in the m/z = 1503 parent ion mixture. Furthermore, in combination with the diagnostic ion at m/z = 794, the existence of three additional truncated isomeric structures could be proposed (if we restrict our analysis to the reported HMO structures, up to lacto-*N*-decaose), but even more isomeric structures are possible, when Type I and Type II branches are considered [19]. Therefore, at least five novel structure types could be added to the list of possible isomeric octasaccharide structures present in the parent ion at m/z = 1503, most of them as truncated isomers.

2.3. Analysis of Branched Higher Molecular Weight Oligosaccharides

In Figures 2 and 3, we showed how the existence of truncated structures can exponentially increase the number of theoretically possible isomers present in the isobaric parent ion mixture. In Figure 4, the MALDI Q-TOF CID spectrum of the parent ion at m/z = 1665 and the conditional diagnostic ion at m/z = 408 was not present, but the positive structural evidence was obtained by two important fragment ions, the conditional diagnostic ion at m/z = 773 and the diagnostic ion at m/z = 794. The diagnostic ion at m/z = 367 together with the conditional diagnostic ion at m/z = 773 provides evidence of branched nonasaccharide structures with a branching point at Hex_2. The diagnostic ion at m/z = 794 indicates the presence of branched nonasaccharide structures with a branching point at Hex_4. By applying biosynthetic rules and restricting our analysis to the reported HMO structures (up to lacto-*N*-decaose), only three out of 13 theoretical branched structure types remain possible.

Figure 4. Fragmentation spectrum of the parent ion at m/z = 1665.59. The analysis of fragment ions revealed the presence of isobaric nonasaccharide alditols.

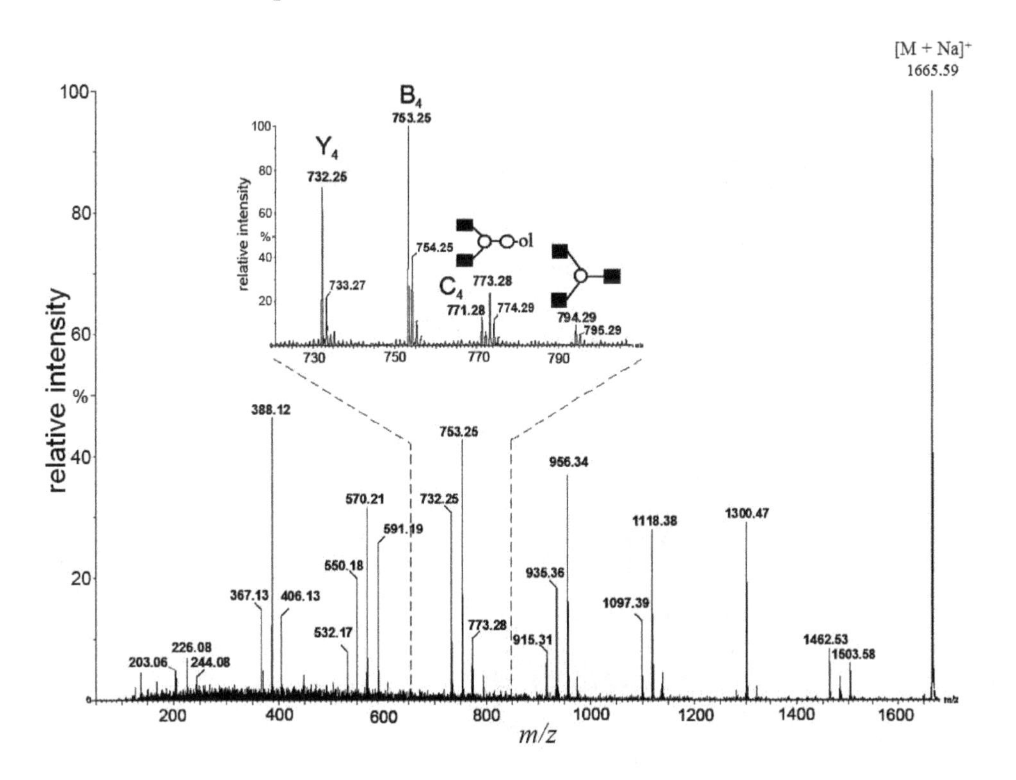

2.4. MALDI Q-TOF CID of Low-Intensity and High Molecular Weight Parent Ions

The ability to sequence high molecular weight and low-intensity precursor ions with MALDI Q-TOF CID is illustrated in Figure 5, where the CID spectra of parent ions at m/z = 1608 (Figure 5A) and 2338 (Figure 5B) are shown. The precursor ion at m/z = 1608 is of low abundance, and the precursor ion at m/z = 2338 is the highest m/z ion in the primary mixture (Figure 1); both are fucosylated. Both low abundant precursor ions delivered on the AB SCIEX QSTAR® Pulsar i, with the "enhance all" function enabled, good evidence and sufficient sequence information to deduce linear octasaccharide and dodecasaccharide HMO chains. Furthermore, a significant amount of fucosylated product ions were detected. We were able to clearly assign the fucosylation site to Hex_4 in the case of fucosylated octasaccharide (Figure 5A). In the case of fucosylated dodecasaccharide (Figure 5B), fragment ions corresponding to either Y_5 or Y_5 + F ions were not detectable, while the $Y_{6\text{-}8}$ + F ion series was visible. Accordingly, the position of the fucose residue in the dodecasaccharide HMO structure was not assigned, indicating the limits of sensitivity at sequencing low intensity, high molecular weight oligosaccharides.

Figure 5. The potentials and limits of fragmentation data using low abundant ions at high m/z values. (**A**) Parent ion at m/z = 1608.43; and (**B**) parent ion at m/z = 2338.73. The spectra were acquired on the AB SCIEX QSTAR® Pulsar i (see the Experimental Section).

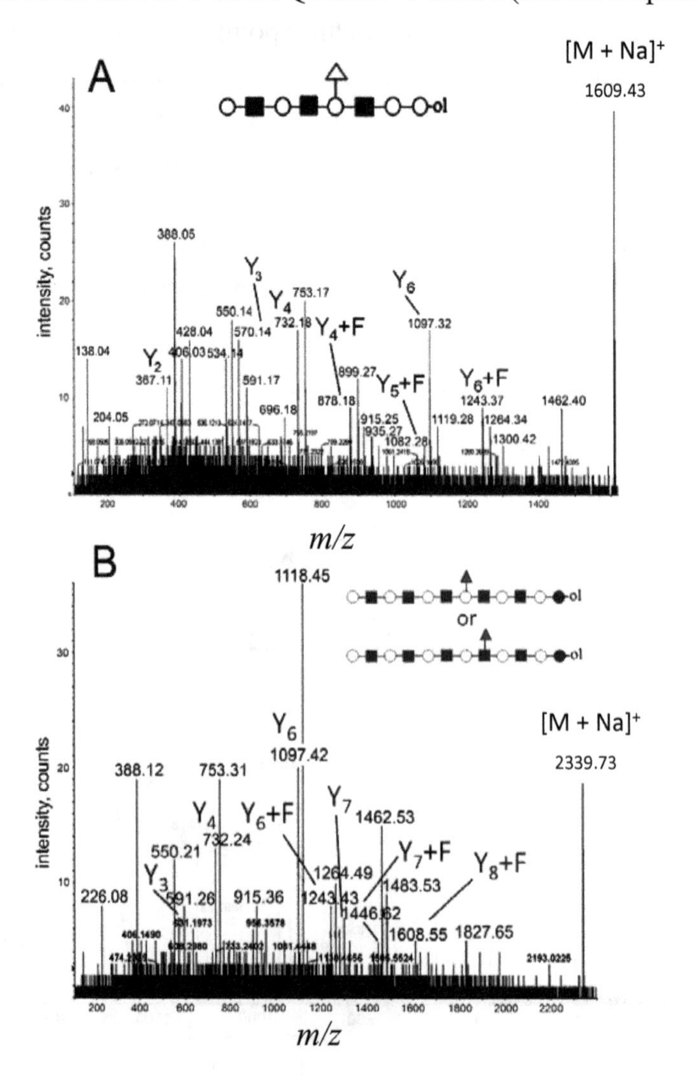

2.5. Analysis of HMO Truncated Structures

A number of oligosaccharides have been identified as truncated structures by their MS/MS patterns. In Figure 6, the possible mechanism of the generation of such structures is presented. The proposed reaction schemes in five practical categories (I–V) for linear and branched species is illustrated, following single or multiple truncation, based on fragmentation data and the biosynthetic considerations of the elongation of the lactose core by *N*-acetyllactosamine units in a linear or branched fashion. Most evident are oligosaccharide structures with an odd monosaccharide composition, which could possibly arise also by the exoglycosidase cleavages [6].

Figure 6. The formation of truncated oligosaccharide structures in the human milk oligosaccharide (HMO) fraction. ⚡ and ⚡ depict various possible hydrolysis events, each of which, alone (**A**–**D** right, **E**) or combined (**D** left), yield different oligosaccharide structure types shown in the figure. ■ denotes *N*-acetylglucosamine, ○ galactose and ● glucose.

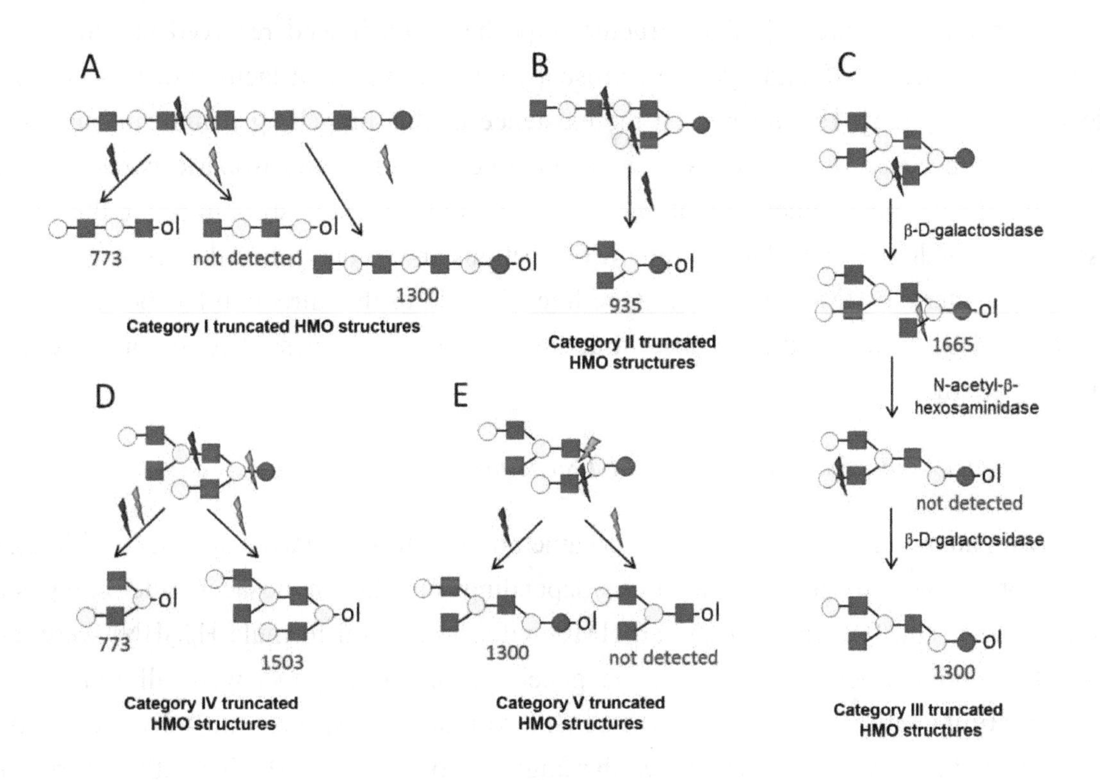

Category I (Figure 6A) is the simplest type of hydrolysis. The resulting oligosaccharides would give rise to a fragment ion at m/z = 408. Furthermore, these hydrolytic products could all give rise to odd-numbered linear oligosaccharide structures. Similar to Category I, the same hydrolytic events could occur on branched oligosaccharide structures, giving rise to odd-numbered Category II branched oligosaccharide structures (Figure 6B). Category III represents possible exoglycosidase events, described previously [6]. The sequential trimming of monosaccharide units on the nonreducing termini can yield structures, such as H_5HN_4 (m/z = 1665), which is the most abundant component in our sample (see Figure 1). In the case of the nonasaccharide, the trimming of another terminal galactose residue seems favourable, leading to respective H_4HN_4 structures, while the trimming of a *N*-acetylglucosamine (GlcNAc) residue seems unfavourable or leads to a quickly degradable product. Category IV hydrolysis depicts the type of structures arising from hydrolytic events on the nonreducing end of glucose. All parent ion mixtures that gave rise to the fragment ion at m/z = 611 could theoretically contain these type of structures (parent ions at m/z = 773, 1138 and 1503), although they seem to be present in very low amounts. Category V hydrolysis shows the remaining two theoretical hydrolytic fragments. We detected some of these types of fragments unambiguously in parent ions at m/z = 1138 and 1300, albeit they were present only near the detection limit in the latter.

2.6. Fragmentation of the Ion at m/z = 1341

The assignment of the parent ion at m/z = 1341.51 is based on the fragment ions at m/z = 794 and 773 (spectrum not shown). This type of branching pattern has been described before, but is still considered novel [2,24], perhaps due to its incomplete characterization and rare observance in the published literature. One novel and five additional possible HMO structures were postulated recently to be based on this structure [2]. This structure type has been indeed resolved previously by FAB (fast atom bombardment) MS and NMR spectroscopy on an example of lacto-*N*-decaose in 1988 [20], as reviewed recently [25]. Having proven the existence of this branching pattern in our sample and HMO mixtures, in general, it is indicated as a possible isomer in the isobaric parent ion mixtures containing the diagnostic fragment ion at m/z = 794. It would be interesting to prove the existence of these isomers in the high molecular weight HMO fractions and to determine the relative abundance of this branching pattern in HMO in general. One line of study in this direction has been carried out by Amano *et al.* [19], in which the decaose structure has been best described so far in terms of its fucosylation patterns.

2.7. General Conclusions about Identified HMO Structures

After having analysed all the CID data, it became possible to summarize the data (Table 2), as well as to categorize the postulated structures, depending on the monosaccharide composition of oligosaccharides (Table 3). Accordingly, all HMO with the general formula $H_{n+2}HN_n$ were identified as linear. The odd-numbered HMO with the general formula $H_{n+1}HN_n$ were all truncated on the nonreducing termini only. The HMO with the general formula H_nHN_n were most diverse. Unlike in the first two groups, this group always contained the fragment ion at m/z = 408, indicative of truncation on the reducing termini. They also had multiple truncations on the non-reducing termini. Surprisingly, they are not the most abundant structures in the spectrum and are only about half as abundant as the previous two groups. Finally, the HMO of the general formula $H_{n-1}HN_n$ were of very low abundance. They also yielded the m/z = 408 fragment ion, although it could not be detected for H_3HN_4, most likely due to the low intensity of fragment ions obtained in the CID spectrum. Monofucosylated parent ions corresponding to all these groups were detected, except for the $H_{n-1}HN_n$ group. Further and more detailed studies will be carried out to study these phenomena. In particular, quantitative analysis should provide important insight about the abundance of individual HMO isomers.

Table 3. Categorization of HMO structures identified in this study based on their monosaccharide composition. ■ denotes *N*-acetylglucosamine, ○ galactose and ● glucose. * structures that could be neither confirmed nor excluded. * present in very small amounts.

Monosaccharide composition	HMO structure types identified
$H_{n+2}HN_n$	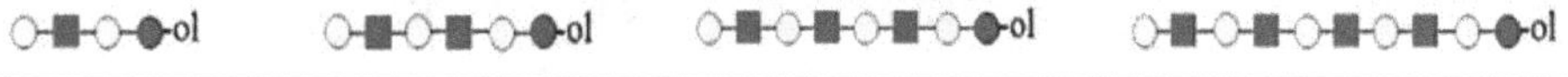
$H_{n+1}HN_n$	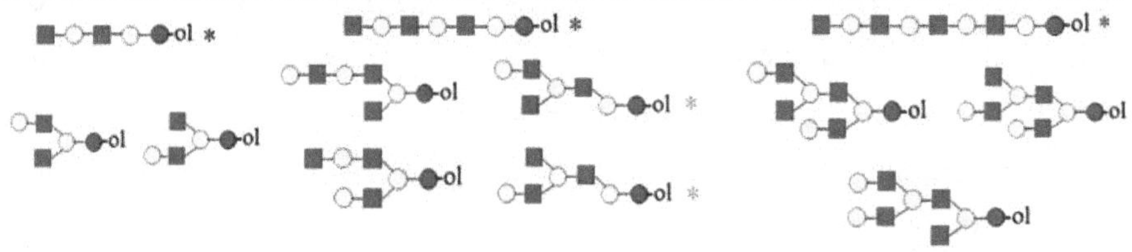
H_nHN_n	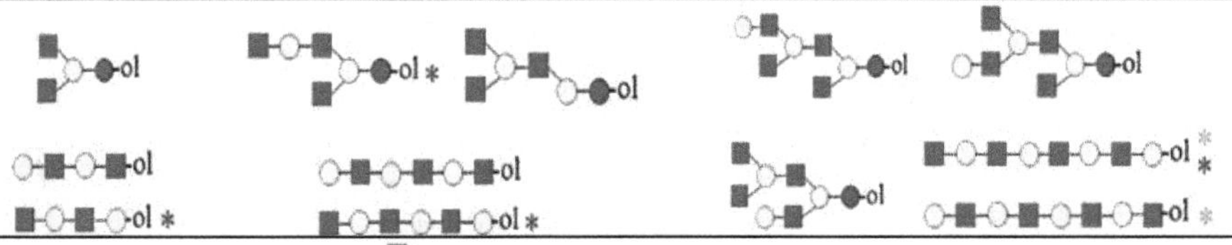
$H_{n-1}HN_n$	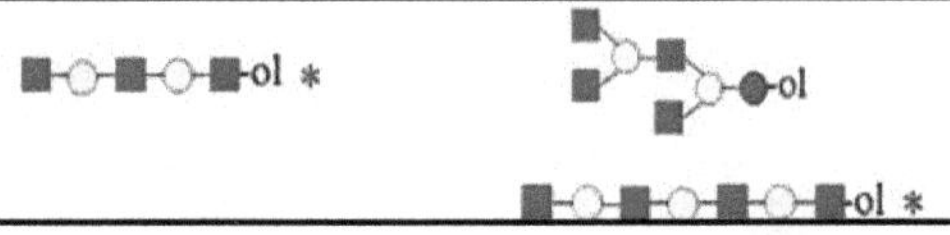

3. Experimental Section

A pool of human milk was obtained from Donor 0, a Le[a] non-secretor. The high molecular weight chromatographic fraction was obtained by gel permeation Biogel P-4 chromatography and submitted to mild hydrolysis to remove fucose and sialic acid, as described in Bruntz *et al.* [20]. The oligosaccharide fraction was reduced with $NaBH_4$ and analysed using MALDI MS in the positive ion mode.

2,5-Dihydroxybenzoic acid (DHB) (Sigma, Steinheim, Germany) was used as a MALDI matrix. For MALDI TOF analysis, γ = 20 mg/mL, dissolved in H_2O/ACN (acetonitrile) 1:1 (*v*/*v*), was used. For MALDI Q-TOF analysis, γ = 80 mg/mL in H_2O/ACN 70%:30% (*v*/*v*) was used. The oligosaccharide sample was dissolved in ddH_2O, and 1 μL of the sample was mixed with 1 μL of matrix solution on the MALDI target and allowed to dry. MALDI TOF MS experiment was performed on a TofSpec 2E instrument (Micromass, Manchester, UK), equipped with a LeCroy (Teledyne Lecroy Inc., New York, NY, USA) digitizer LSA1000 with a 2 GHz acquisition rate. The majority of the MALDI Q-TOF CID experiments were conducted on a prototype Micromass instrument. Additional MALDI Q-TOF CID experiments were performed on a commercial instrument, AB SCIEX QSTAR® Pulsar i.

In general, there are differences in experimental parameters between these three instruments, as axial (MALDI TOF) and orthogonal (MALDI Q-TOF) instruments have different modes of operation. For MALDI TOF MS, the minimum laser fluence that allowed stable signal collection was used, in order to increase the spectral resolution, to minimize possible detector saturation and to minimize in-source and post-source decay (ISD and PSD). For the prototype MALDI Q-TOF instrument from Micromass, as well as the commercial AB SCIEX QSTAR® Pulsar i, no sample-to-sample signal optimization was required, as the instrument default method was robust enough. The practical difference in the mode of operation between MALDI TOF and MALDI Q-TOF instruments is the high laser fluence requirement for MALDI Q-TOF instruments. This is technically necessary because orthogonal instruments are less efficient in the ion transport than axial instruments, and therefore, more ions need to be produced within the oMALDI (orthogonal MALDI) source. Two practical consequences of higher laser fluence used in our hands were: (i) the ability to use higher γ(DHB) as a matrix solution on MALDI Q-TOF instruments; and (ii) the use of DHB as a matrix was required, as other matrices used in our laboratory did not give any signal on a MALDI Q-TOF instrument. Despite these technical and experimental differences, the MS1 spectra on all three instruments were very similar. However, in our hands, the MALDI TOF oligosaccharide map of the analysed chromatographic fraction acquired on the Micromass TofSpec 2E was of slightly higher quality in terms of signal intensity and is, thus, presented here.

MS/MS spectra were obtained by collisionally-induced dissociation (CID) using Ar as a collision gas on MALDI Q-TOF instruments under conditions like low-energy CID. The collision energy that produced the optimal fragmentation pattern across the whole *m/z* range for a given parent ion was adjusted to be proportional to the parent ion *m/z* value. The AB SCIEX QSTAR® Pulsar i instrument has a unique additional mode of operation in MS/MS mode, in which it can amplify the intensity of the detected fragment ions by using a "pulse" function option, which allows the user to select which *m/z* range in the CID spectrum needs to be "enhanced" (in terms of fragment ion intensity). When this option is chosen, the built-in control software automatically recalculates "pulsing" windows, which determine which fragment ion packets are sent to the TOF analyser more often, thus increasing their relative intensities in the final CID spectrum. The "enhance all" option in the control software allows enhanced product ion detection over the whole *m/z* range of the CID spectrum. This option in the recording of the CID spectra of low-intensity parent ions was functional (Figure 5), as well as in the recording of diagnostic low-intensity fragment ions (Figure 3B).

4. Conclusions

De novo analysis of tandem MS data is described as an option for an in-depth view on already known or new structures. Since many molecular ions represented isomeric mixtures of oligosaccharides, a rationale for the assignment based on the biosynthetic pathway was applied. Oligosaccharide structures not found in databases were proposed for assignment according to specific diagnostic ions detected in fragmentation spectra. In the mixture of linear and branched HMO structures within a single precursor ion *m/z* value, the isomeric components were assigned using a diagnostic ion analysis technique. A minor amount of fucosylated oligosaccharides was found, where fucosylation occurred on a single site. In summary, four non-fucosylated linear HMO and three monofucosylated linear HMO were characterized. About twenty branched structure types were proposed, as in agreement with MS/MS data, known HMO structures and hydrolytic patterns. These structures are furthermore isomeric with many possible biosynthetic HMOs and, when taking into account the 1-3 and 1-4 galactose linkage isomers, over a hundred HMO structures can be assumed to be present in the analysed chromatographic fraction. The summarized results are depicted in Tables 2 and 3. Truncated structures without glucose are omitted for clarity.

Although we were able to determine the presence of isomeric structures in our sample, our data also illustrate the limitations of MALDI Q-TOF CID technique in their full characterization. It is important to note that mass spectrometry in general can overcome these limitations in several ways: (i) by using an additional liquid chromatography (LC) or capillary electrophoresis (CE) separation step prior to MS analysis; (ii) by using high-energy CID to generate cross-ring cleavages; and (iii) by using MS^n fragmentation techniques. Additional LC separation step prior to MALDI Q-TOF MS/MS analysis can be done in off-line mode [26]. In on-line mode, the applicability of LC-ESI-MS/MS [2,3,17] and CE-LIF (laser induced fluorescence)-MS^n [16] in the analysis of HMO has been well documented. Unlike low-energy CID, used in MALDI Q-TOF instruments, high-energy CID used in MALDI TOF-TOF instruments typically yields cross-ring cleavages, which can carry information about the branching pattern of the parent ion [14,15]. Finally, the MS^n technique has been successfully applied to the analysis of HMO [19]. In general, different MS instrumentation setups are required for obtaining the maximum amount of structural information about the analysed sample and these often require either more time for analysis (LC-MS, CE-MS, MS^n), a derivatization step prior to MS analysis (CE-MS, MS^n) [16,19,27] and more time or additional software tools to analyse the large amount of data these techniques yield [2,18]. Traditionally, MALDI MS has been advantageous in the high-throughput MS1 screening of biological samples, and with modern instrumentation allowing the acquisition of high-quality MALDI MS/MS data, this property of MALDI MS can be extended to the high-throughput screening of tandem MS data, as well.

The advantages of MALDI Q-TOF CID for structural elucidation of complex oligosaccharides are the precursor ion selection properties, the ability to sequence very low abundant parent ions and the speed of the MS/MS data acquisition. The majority of MS1 signals could be sequenced with the prototype Micromass MALDI Q-TOF instrument, and all detectable signals could be sequenced with the commercial AB SCIEX MALDI Q-TOF instrument. This technical advancement alone allows for a huge jump in the data throughput of a glycomic research laboratory, where MALDI MS is the method of choice for all preliminary experiments related to the structural characterization of complex oligosaccharide mixtures. Finally, the advantages of the MALDI Q-TOF MS/MS indicate its applicability to high-throughput sample analysis, requiring new software tools for faster data analysis [18,28].

Future studies on high molecular weight HMO should provide better insight on biologically interesting data, like the quantitation of isomers and the reexamination of sequences, which indicate the biosynthetic pathways involved. Quantitative studies are necessary in order to determine the relative abundance of particular isomers, *i.e.*, are they all present in equal, random or more selective amounts, where the presence of certain isomers can be neglected.

Acknowledgments

This work was partially supported by IZKF (Interdisciplinary Centre for Clinical Research of University of Muenster, Germany) project "Molecular Analysis ZP6", awarded to J.P.K. The authors also thank the Mass Spectrometry Core Facility at the Institute of Molecular Medicine and Genetics, Medical College of Georgia, Augusta, GA, USA, for the use of the QStar Pulsar i instrument.

Author Contributions

M.J. performed experiments on the TofSpec 2E and QStar Pulsar i instruments, analysed the data and wrote the manuscript. R.T.W. designed the Micromass prototype MALDI Q-TOF instrument and performed the related experiments in this work. He also contributed to the Experimental Section. G.P.

developed the isolation procedure for HMO and prepared the sample for analysis. J.P.K. designed the experiment and wrote the manuscript.

References

1. Coppa, G.V.; Gabrielli, O.; Pierani, P.; Catassi, C.; Carlucci, A.; Giorgi, P.L. Changes in carbohydrate composition in human milk over 4 months of lactation. *Pediatrics* **1993**, *91*, 637–641.
2. Wu, S.; Tao, N.; German, J.B.; Grimm, R.; Lebrilla, C.B. Development of an annotated library of neutral human milk oligosaccharides. *J. Proteome Res.* **2010**, *9*, 4138–4151.
3. Wu, S.; Grimm, R.; German, J.B.; Lebrilla, C.B. Annotation and structural analysis of sialylated human milk oligosaccharides. *J. Proteome Res.* **2011**, *10*, 856–868.
4. Marino, K.; Lane, J.A.; Abrahams, J.L.; Struwe, W.B.; Harvey, D.J.; Marotta, M.; Hickey, R.M.; Rudd, P.M. Method for milk oligosaccharide profiling by 2-aminobenzamide labeling and hydrophilic interaction chromatography. *Glycobiology* **2011**, *21*, 1317–1330.
5. Zivkovic, A.M.; German, J.B.; Lebrilla, C.B.; Mills, D.A. Human milk glycobiome and its impact on the infant gastrointestinal microbiota. *Proc. Natl. Acad. Sci. USA* **2011**, *108*, 4653–4658.
6. Newburg, D.S.; Ruiz-Palacios, G.M.; Morrow, A.L. Human milk glycans protect infants against enteric pathogens. *Annu. Rev. Nutr.* **2005**, *25*, 37–58.
7. O'Hara, A.M.; Shanahan, F. The gut flora as a forgotten organ. *EMBO Rep.* **2006**, *7*, 688–693.
8. Wang, B.; Bing, Y.; Karim, M.; Sun, Y.; McGreevy, P.; Petocz, P.; Held, S.; Brand-Miller, J. Dietary sialic acid supplementation improves learning and memory in piglets. *Am. J. Clin. Nutr.* **2007**, *85*, 561–569.
9. Wang, B. Molecular mechanism underlying sialic acid as an essential nutrient for brain development and cognition. *Adv. Nutr.* **2012**, *3*, 465S–472S.
10. De Leoz, M.L.; Gaerlan, S.C.; Strum, J.S.; Dimapasoc, L.M.; Mirmiran, M.; Tancredi, D.J.; Smilowitz, J.T.; Kalanetra, K.M.; Mills, D.A.; German, J.B.; *et al.* Lacto-*N*-tetraose, fucosylation, and secretor status are highly variable in human milk oligosaccharides from women delivering preterm. *J. Prot. Res.* **2012**, *11*, 4662–4672.
11. Stahl, B.; Thurl, S.; Zeng, J.; Karas, M.; Hillenkamp, F.; Steup, M.; Sawatzki, G. Oligosaccharides from human milk as revealed by matrix-assisted laser desorption/ionization mass spectrometry. *Anal. Biochem.* **1994**, *223*, 218–226.
12. Finke, B.; Stahl, B.; Pfenninger, A.; Karas, M.; Daniel, H.; Sawatzki, G. Analysis of high-molecular-weight oligosaccharides from human milk by liquid chromatography and MALDI-MS. *Anal. Chem.* **1999**, *71*, 3755–3762.
13. Xie, Y.; Lebrilla, C.B. Infrared multiphoton dissociation of alkali metal-coordinated oligosaccharides. *Anal. Chem.* **2003**, *75*, 1590–1598.
14. Mechref, Y.; Novotny, M.V. Structural characterization of oligosaccharides using MALDI-TOF/TOF tandem mass spectrometry. *Anal. Chem.* **2003**, *75*, 4895–4903.
15. Lewandrowski, U.; Resemann, A.; Sickmann, A. Laser-induced dissociation/high-energy collision-induced dissociation fragmentation using MALDI-TOF/TOF-MS instrumentation for the analysis of neutral and acidic oligosaccharides. *Anal. Chem.* **2005**, *77*, 3274–3283.

16. Albrecht, S.; Schols, H.A.; van den Heuvel, E.G.; Voragen, A.G.; Gruppen, H. CE-LIF-MS n profiling of oligosaccharides in human milk and feces of breast-fed babies. *Electrophoresis* **2010**, *31*, 1264–1273.
17. Ninonuevo, M.; An, H.; Yin, H.; Killeen, K.; Grimm, R.; Ward, R.; German, B.; Lebrilla, C. Nanoliquid chromatography-mass spectrometry of oligosaccharides employing graphitized carbon chromatography on microchip with a high-accuracy mass analyzer. *Electrophoresis* **2005**, *26*, 3641–3649.
18. Wu, S.; Salcedo, J.; Tang, N.; Waddell, K.; Grimm, R.; German, J.B.; Lebrilla, C.B. Employment of tandem mass spectrometry for the accurate and specific identification of oligosaccharide structures. *Anal. Chem.* **2012**, *84*, 7456–7462.
19. Amano, J.; Osanai, M.; Orita, T.; Sugahara, D.; Osumi, K. Structural determination by negative-ion MALDI-QIT-TOFMSn after pyrene derivatization of variously fucosylated oligosaccharides with branched decaose cores from human milk. *Glycobiology* **2009**, *19*, 601–614.
20. Bruntz, R.; Dabrowski, U.; Dabrowski, J.; Ebersold, A.; Peter-Katalinić, J.; Egge, H. Fucose-containing oligosaccharides from human milk from a donor of blood group 0 Le^a nonsecretor. *Biol. Chem. Hoppe Seyler* **1988**, *369*, 257–273.
21. Wiederschain, G.Y.; Newburg, D.S. Glycosidase activities and sugar release in human milk. *Adv. Exp. Med. Biol.* **2001**, *501*, 573–577.
22. Perreault, H.; Lattová, E.; Šagi, D.; Peter-Katalinić, J. MALDI-MS of glycans and glycoconjugates. In *MALDI MS: A Practical Guide to Instrumentation, Methods and Applications*, 2nd ed.; Hillenkamp, F., Peter-Katalinić, J., Eds.; Wiley-VCH: Weinheim, Germany, 2013; pp. 239–272.
23. Domon, B.; Costello, C.E. A systematic nomenclature for carbohydrate fragmentation in FAB-MS/MS spectra of glycoconjugates. *Glycoconj. J.* **1988**, *5*, 397–409.
24. Chai, W.; Piskarev, V.E.; Zhang, Y.; Lawson, A.M.; Kogelberg, H. Structural determination of novel lacto-*N*-decaose and its monofucosylated analogue from human milk by electrospray tandem mass spectrometry and ^{1}H NMR spectroscopy. *Arch. Biochem. Biophys.* **2005**, *434*, 116–127.
25. Kobata, A. Structures and application of oligosaccharides in human milk. *Proc. Jpn. Acad. Ser. B Phys. Biol. Sci.* **2010**, *86*, 731–747.
26. Ericson, C.; Phung, Q.T.; Horn, D.M.; Peters, E.C.; Fitchett, J.R.; Ficarro, S.B.; Salomon, A.R.; Brill, L.M.; Brock, A. An automated noncontact deposition interface for liquid chromatography matrix-assisted laser desorption/ionization mass spectrometry. *Anal. Chem.* **2003**, *75*, 2309–2315.
27. Amano, J.; Sugahara, D.; Osumi, K.; Tanaka, K. Negative-ion MALDI-QIT-$TOFMS^n$ for structural determination of fucosylated and sialylated oligosaccharides labeled with a pyrene derivative. *Glycobiology* **2009**, *19*, 592–600.
28. Vakhrushev, S.Y.; Dadimov, D.; Peter-Katalinić, J. Software platform for high-throughput glycomics. *Anal. Chem.* **2009**, *81*, 3252–3260.

13

Direct Analysis of hCGβcf Glycosylation in Normal and Aberrant Pregnancy by Matrix-Assisted Laser Desorption/Ionization Time-of-Flight Mass Spectrometry

Ray K. Iles [1,2,3,*], Laurence A. Cole [4] and Stephen A. Butler [1,3,4]

[1] Williamson Laboratory for Molecular Oncology, St Bartholomews Hospital, London EC1A 7BE, UK; E-Mail: Butlersa1@googlemail.com

[2] ELK Foundation for Health Research, An Scoil Monzaird, Crieff PH7 4JT, UK

[3] MAP Diagnostics Ltd., Ely, Cambridgeshire CB6 3FQ, UK

[4] USA hCG Reference Service, Angel Fire, NM 87710, USA; E-Mail: larry@hcglab.com

* Author to whom correspondence should be addressed; E-Mail: ray@iles.net

---

Abstract: The analysis of human chorionic gonadotropin (hCG) in clinical chemistry laboratories by specific immunoassay is well established. However, changes in glycosylation are not as easily assayed and yet alterations in hCG glycosylation is associated with abnormal pregnancy. hCGβ-core fragment (hCGβcf) was isolated from the urine of women, pregnant with normal, molar and hyperemesis gravidarum pregnancies. Each sample was subjected to matrix-assisted laser desorption/ionization time-of-flight mass spectrometry (MALDI TOF MS) analysis following dithiothreitol (DTT) reduction and fingerprint spectra of peptide hCGβ 6–40 were analyzed. Samples were variably glycosylated, where most structures were small, core and largely mono-antennary. Larger single bi-antennary and mixtures of larger mono-antennary and bi-antennary moieties were also observed in some samples. Larger glycoforms were more abundant in the abnormal pregnancies and tri-antennary carbohydrate moieties were only observed in the samples from molar and hyperemesis gravidarum pregnancies. Given that such spectral profiling differences may be characteristic, development of small sample preparation for mass spectral analysis of hCG may lead to a simpler and faster approach to glycostructural analysis and potentially a novel clinical diagnostic test.

Keywords: hCG; hCGβcf; MALDI TOF MS; pregnancy; hydatidiform mole; hyperemesis gravidarum; glycosylation

---

1. Introduction

In a post-genomic era the importance of proteoforms has come to the fore [1], and it is the subtleties of the proteoform that underlay many pathologies not yet characterized at a genetic level. This is not simply splice variants but the form a protein takes within a functional cellular and physiological system. Critical to clinical functionality of a coded protein are its post translational modifications, e.g., pre- and pro-peptide cleavage, phosphorylation and glycosylation. Detection and relative quantification of particular proteoforms will form the bases of new biomarker discovery and not necessarily simple measurement of any given mass of protein [2].

The detection of human chorionic gonadotropin (hCG) is used extensively in obstetrics and gynecology for the detection and monitoring of pregnancy. The hormone is an αβ hetero-dimeric glycoprotein with eight glycosylation sites, comprising four *N*-linked oligosaccharides and four *O*-linked oligosaccharides. Two *N*-linked oligosaccharides are attached to each of the subunit polypeptide chains by β-*N*-glycosidic bonds to asparagine residues. These moieties share the same basic structural characteristics: *N*-acetylglucosamine (GlcNAc) is attached to an asparagine residue followed by another GlcNAc, mannose, and two more branches of mannose. This is the mono-antennary pentasaccharide core with the remaining components being variable [3–5]. The *O*-linked oligosaccharides are attached by α-*O*-glycosidic bonds onto serine residues of the β-subunit carboxyl terminal peptide [6–9].

Carbohydrate heterogeneity has been extensively reported for the free β-subunit of hCG (hCGβ) with variable mono-, bi-, and tri-antennary carbohydrate structures being found in normal and abnormal pregnancies, as well as in gestational trophoblastic disease and in particular choriocarcinoma and early pregnancy [10–14]. In general, a greater proportion of tri-antennary oligosaccharide structures are usually indicative of abnormalities in pregnancy, while bi-antennary forms account for the majority of structures found in normal pregnancy [13].

hCG is excreted intact into the urine, as documented by extensive implementation of urinary pregnancy testing. However, hCG is also degraded in liver and kidneys and a large proportion of immunoreactive hCG in the urine is attributed to this urinary degradation product of the hCGβ subunit hCG β-core fragment (hCGβcf). The carbohydrate structures of the hCGβcf have been studied independently [15,16] and the molecule is composed of peptides, β 6–40 and β 55–92, connected by four disulfide bridges. It retains many of the antigenic determinants of the original hCGβ molecule prior to metabolism, which occurs primarily in the kidney [17]. The β 6–40 polypeptide chain contains the two hCGβ *N*-linked carbohydrate moieties, although the oligosaccharides are truncated due to metabolism. Urinary hCGβcf can be isolated with relatively straightforward procedures [15] from a simple urine sample and offers a convenient way of providing insights into glycosylation of the hCGβ subunit and therefore the hCG from which it was derived [18]. This presents an opportunity to indirectly study pregnancy disorders known to exhibit glycoform variants of hCG.

Matrix-assisted laser desorption/ionization time-of-flight mass spectrometry (MALDI TOF MS) is a technique that can be used for the determination of the mass of macromolecules, originally developed by Karas & Hillenkamp [19]. MALDI TOF MS can be used in the characterization of glycopeptides [20] and/or oligosaccharides that are released from glycoproteins with the use of enzymatic digestion [21–23]. Dithiothreitol (DTT) can also be used in situations where disulfide linkages are present and can reduce the mass of peptides bringing them into relatively optimum resolution for this mass spectrometer.

In the case of hCGβcf, the amino acids β 55–92 are linked to β 6–40 from the original β-subunit in hCG. After disulfide reduction, these two peptides along with glycosylation moieties can be analyzed by MALDI TOF MS and oligosaccharide masses calculated by subtraction of the peptide mass of the β 6–40 chain from the observed peak mass of each glycoform. Carbohydrate heterogeneity has been reported on hCGβcf and a population of mono- and bi-antennary structures has been proposed by various studies [24–27]. Using a MALDI TOF MS technique we have previously shown that the remaining oligosaccharide structures found on hCGβcf do not possess sialic acid and the extent to which those structures are truncated prior to urinary excretion as hCGβcf [28]. This made it possible to analyze glycosylation moieties whilst still attached to the peptide, thus eliminating the need for glycosidase digestion. However, this previous work was conducted on a pooled sample preparation and there has, as yet, been no report of hCGβcf glycosylation patterns from individual patients. In order to provide some insights in hCGβcf glycosylation in aberrant pregnancies, we used the same MALDI TOF MS technique to analyze hCGβcf isolated from individual patient samples with normal pregnancy or conditions such as molar pregnancy and hyperemesis gravidarum.

2. Results and Discussion

2.1. Mass Spectral Profiles

hCGβcf purified from pregnancy urine samples (normal, molar and hyperemesis gravidarium) subjected to MALDI TOF MS generated mass spectra for hCGβcf displaying a broad peak between *m/z* 8700 and 10,700, as published previously [28]. On reduction of the disulfide linkages using DTT, this broad peak was replaced by a set of lower molecular weight peaks (Figure 1). A peak at *m/z* 3950 was seen in the spectra from hCGβcf samples N2βcf and HGβcf (Figure 2b,e). Common to all samples was the peak at *m/z* 4156.8, corresponding to the non-glycosylated hCGβcf peptide β 55–92 (Figure 1).

2.2. Determination of Glycostructures

Prediction of the glyco-structures that resulted in the remaining peaks was achieved by the subtraction of the corresponding mass of the primary amino acid sequence of β 6–40 from the observed *m/z* values corresponding to the glycosylated isoforms (Figure 2 and Table 1). Despite the fact that the exact predicted mass of the hCGβcf asparagine-linked carbohydrate moieties were not observed directly, the low percentage errors between the observed and expected mass match of the peaks acquired show that it is likely that these glycoforms were detected. The proposed carbohydrate moieties identified from the mass spectra are shown in Figure 2. Each of the five pregnancy samples contained between 8 and 11 out of the 25 glycosylated forms of β 6–40 identified in this set of samples (Figure 2 and Table 1).

Figure 1. Matrix-assisted laser desorption/ionization time-of-flight mass spectrometry (MALDI TOF MS) of human chorionic gonadotropin β-core fragment (hCGβcf) treated with dithiothreitol (DTT). hCGβcf purified from pregnancy urine samples; Normal (**A**,**B**), Molar (**C**,**D**) and Hyperemesis Gravidarium (**E**). Disulfide linkages were reduced using DTT. The indicated peak at *m/z* 4156.8 (β 55–92) appears in all samples and represents the unglycosylated peptide of beta-core. Arrowed peak (↓) only appears in samples N2βcf and HGβcf and indicates a fragment smaller than β 55–92 and as such is likely to be β 6–40 with minimal or no-glycosylation. All remaining peaks are attributed β 6–40 glycopeptides and described in Table 1.

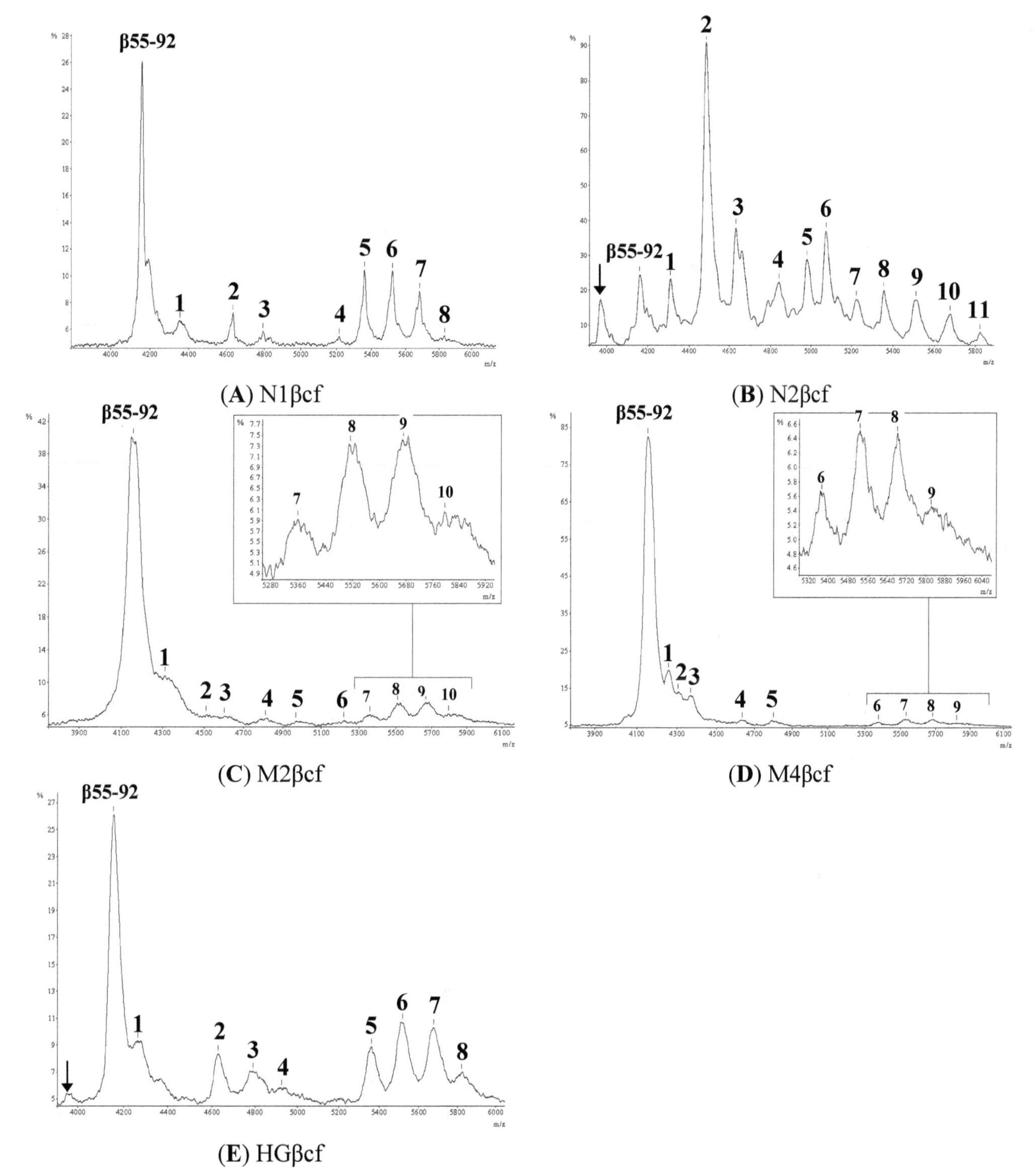

Figure 2. Oligosaccharide structures of hCGβcf. Structures identified in samples used in this study. The information for each structure includes; structure letter, schematic and molecular weight (Da). ■, GlcNAc (221.2 Da); ○, mannose (180.2 Da); Δ, Fucose (164.2 Da); ●, Galactose (180.2 Da).

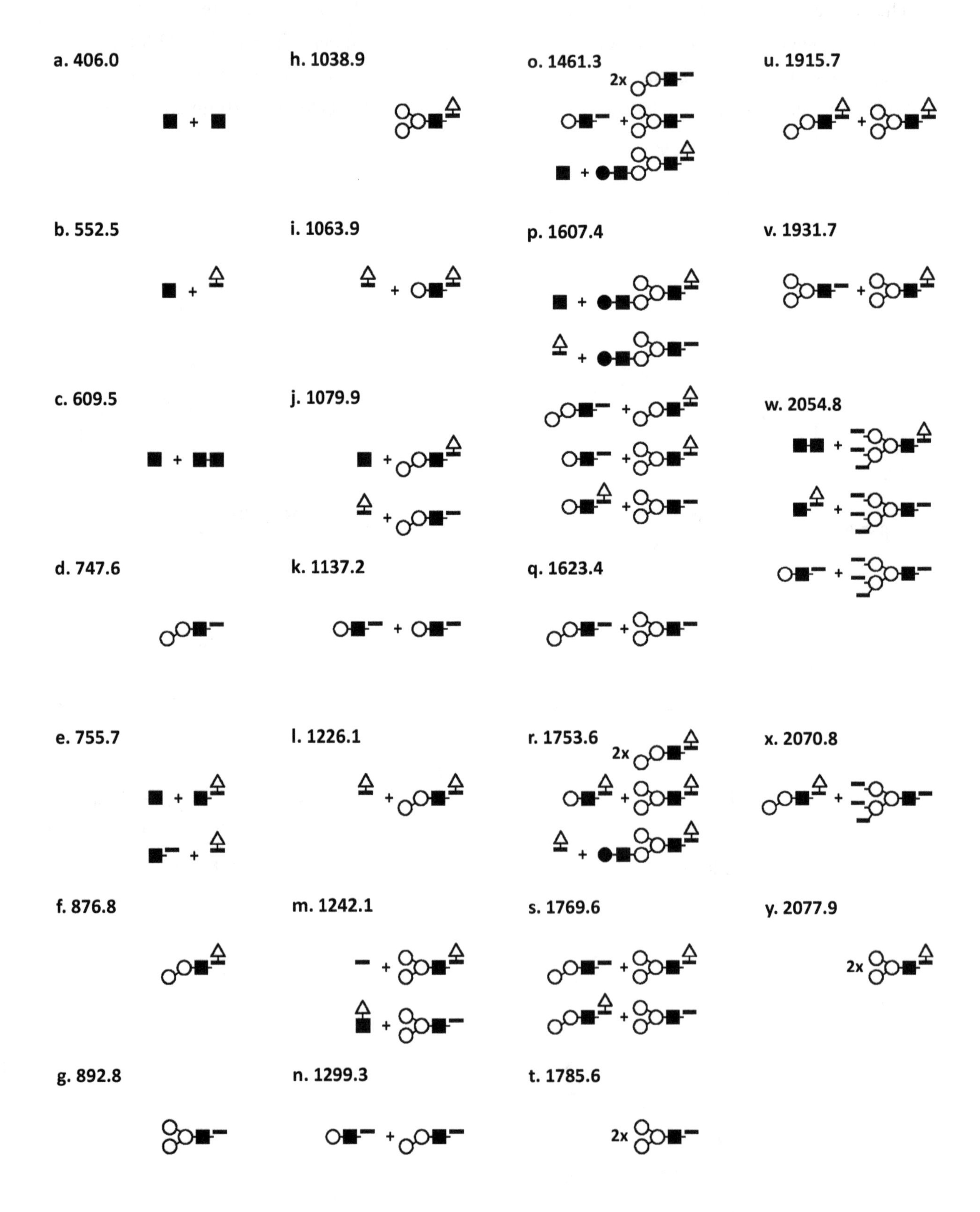

Table 1. Identifying MALDI TOF MS peaks. For each peak in each sample; an inferred oligosaccharide (CHO) mass was calculated and best fit structure assigned (Figure 2). The theoretical mass of the glycopeptides (β 6–40 plus CHO moiety) was then calculated as percentage fit (mass match) to the observed peak mass. The calculated relative abundance of each observed peak represents the proportion of area under the curve for the mass spectral range (*m/z* 4200–6000) for that peak/glycopeptides (% abundance).

Peak	Observed Mass (*m/z*) $[M + H]^+$	Predicted Carbohydrate Structure (Figure 2)	Predicted Mass (Da) of Glycopeptide	% Mass Match	% Abundance
			N1bcf		
1	4353.8	c	4361.9	0.9982	16.2
2	4634.7	f	4629.2	1.0012	5.1
3	4798.8	h	4790.8	1.0017	4.4
4	5220.8	o	5213.7	1.0014	3.2
5	5366.7	p	5359.8	1.0013	18.1
6	5529.1	s	5522.0	1.0013	22.8
7	5689.8	v	5684.1	1.0010	17.8
8	5840.6	y	5830.3	1.0018	12.4
			N2bcf		
1	4307.7	b	4277.9	1.007	4.7
2	4477.6	d	4483.0	1.0012	35.2
3	4630.9	g	4630.9	1.0030	17.3
4	4837.4	j	4832.3	0.9989	6.6
5	4976.6	l	4978.5	1.0004	5.3
6	5074.3	n	5051.4	0.9955	7.5
7	5219.7	o	5213.7	0.9899	3.3
8	5352.6	p	5359.8	1.0013	5.9
9	5504.9	r	5506.0	1.0002	7.3
10	5672.8	u	5568.1	0.9992	5.3
11	5820.9	x	5823.2	1.0004	1.6
			M2bcf		
1	4310	b	4304.9	0.9988	24
2	4515.2	e	4508.1	0.9984	0.3
3	4601.7	f	4628.4	1.0058	0.6
4	4805.8	i	4816.3	1.0022	13.1
5	4966.4	m	4994.5	1.0056	9.8
6	5219.1	o	5213.7	0.9970	1.7
7	5359.5	p	5359.8	1.0001	9.8
8	5518.9	s	5522.0	1.0006	13.4
9	5673	u	5668.1	0.9991	15.3
10	5796.1	w	5807.2	1.0019	12

Table 1. *Cont.*

Peak	Observed Mass (m/z) $[M + H]^+$	Predicted Carbohydrate Structure (Figure 2)	Predicted Mass (Da) of Glycopeptide	% Mass Match	% Abundance
			M4bcf		
1	4256.3	a	4158.4	0.9770	25.6
2	4305.2	b	4304.9	0.9999	16.2
3	4369	c	4361.9	0.9984	24.5
4	4639.1	f	4629.2	0.9979	6.1
5	4801.5	h	4791.3	0.9979	8.1
6	5378.9	q	5375.8	0.9994	4.1
7	5535.7	t	5538.0	1.0004	7.3
8	5686.9	v	5684.1	0.9995	5.7
9	5824.9	x	5823.2	0.9997	2.4
			HGbcf		
1	4265.4	b	4304.4	1.0091	12.2
2	4628.3	f	4629.2	1.0002	9.2
3	4790.2	h	4791.3	1.0002	8.9
4	4929.4	k	4889.6	0.9919	3.1
5	5361.6	p	5359.8	0.9997	13.5
6	5516.8	s	5522.0	1.0079	21.7
7	5675.7	u	5568.1	0.9987	14.9
8	5820.2	x	5823.2	1.0005	16.5

2.3. Relative Abundance of Glycoforms

The most commonly detected glycol-structure found in 4 of 5 of the samples were Figure 2 structures b (m/z 552.5), f (m/z 876.8), p (m/z 1607.4); and in 3 of 5 samples h (m/z 1038.9), o (m/z 1461.3), s (m/z 1769.6) and v (m/z 1915.7). Collectively structures b (m/z 552.5), p (m/z 1607.4), and s (m/z 1769.6) represent a third of the peak abundance of all the spectra.

The incidence of the remaining glyco-structures was low as was the abundance of the mass spectra generated for the urine samples from normal pregnancies; sample N1βcf had one unique peak at m/z 2077.9 (carbohydrate structure y in Figure 2) and sample N2βcf four- structures: d (m/z 747.6), g (m/z 892.8), j (m/z 1079.9) and l (m/z 1226.1). Peaks corresponding to structures e (m/z 755.7), i (m/z 1063.9), m (m/z 1242.1) and w (m/z 2054.8) were present only in the hCGβcf purified from M2βcf, whilst the spectra for the second molar pregnancy urine sample M4βcf displayed peaks representing structure q (m/z 1623.4) and t (m/z 1769.6). Interestingly the hCGβcf preparation from the hyperemesis gravidarium pregnancy urine did not reveal any unique glycoforms.

Fucose at 1–6 of the basal GlcNac was a common retained feature of the residual glycosylation moieties, occurring in 16 of the 25 identified structures and in terms of abundance could be accounted for in 76% of the peak areas of the combined samples.

The glyco-structures that contributed to the greatest proportion of samples are; N1βcf—s (m/z 1769.6) (22.8%); N2βcf—d (m/z 747.6) (35.2%), M2βcf—b (m/z 552.5) (24%), M4βcf—a (m/z 406.0) (24.5%) and HGβcf—s (m/z 1769.6) (21.7%). Mono-antennary structures (m/z 406–1226.1) and bi-antennary structures (m/z 892.8–2077.9) were found in all samples. Tri-antennary carbohydrate moieties w (*m*/*z* 2054.8) and x (*m*/*z* 2070.8) were only detected in molar pregnancy-M2βcf and Hyperemesis gravidarum-HGβcf samples.

2.4. Discussion

HCG is produced by placental trophoblast cells and is a glycoprotein hormone in the diagnosis of pregnancy testing and in the detection of cancer. It would be a significant improvement on current methods to develop a rapid and reliable analytical technique for the characterisation of peptide and carbohydrate portions of hCG rather than a simple quantification of serum or urine levels. By differentiating between those hCG moieties present and with the development of analytical peptide standards for hCG, progress can really be made in identifying hCG glyco-variants as specific clinical diagnostic markers. As such the utilisation of mass spectrometry for the detection and characterisation of hCG would provide an additional diagnostic tool in both the monitoring of pregnancy and cancer.

This study examined the structural heterogeneity of hCGβcf from individuals with normal pregnancy, hydatidiform mole and hyperemesis gravidarum. MALDI TOF MS was used to analyse hCGβcf isolated from individual pregnancy urine samples. Reduction of hCGβcf purified from normal pregnancy urine resulted in the separation of the two peptides; non-glycosylated (β 55–92) and glycosylated (β 6–40) chains corresponding to the mass spectral peaks at *m*/*z* 4156.8 and 5840.6 respectively (See Figure 1). In addition to the peaks attributed to the non-glycosylated and the glycosylated peptide and its glycoforms (discussed extensively below) hCGβcf mass spectra from samples N2βcf and HGβcf displayed a peak at *m*/*z* 3950. This peak was not detected in the pooled urine samples from multiple pregnancies from our previous study. The *m*/*z* value of this species is too high to attribute the peak to a non-glycosylated β 6–40, we speculate therefore that this is a glycoform of β 6–40 with a carbohydrate moiety of approximate molecular mass 197.6 Da. Studies of the biosynthesis of *N*-linked sugar chains have demonstrated that a common core of 3 mannose residues forming two branches of $GlcNAc_2$ ($Man_3GlcNAc_2$) is transferred *en bloc* to the polypeptide chain and that the removal of portions of this unit and addition of other sugar residues occur during subsequent processing [29,30]. This processing of the *N*-linked oligosaccharide and also the attachment of the sugars to specific serine or threonine residues takes place in the Golgi apparatus [31]. In the first instance it is possible to suggest that this peak is due solely to the attachment of either galactose or mannose directly on the β 6–40, as there molecular weights are both 180.2 Da. However, in line with the mechanism by which *N*-linked carbohydrates are processed, it may be that this peak represents β 6–40 with one GlcNAc (*m*/*z* 221.21) suggesting that this oligosaccharide may have been removed during processing and modified no further.

The remaining mass spectral peaks are attributed to the multiple glycosylated forms of the peptide β 6–40. Absolute quantification of the relative amounts of each carbohydrate moiety was not possible using this method, one of the perceived restrictions of MALDI TOF MS is its inability to quantitate from spectra. However, we applied a semi-quantitative approach by determining the areas under the peaks of the reduced peptides, similar to that used for data generated by HPLC. These results suggest that hCG is *N*-linked hyperglycosylated to a greater extent in disease and abnormal pregnancy as has been previously described [13,14] and that these glycosylation moiety variation structures are reflected through to the pattern and abundance of urinary metabolite hCGβcf glycoforms. This combined finding suggests a possible use of hCGβcf glycoform analysis by MALDI TOF MS or other methodologies as a novel marker of these diseases.

However, before the use of MALDI TOF MS, as described here, is a clinical reality several technical problems need to be overcome: The first is that we examined purified hCGβcf originating from large pools of collected urine. This volume collection alone renders this approach in-practical for routine clinical analysis purposes. Micro-scale enrichment columns (akin to Zip Tips™) may be needed to process, in both terms of analyte concentration and purity, the much smaller volume urine samples available/collected for large sample sets of clinical samples to be logistically (and economical) analysed by this proposed approach. Secondly, and as referred to above, a major criticism of MALDI TOF MS is that it is not quantifiable. That is the *y*-axis is a relative intensity within a profile and not directly proportional to the various amounts of given molecules present in the sample, *i.e.*, the molecules that ionize easily give more intense signals compared to molecules that might be more abundant but do not ionise easily, and therefore give weaker intensity signals. This reduces the value of MALDI TOF MS spectral data; but to partially overcome this we have adopted a normalisation approach in order to render peak intensities axis comparable between sample spectra. Thus, we transformed the *y*-axis values to a percentage of the spectral region being compared. It has yet to be seen if such a simple processing approach is sufficiently robust to be reproducible when comparing large numbers of samples in a clinical diagnostic situation.

There is some debate as to whether the carbohydrate composition of hCGβcf in pregnancy urine can be directly correlated to that of the parental hCGβ subunit. The results from this study are in line with literature that suggests that carbohydrate heterogeneity has been found in hCGβ in both normal and abnormal pregnancies and that this remains in the terminal urinary degradation product hCGβcf [13,28]. Other studies suggest that hCGβcf glycoforms are very different from that of the hCGβ subunit, proposing the presence of shortened asparagine-linked oligosaccharides on hCGβcf that had generally been metabolised to their pentasaccharide cores as well as smaller sugars [24–27]. One such study reported that 22%–44% of the hCGβcf failed to show binding ability to Concavalin A, which according to the authors is as a consequence of having no sugar molecules [27]. In the previous study, our group have shown that the hCGβcf glycosylated peptide β 6–40 is never completely trimmed of oligosaccharides and that there is only one non-glycosylated hCGβcf peptide, β 55–92 at *m*/*z* 4156.8 after reduction of hCGβcf with DTT [28]. This is also true for the samples in the current study. This discrepancy in the literature may be due to the difference in hCGβcf preparations; the source or the methods used for its purification and characterisation.

In our earlier study [28] hCGβcf was purified from pooled normal pregnancy urine samples and isolated by sequential size exclusion. In the present study the method of purification was ion exchange chromatography and in this case each sample; normal, molar and hyperemesis gravidarum was processed and analysed individually similar to that performed by Elliott *et al.* [13]. In fact some of the preparations used here were prepared alongside this study and as such can be compared directly to M4 and M2 hCG described therein. The largest oligosaccharides (*m*/*z* 2420.3 and 2598.4) detected in the pooled urine hCGβcf population previously were not detected in this cohort of patients in which the largest carbohydrate moiety was identified at *m*/*z* 2077.9. Previous studies in normal pregnancy suggest that by the 10th week of gestation some tri-antennary forms fall to less than 10% of total hCG, indicating that hCG tri-antennary glycoforms, including hyperglycosylated hCG are only seen in significant proportions earlier in pregnancy [13,32,33]. As the samples collected for this study were from the gestational period 7 to 13 weeks, it is possible that the proportion of hCGβcf composed of tri-antennary sugars in the urine samples from the weeks before gestation week 10 were diluted or cleared and in each of the samples this structure occurs in such low concentrations as to be undetectable by this technique.

Our previous study [28] proposed that there was a general absence of galactose with the pooled samples even though in one of the spectra a structure was observed that correlated to a single peak (*m*/*z* 1607.5) [28]. Peaks corresponding to this same glycoform (structure l, Figure 2) have been identified in the present study in all samples except M1βcf. Carbohydrate structures j, k, l, m and n (Figure 2) which have been attributed to peaks in the samples analysed for this study contain, in some isoforms, a galactose residue. In the literature the galactose content of hCGβcf has been reported differently. In some studies involving carbohydrate analysis after acid hydrolysis or the conversion of sugars to glycamines, small amounts of galactose has been detected [15,27]. In contrast to this, other groups have found hCGβcf *N*-linked sugars lacking galactose [24,26]. MALDI TOF MS analysis of the samples in this study has highlighted peaks which contribute significantly to the overall spectrum that cannot be correlated directly with currently identified carbohydrate moieties. It is tempting to speculate about the presence of additional hCGβcf peptide variants as has been suggested previously and their potential involvement in pregnancy and pregnancy associated disorders [14].

3. Materials and Methods

3.1. Biological Samples

Urine samples from five individual pregnancies were used in this study: two were complete molar pregnancies (M2βcf, M4βcf; *i.e.*, moles existed *in utero* when the urine sample was taken), one hyperemesis gravidarum (HGβcf), and two from apparently normal uncomplicated pregnancies (N1βcf, N2βcf). Because hCG reaches its highest levels in urine during the 10th week of pregnancy, all samples were obtained between the 7th and the 13th week of gestation, therefore allowing a 3-week window on either side of the hCG peak; 3 to 5 L of urine were collected continuously from each individual over several days. M2βcf and M4βcf were collected and stored (−80 °C) previously (and intact hCG extracted, the structure of which was reported earlier) [11]. The other samples were collected and purified at the University Of New Mexico School Of Medicine (Albuquerque, MN, USA) following full consent from pregnant women and ethical approval for the study was granted by the OB/GYN departmental research ethics committee.

3.2. Sample Treatments

Proteins were precipitated from urine, initially with acetone (acetone:urine = 2:1 (*v*:*v*)) (Merck, Nottingham, UK) overnight at 4 °C according to methods described previously [34]. The precipitate was collected by centrifugation, and re-dissolved in a minimum amount of distilled-deionized water and re-precipitated with ethanol (ethanol:sample = 9:1 (*v*:*v*)) (Merck) overnight at 4 °C. The resulting precipitate was collected by centrifugation, air-dried to remove excess ethanol, re-dissolved in a minimum amount of distilled-deionized water, and dialyzed against 0.05 M ammonium bicarbonate.

Samples M2βcf and M4βcf were initially fractionated by size exclusion chromatography on an S-200 Sephacryl column (Pharmacia, Piscataway, NJ, USA). The hCGβcf content of each fraction was then determined by specific immunoassay [35]. These samples were co-purified along with intact hCG, some of which were later characterized [13]. The hCGβcf fractions were lyophilized and stored at −80 °C.

All samples were then fractionated on a DEAE-Sepharose ion exchange column [36]. One hundred and thirty milliliter of DEAE-Sepharose CL-6B (Pharmacia, Piscataway, NJ, USA) was packed into an XK26 column (Pharmacia) (26 × 245 mm) at a flow rate of 1.5 mL/min. The void volume (V_0) was calculated by detection of changes in salt concentration using silver nitrate precipitation, after the elution buffer was changed from 0.1 to 1 M ammonium bicarbonate (V_0 = 144 mL). The column was then equilibrated with 2 L of 0.1 M ammonium bicarbonate buffer (pH 7.1) (Sigma-Aldrich, St. Louis, MO, USA) and kept at 4 °C at all times in order to prevent protein degradation by bacterial enzymes. Individual samples were loaded onto the column and the column was eluted with 200 mL of stepwise increases in ammonium bicarbonate buffer (pH 7.1) starting with 0.1 M at a flow rate of 1.5 mL/min, followed by 200 mL of 0.15 M, 200 mL of 0.2 M, 200 mL of 0.25 M and finally 200 mL of 1 M ammonium bicarbonate buffer. The eluent was collected as 9 mL fractions and the concentration of hCGβcf in each fraction was determined by enzyme-linked immunosorbent assay (ELISA).

3.3. hCGβcf Enzyme-Linked Immunosorbent Assay (ELISA)

The assay utilized a monoclonal antibody INN-hCG-106 against the β 11 epitope on hCGβcf as the capture antibody [37]. The S504 polyclonal antibody [38] was used as a primary detection antibody and a donkey-anti-sheep-HRP monoclonal (Jackson Immunoresearch Inc., West Grove, PA, USA) was used as a secondary detection antibody. All fractions with hCGβcf immunoreactivity were pooled and their hCGβcf levels were determined once again.

3.4. Matrix-Assisted Laser Desorption/Ionization Time-of-Flight Mass Spectrometry (MALDI TOF MS)

Post DEAE fractionation, samples were lyophilized against liquid nitrogen in order to remove buffers prior to mass spectrometric analysis. After two freeze-dry/rehydration cycles, the protein was re-dissolved in a minimum amount of distilled-deionized water.

3.5. Whole Molecule hCGβcf Analysis (Non-Reduced)

One micro-liter of sample was applied to a stainless steel MALDI TOF MS target and allowed to dry and crystallize at room temperature. 0.6 μL of sinapinic acid (20 mg/mL^{-1}) (Sigma-Aldrich) in acetonitrile (Merck) and 0.1% trifluoroacetic acid (Merck) was applied on top of the sample and allowed to dry prior to mass spectrometric analysis.

3.6. Dithiothreitol (DTT)-Treated hCGβcf Analysis (Reduced)

Five micro-liters of neat sample was incubated with 5 μL of 100 mM DTT (Sigma-Aldrich) in 100 mM ammonium bicarbonate for 1 h at room temperature. Sample and matrix were then applied on the MALDI TOF MS target as described above.

A pulsed nitrogen laser (λ_{max} = 337 nm) was used to desorb ions from the sample, which were accelerated by a 20 kV electrical field down a 0.5 m linear tube and detected by a micro-channel plate detector. The detector was digitized at a sampling rate of 500 MHz. Spectra were generated by summing 20–30 laser shots by using a Finnigan LASERMAT 2000 instrument (Thermo-Finnigan, Waltham, MA, USA).

Mass calibration was assigned using horse heart cytochrome C (20 pmol/μL "on target") as an external calibrant (two point calibration at $[M + H]^+$ = 12,361 Da and $[M + 2H]^{2+}$ = 6181 Da) for spectral analysis of whole hCGβcf. For spectra analysis of DTT-reduced hCGβcf, the non-glycosylated peptide of hCGβcf (β 55–92) was used as an internal calibrant (one point calibration at $[M + H]^+$ = 4156.8 Da; calculated from its given primary sequence). A 0.5% error during peak mass allocation was allowed for, as this was typical in the linear mode for the MALDI instrument used.

3.7. Treatment of Spectra

In order to determine the masses of the carbohydrate moieties, a previously described method was used [28,39]: reduced hCGβcf spectra were calibrated by using the β 55–92 non-glycosylated peptide as described above. The inferred masses were determined by subtracting the mass of the glycosylated peptide of hCGβcf (β 6–40), which was calculated from its given primary sequence at a mass of 3752.4 Da. The carbohydrate content of each peak was then determined by sequential subtraction of the masses of individual sugar residues [28]. An error ≤0.25% was allowed between observed and predicted carbohydrate masses.

The percentage represented by each of the peaks in individual spectra was also calculated by using the following formula:

%Area = [(Peak height from baseline × Peak width at ½ height) × 100] ÷ Σ of Spectrum Peak Area

4. Conclusions

In conclusion, hCGβcf hyperglycosylation due to tri-antennary glycoforms was found to be the highest in the urine from women with molar and hyperemesis gravidarum pregnancies compared to the samples from normal pregnancy. Although such molecules are subject to metabolic processing, this supports previously published data from Elliott *et al.*, which has shown that hCG is *N*-linked hyperglycosylated to a greater extent in disease and abnormal pregnancy. Although a very high percentage of tri-antennary glycoforms were seen on the hCGβ subunit in abnormal pregnancy in that study [13], such distinct hyperglycosylation has not previously been seen as clearly in hCGβcf. The MALDI TOF MS technique described here, although not definitive is considerably simpler and faster than conventional approaches to glycostructural analysis and presents a potential novel approach to provide additional clinical information. Chromatographic purification prior to MALDI TOF MS analysis is still laborious; however, this may become unnecessary when coupled with affinity capture MALDI techniques as described by Neubert *et al.* [40] and in turn this may lead to more rapid analysis of multiple patients from spot urine samples.

The application of mass spectrometry in the analysis of glycosylation proteforms is developing rapidly. Glycomics, as demonstrated in Manfred Wuhrer's recent review, is now entering the clinical diagnostic arena [41] and, as a result, international searchable databases specifically addressing glycosylation patterns are emerging [42].

Acknowledgments

The authors gratefully acknowledge Peter Berger of the University of Innsbruck (Innsbruck, Austria) for providing the antibody against hCGβcf (INN-hCG-106). The authors also acknowledge Sotiris Malatos for conducting the mass spectrometry work and contributing initial comments for the manuscript. This work was supported by the Joint Research Board of St Bartholomew's Hospital; NIH grants HD35654 & CA44131 to Larry Cole and also the support of the ELK Foundation.

Author Contributions

S.A.B. and L.A.C. carried out sample preparation and purification. R.K.I. analyzed the MS data. S.A.B. and L.A.C. prepared the clinical sample information. R.K.I. and S.A.B. designed the experiment and drafted the manuscript. All authors read and approved the final manuscript.

References

1. Smith, L.M.; Kelleher, N.L.; Proteomics consortium down. Proteoform: A single term describing protein complexity. *Nat. Methods* **2013**, *10*, 186–187.
2. Liu, H.; Zhang, N.; Wan, D.; Cui, M.; Liu, Z.; Liu, S. Mass spectrometry-based analysis of glycoproteins and its clinical applications in cancer biomarker discovery. *Clin. Proteomics* **2014**, *11*, 14.
3. Kessler, M.J.; Reddy, M.S.; Shah, R.H.; Bahl, O.P. Structures of *N*-glycosidic carbohydrate units of human chorionic gonadotrophin. *J. Biol. Chem.* **1979**, *254*, 7901–7908.
4. Boime, I.; Ben-Menahem, D.; Olijve, W. Studies of recombinant gonadotropins: Intersection of basic science and therapeutics. In *Molecular Biology in Reproductive Medicine*; Fauser, B.C.J.M., Rutherford, A.J., Strauss, J.F., van Steirteghem, A., Eds.; The Parthenon Publishing Group: London, UK, 1999; pp. 148–164.
5. Kobata, A.; Takeuchi, M. Structure, pathology and function of the *N*-linked sugar chains of human chorionic gonadotropin. *Biochim. Biophys. Acta* **1999**, *1455*, 315–326.
6. Kessler, M.J.; Mise, T.; Ghai, R.D.; Bahl, O.P. Structure and location of the *O*-glycosidic carbohydrate units of human chorionic gonadotrophin. *J. Biol. Chem.* **1979**, *254*, 7909–7914.
7. Cole, L.A.; Birken, S.; Perini, F. The structures of the serine-linked sugar chains on human chorionic gonadotropin. *Biochem. Biophys. Res. Commun.* **1985**, *126*, 333–339.
8. Cole, L.A. Distribution of *O*-linked sugar units on hCG and its free α-subunit. *Mol. Cell. Endocrinol.* **1987**, *50*, 45–57.
9. Amano, J.; Nishimura, R.; Mochizuki, M.; Kobata, A. Comparative study of the mucin-type sugar chains of human chorionic gonadotropin present in the urine of patients with trophoblastic diseases and healthy pregnant women. *J. Biol. Chem.* **1988**, *263*, 1157–1165.
10. Yoshimoto, Y.; Wolfsen, A.; Odell, W.D. Glycosylation, a variable in the production of hCG by cancers. *Am. J. Med.* **1979**, *67*, 414–420.

11. Mizuochi, T.; Nishimura, R.; Derappe, C.; Taniguchi, T.; Hamamoto, T.; Mochizuki, M.; Kobata, A. Structures of the asparagine-linked sugar chains of human chorionic gonadotropin produced in choriocarcinoma. *J. Biol. Chem.* **1983**, *258*, 14126–14129.
12. Damm, J.B.L.; Voshol, H.; Hard, K.; Kamerling, J.P.; van Dedem, G.W.K.; Vliegenthart, J.F.G. The β-subunit of human chorionic gonadotropin contains *N*-glycosidictrisialo tri-antennary and tri-antennary carbohydrate chains. *Glycoconj. J.* **1988**, *5*, 221–233.
13. Elliot, M.M.; Kardana, A.; Lustbader, J.W.; Cole, L.A. Carbohydrate and peptide structure of the α- and β-subunits of human chorionic gonadotropin from normal and aberrant pregnancy and choriocarcinoma. *Endocrine* **1997**, *7*, 15–32.
14. Cole, L.A.; Cermik, D.; Bahado-Singh, R. Oligosaccharide variants of hCG-related molecules: Potential screening markers for Down syndrome. *Prenat. Diagn.* **1997**, *17*, 1187–1190.
15. Birken, S.; Armstrong, E.G.; Kolks, M.A.; Cole, L.A.; Agosto, G.M.; Krichevsky, A.; Vaitukaitis, J.L.; Canfield, R.E. Structure of the human chorionic gonadotrophin β-subunit fragment from pregnancy urine. *Endocrinology* **1988**, *123*, 572–583.
16. Cole, L.A.; Tanaka, A.; Kim, G.S.; Park, S.Y.; Koh, M.W.; Schwartz, P.E.; Chambers, J.T.; Nam, J.H. β-Core fragment (β-core, UGP or UGF), a tumor marker: A 7-year report. *Gynecol. Oncol.* **1996**, *60*, 264–270.
17. Nisula, B.C.; Blithe, D.L.; Akar, A.; Lefort, G.; Wehmann, R.E. Metabolic fate of human choriogonadotropin. *J. Steroid Biochem.* **1989**, *33*, 733–737.
18. Okamoto, T.; Matsuo, K.; Niu, R.; Osawa, M.; Suzuki, H. Human chorionic gonadotropin (hCG) β-core fragment is produced by degradation of hCG or free hCGβ in gestational trophoblastic tumors: A possible marker for early detection of persistent post molar gestational trophoblastic disease. *J. Endocrinol.* **2001**, *171*, 435–443.
19. Karas, M.; Hillenkamp, F. Laser desorption ionisation of proteins with molecular masses exceeding 10,000 daltons. *Anal. Chem.* **1988**, *60*, 2299–2301.
20. Dell, A.; Morris, H.R. Glycoprotein structure determination by mass spectrometry. *Science* **2001**, *291*, 2351–2356.
21. Sutton, C.W.; O'Neil, J.A.; Cottrell, J.S. Site-specific characterisation of glycoprotein carbohydrates by exoglycosidase digestion and laser desorption mass spectrometry. *Anal. Biochem.* **1994**, *218*, 34–46.
22. Yang, Y.; Orlando, R. Identifying the glycosylation sites and site-specific carbohydrate heterogeneity of glycoproteins by matrix-assisted laser desorption/ionization mass spectrometry. *Rapid Commun. Mass Spectrom.* **1996**, *10*, 932–936.
23. Iwase, H.; Tanaka, A.; Hiki, Y.; Kokubo, T.; Ishii-Karakasa, I.; Nishikido, J.; Kobayashi, Y.; Hotta, K. Application of matrix-assisted laser desorption ionization time-of-flight mass spectrometry to the analysis of glycopeptide-containing multiple *O*-linked oligosaccharides. *J. Chromatogr. B* **1998**, *709*, 145–149.
24. Blithe, D.L.; Akar, A.H.; Wehmann, R.E.; Nisula, B.C. Purification of β-core fragment from pregnancy urine and demonstration that its carbohydrate moieties differ from those of native human chorionic gonadotropin-β. *Endocrinology* **1988**, *122*, 173–180.
25. Blithe, D.L.; Wehmann, R.E.; Nisula, B.C. Carbohydrate composition of β-core. *Endocrinology* **1989**, *125*, 2267–2272.

26. Endo, T.; Nishimura, R.; Saito, S.; Kanazawa, K.; Nomura, K.; Katsuno, M.; Shii, K.; Mukhopadhyay, K.; Baba, S.; Kobata, A. Carbohydrate structures of β-core fragment of human chorionic gonadotropin isolated from a pregnant individual. *Endocrinology* **1992**, *130*, 2052–2058.
27. De Medeiros, S.F.; Amato, F.; Matthews, C.D.; Norman, R.J. Molecular heterogeneity of the β-core fragment of human chorionic gonadotrophin. *J. Endocrinol.* **1993**, *139*, 519–532.
28. Jacoby, E.S.; Kicman, A.T.; Laidler, P.; Iles, R.K. Determination of the glycoforms of human chorionic gonadotrophin β-core fragment by matrix-assisted laser desorption/ionization time-of-flight mass spectrometry. *Clin. Chem.* **2000**, *46*, 1796–1803.
29. Waechter, C.J.; Lennarz, W.J. The role of polyprenol-linked sugars in glycoprotein synthesis. *Annu. Rev. Biochem.* **1976**, *45*, 95–112.
30. Kornfeld, S.; Li, E.; Tabas, I. The synthesis of complex-type oligosaccharides. II. Characterization of the processing intermediates in the synthesis of the complex oligosaccharide units of the vesicular stomatitis virus G protein. *J. Biol. Chem.* **1978**, *253*, 7771–7778.
31. Hirschberg, C.B.; Snider, M.D. Topography of glycosylation in the rough endoplasmic reticulum and Golgi apparatus. *Annu. Rev. Biochem.* **1987**, *56*, 63–87.
32. Butler, S.A.; Khanlian, S.A.; Cole, L.A. Detection of early pregnancy forms of human chorionic gonadotropin by home pregnancy test devices. *Clin. Chem.* **2001**, *47*, 2131–2136.
33. Cole, L.A.; Khanlian, S.A.; Sutton, J.M.; Davies, S.; Stephens, N.D. Hyperglycosylated hCG (invasive trophoblast antigen, ITA) a key antigen for early pregnancy detection. *Clin. Biochem.* **2003**, *36*, 647–655.
34. Birken, S.; Berger, P.; Bidart, J.M.; Weber, M.; Bristow, A.; Norman, R.; Sturgeon, C.; Stenman, U.H. Preparation and characterization of new WHO reference reagents for human chorionic gonadotropin and metabolites. *Clin. Chem.* **2003**, *49*, 144–154.
35. Berger, P.; Sturgeon, C.; Bidart, J.M.; Paus, E.; Gerth, R.; Niang, M.; Bristow, A.; Birken, S.; Stenman, U.H.; The ISOBM TD-7 Workshop on hCG and related molecules. Towards user-oriented standardization of pregnancy and tumor diagnosis: Assignment of epitopes to the three-dimensional structure of diagnostically and commercially relevant monoclonal antibodies directed against human chorionic gonadotropin and derivatives. *Tumor Biol.* **2002**, *23*, 1–38.
36. Canfield, R.E.; Ross, G.T. A new reference preparation of human chorionic gonadotrophin and its subunits. *Bull. World Health Organ.* **1976**, *54*, 463–472.
37. Berger, P.; Bidart, J.M.; Delves, P.S.; Dirnhofer, S.; Hoermann, R.; Isaacs, N.; Jackson, A.; Klonisch, T.; Lapthorn, A.; Lund, T.; *et al.* Immunochemical mapping of gonadotropins. *Mol. Cell. Endocrinol.* **1996**, *125*, 33–43.
38. Lee, C.L.; Iles, R.K.; Shepherd, J.H.; Hudson, C.N.; Chard, T. The purification and development of a radioimmunoassay for β-core fragment of human chorionic gonadotrophin in urine: Application as a marker of gynaecological cancer in premenopausal and postmenopausal women. *J. Endocrinol.* **1991**, *130*, 481–489.
39. Jacoby, E.S.; Kicman, A.T.; Iles, R.K. Identification of post-translational modifications resulting from LHβ polymorphisms by matrix-assisted laser desorption time-of-flight mass spectrometric analysis of pituitary LHβ core fragment. *J. Mol. Endocrinol.* **2003**, *30*, 239–252.

40. Neubert, H.; Jacoby, E.S.; Bansal, S.S.; Iles, R.K.; Cowan, D.A.; Kicman, A.T. Enhanced affinity capture MALDI TOF MS: Orientation of an immunoglobulin G using recombinant protein G. *Anal. Chem.* **2002**, *74*, 3677–3683.
41. Wuhrer, M. Glycomics using mass spectrometry. *Glycoconj. J.* **2013**, *30*, 11–22.
42. Baycin Hizal, D.; Wolozny, D.; Colao, J.; Jacobson, E.; Tian, Y.; Krag, S.S.; Betenbaugh, M.J.; Zhang, H. Glycoproteomic and glycomic databases. *Clin. Proteomics* **2014**, *11*, 15.

14

Interleukin-6 Receptor rs7529229 T/C Polymorphism is Associated with Left Main Coronary Artery Disease Phenotype in a Chinese Population

Feng He [1], Xiao Teng [1], Haiyong Gu [2], Hanning Liu [1], Zhou Zhou [1], Yan Zhao [1], Shengshou Hu [1,*] and Zhe Zheng [1,*]

[1] State Key Laboratory of Cardiovascular Disease, Fuwai Hospital, National Center for Cardiovascular Diseases, Chinese Academy of Medical Sciences and Peking Union Medical College, Beijing 100037, China; E-Mails: drhefeng@gmail.com (F.H.); tengxiao7@163.com (X.T.); zsylhn@gmail.com (H.L.); contyzhou@msn.com (Z.Z.); zhaoyan19791025@aliyun.com (Y.Z.)

[2] Department of Cardiothoracic Surgery, Affiliated People's Hospital of Jiangsu University, Zhenjiang 212002, China; E-Mail: haiyong_gu@hotmail.com

* Authors to whom correspondence should be addressed; E-Mails: shengshouhu@yahoo.com (S.H.); zhengzhe@fuwai.com (Z.Z.)

Abstract: Left main coronary artery disease (LMCAD) is a particular severe phenotype of coronary artery disease (CAD) and heritability. Interleukin (IL) may play important roles in the pathogenesis of CAD. Although several single nucleotide polymorphisms (SNPs) identified in *IL* related genes have been evaluated for their roles in inflammatory diseases and CAD predisposition, the investigations between genetic variants and CAD phenotype are limited. We hypothesized that some of these gene SNPs may contribute to LMCAD phenotype susceptibility compared with more peripheral coronary artery disease (MPCAD). In a hospital-based case-only study, we studied *IL-1A* rs1800587 C/T, *IL-1B* rs16944 G/A, *IL-6* rs1800796 C/G, *IL-6R* rs7529229 T/C, *IL-8* rs4073 T/A, *IL-10* rs1800872 A/C, and *IL-10* rs1800896 A/G SNPs in 402 LMCAD patients and 804 MPCAD patients in a Chinese population. Genotyping was done using matrix-assisted laser desorption/ionization time-of-flight mass spectrometry (MALDI-TOF MS) and ligation detection reaction (LDR) method. When the *IL-6R* rs7529229 TT homozygote genotype was used as the reference group, the CC or TC/CC genotypes were associated with the increased risk for LMCAD (CC *vs.* TT, adjusted odds ratio(OR) = 1.46,

95% confidence interval (CI) = 1.02–2.11, $p = 0.042$; CC + TC *vs*. TT, adjusted OR = 1.31, 95% CI = 1.02–1.69, $p = 0.037$). None of the other six SNPs achieved any significant differences between LMCAD and MPCAD. The present study suggests that *IL-6R* rs7529229 T/C functional SNP may contribute to the risk of LMCAD in a Chinese population. However, our results were limited. Validation by a larger study from a more diverse ethnic population is needed.

Keywords: *IL-6R*; polymorphisms; left main artery; coronary artery disease; mass spectrometry

1. Introduction

Coronary artery disease (CAD) is the leading cause of mortality and morbidity in many countries including China [1,2]. Multifaceted phenotypes of CAD (number of involved vessels, location of lesions, severity of diameter narrowing, and morphology of lesions) have different mechanisms. Left main coronary artery disease (LMCAD) is a particular severe phenotype of CAD because it is associated with higher risk of fatal cardiovascular events [3]. Thus, any information in detecting the asymptomatic relatives of these patients can be useful for primary prevention.

LMCAD was heritability and frequently shared by siblings with CAD [4]. Fischer *et al.* [5] reported a stronger genetic component in LMCAD phenotype. However, Kolovou *et al.* [6] found no significant difference in cholesteryl ester transfer protein (CETP) allele frequency or genotype distribution among LMCAD and more peripheral coronary artery disease (MPCAD) patients. Thus, more investigations are needed in detecting association between genetic variants and CAD phenotype.

Inflammation is important in the initiation, progression, and clinical outcome of CAD [7,8]. Interleukin (IL) plays a key role in the inflammatory response, immune regulation and development of CAD. IL-1 family (including IL-1A, IL-1B *et al.*) plays a major role in inflammation by affecting antigen recognition patterns and lymphocyte function [9]. IL-6 is a multi-functional cytokine involved in various contradictory processes, a high circulating concentration of IL-6 is associated with increased risk of CAD [10]. IL-6 binds to IL-6R and then activates an intracellular signaling cascade leading to the inflammatory response [11]. IL-8 plays a key role in the development of atherosclerotic plaques [12]. IL-10 has anti-inflammatory and immunosuppressive effects by decreasing the production of pro-inflammatory mediators, exerts important protective effects on atherosclerotic lesion development [13].

Recently, a link between CAD genetic susceptibility and the response to inflammatory signaling has been established [8,14]. Growing evidence for heritability of pro-inflammatory state suggests that individual genetic background also modulate the development of CAD and its magnitude [15,16]. Despite well-described associations between inflammatory-related genetic variation and susceptibility to CAD, there is a paucity of data regarding to the phenotype of CAD.

The identification of novel genetic variants for use in assessing early risk of LMCAD have potentially important clinical implications, such as identifying high-risk individuals and adapting therapeutic management to the individual's genetic make-up. Single nucleotide polymorphisms (SNPs) strongly influence the plasma levels and biological activity of the corresponding proteins. On the basis of the biological and pathologic significance of IL, we performed a genetic association study on the SNPs of *IL-1A*, *IL-1B*, *IL-6*, *IL-6R*, *IL-8* and *IL-10*.

The objective of this investigation was to evaluate the association between *IL* SNPs and LMCAD susceptibility compare with MPCAD in a hospital-based case-only study. We performed genotyping and analyses for the seven SNPs in a cohort of 1206 CAD patients (402 LMCAD patients and 804 MPCAD patients) in a Chinese population.

2. Results and Discussion

2.1. Characteristics of the Study Population

The demographic and clinical characteristics of all subjects are summarized in Table 1. Among 1206 DNA samples, the seven SNPs were successful ranging from 98.59% to 99.92% (Table 2). The mean age for LMCAD patients (62.24 ± 8.66) is significantly higher than for MPCAD patients (60.14 ± 8.96), $p < 0.001$ (Table 1). There are no significant difference for sex and mean body mass index (BMI), family history of CAD, previous smoker, hypertension, hyperlipidemia, diabetes mellitus, mean ejection fraction, circumflex branch of left coronary artery system, right coronary artery system, off-pump coronary artery bypass grafting (OPCAB) or conventional coronary artery bypass grafting (cCABG) for LMCAD patients and MPCAD patients (Table 1). LMCAD patients are more likely to have three disease territories than MPCAD patients (93.3% *vs.* 87.7%, $p = 0.003$) (Table 1).

2.2. Associations between the Seven SNPs and Risk of LMCAD and MPCAD

The genotype frequencies of the *IL-6R* rs7529229 T/C polymorphism were 34.1% (TT), 49.0% (TC) and 16.9% (CC) in LMCAD patients, and 40.4% (TT), 45.9% (TC) and 13.7% (CC) in MPCAD patients ($p = 0.027$) (Table 3). When the *IL-6R* rs7529229 TT homozygote genotype was used as the reference group, the CC genotype was associated with the increased risk for LMCAD (CC *vs.* TT, adjusted OR = 1.46, 95% CI = 1.02–2.11, $p = 0.042$). The TC genotype was not associated with the risk for LMCAD (TC *vs.* TT, adjusted OR = 1.26, 95% CI = 0.97–1.65, $p = 0.088$). In the dominant model, when the *IL-6R* rs7529229 TT genotype was used as the reference group, the TC/CC genotypes were associated with the increased risk for LMCAD (TC/CC *vs.* TT, adjusted OR = 1.31, 95% CI = 1.02–1.69, $p = 0.037$). The polymorphism was not associated with the risk for LMCAD in recessive genetic models (CC *vs.* TT/TC, adjusted OR = 1.29, 95% CI = 0.92–1.79, $p = 0.140$).

None of the other six polymorphisms achieved a significant difference in the genotype distributions between cases and controls. The six polymorphisms were not associated with the risk for LMCAD/MPCAD both in homozygote comparison, heterogeneity comparison, dominant genetic model and recessive genetic models (Table 3).

Table 1. Patient Demographics and Risk Factors with coronary artery disease (CAD) (Left main coronary artery disease (LMCAD) and more peripheral coronary artery disease (MPCAD)).

Variable	LMCAD [b] (*n* = 402)	MPCAD [c] (*n* = 804)	*p*-Value	All CAD (*n* = 1206)
Mean age, *y*	62.24 (±8.66)	60.14 (±8.96)	**<0.001**	60.84 (±8.91)
Woman, %	67 (16.7)	170 (21.1)	0.065	237 (19.7)
Mean BMI [a], kg/m^2	25.55 (±3.16)	25.75 (±3.13)	0.299	25.68 (±3.14)
Family history of CAD, %	21 (5.2)	32 (4.0)	0.323	53 (4.4)
Previous smoker, %	211 (52.5)	422 (52.5)	1.000	633 (52.5)
Hypertension, %	261 (64.9)	530 (65.9)	0.732	791 (65.6)
Hyperlipidemia, %	275 (68.4)	573 (71.3)	0.290	848 (70.4)
Diabetes mellitus, %	125 (31.1)	271 (33.7)	0.363	396 (32.8)
Mean ejection fraction, %	60.25 (±7.96)	59.55 (±8.76)	0.177	59.78 (±8.50)
Disease territories, %				
1–2	27 (6.7)	99 (12.3)	**0.003**	126 (10.4)
3	375 (93.3)	705 (87.7)	**0.003**	1080 (89.6)
LMCAD, %	402 (100)	0 (0)	—	402 (33.3)
Anterior descending artery system, %	386 (96.0)	793 (98.6)	**0.004**	1179 (97.8)
Circumflex branch of left coronary artery system, %	373 (92.8)	744 (92.5)	0.876	1117 (92.6)
Right coronary artery system, %	374 (93.0)	751 (93.4)	0.807	1125 (93.3)
OPCAB/cCABG [d]	209/193	438/366	0.414	647/559

[a] BMI: body mass index; [b] LMCAD: left main coronary artery disease; [c] MPCAD: more peripheral coronary artery disease; Bold values are statistically significant ($p < 0.05$); [d] OPCAB: off-pump coronary artery bypass grafting; cCABG: conventional coronary artery bypass grafting.

Table 2. Primary information for seven genotyped single nucleotide polymorphisms (SNPs).

Genotyped SNPs	Chr [a]	Regulome DB Score [b]	TFBS [c]	Splicing (ESE or ESS)	Location	MAF [d] for Chinese in Database	% Genotyping Value
IL-1A: rs1800587 C/T	2	5	Y	Y	5' UTR	0.073	99.92
IL-1B: rs16944 G/A	2	1f	Y	—	5' near gene	0.453	99.83
IL-6: rs1800796 C/G	7	4	Y	—	ncRNA	0.233	99.75
IL-6R: rs7529229 T/C	1	2b	—	—	intron	0.442	98.59
IL-8: rs4073 T/A	4	2b	Y	—	5' near gene	0.389	98.84
IL-10: rs1800872 A/C	1	5	Y	—	5' near gene	0.238	99.83
IL-10: rs1800896 A/G	1	6	Y	—	5' near gene	0.059	99.83

Table 3. Main effects of SNPs on LMCAD risk.

Genotyped SNPs	Genotyping (AA/AB/BB) [a] LMCAD (n = 402)	MPCAD (n = 804)	AB *vs.* AA Adjusted OR (95% CI); *p*-Value	BB *vs.* AA Adjusted OR (95% CI); *p*-Value	BB *vs.* (AA + AB) Adjusted OR (95% CI); *p*-Value	(BB + AB) *vs.* AA Adjusted OR (95% CI); *p*-Value	*p* Trend
IL-1A: rs1800587 C/T	331/67/3	645/145/14	0.91 (0.66–1.26); 0.570	0.42 (0.12–1.42); 0.175	0.43 (0.12–1.50); 0.183	0.87 (0.64–1.19); 0.374	0.314
IL-1B: rs16944 G/A	104/203/95	212/397/193	1.04 (0.78–1.39); 0.794	1.01 (0.72–1.42); 0.974	0.98 (0.74–1.30); 0.890	1.03 (0.78–1.35); 0.841	0.948
IL-6: rs1800796 C/G	182/174/46	358/348/95	0.97 (0.75–1.25); 0.797	0.96 (0.65–1.43); 0.856	0.98 (0.67–1.43); 0.916	0.97 (0.76–1.23); 0.782	0.970
IL-6R: rs7529229 T/C	135/194/67	320/364/109	1.26 (0.97–1.65); 0.088	**1.46 (1.02–2.11); 0.042**	1.29 (0.92–1.79); 0.140	**1.31 (1.02–1.69); 0.037**	0.080
IL-8: rs4073 T/A	127/198/71	280/375/141	1.18 (0.90–1.56); 0.226	1.14 (0.80–1.62); 0.482	1.03 (0.75–1.41); 0.862	1.17 (0.91–1.52); 0.231	0.545
IL-10: rs1800872 A/C	155/197/49	336/371/96	1.15 (0.89–1.49); 0.277	1.08 (0.73–1.60); 0.716	1.00 (0.69–1.44); 0.982	1.14 (0.89–1.46); 0.305	0.558
IL-10: rs1800896 A/G	330/69/2	639/154/10	0.86 (0.63–1.18); 0.350	0.38 (0.08–1.76); 0.217	0.39 (0.09–1.81); 0.230	0.83 (0.61–1.13); 0.242	0.316

[a] AA/AB/BB means homozygote, heterozygote and mutated homozygote; Bold values are statistically significant ($p < 0.05$); Adjusted for age and sex.

3. Experimental Section

3.1. Ethical Approval of the Study Protocol

The study protocol was approved by the Review Board of Peking Union Medical College (Beijing, China). We have complied with the World Medical Association Declaration of Helsinki regarding ethical conduct of research involving human subjects and/or animals. All patients provided written informed consent to be involved in the study.

3.2. Study Subjects

This study involved 1206 patients with CAD for the purpose undergoing coronary artery bypass grafting (CABG). Patients were consecutively recruited from the Cardiovascular Institute and Fuwai Hospital, Chinese Academy of Medical Sciences and Peking Union Medical College (Beijing, China) between December 2007 and December 2008. All patients were diagnosed using angiography which was scored systematically by an experienced interventional cardiologist. CAD patients were defined as having angiographic coronary stenosis of at least 50% lumen reduction. LMCAD was defined as at least 50% stenosis by visual assessment in the left main (LM) vessel, including ostial stenosis. While lesions compromising the lumen by ≥50% further from LM were defined as MPCAD [6,17].

All subjects were genetically unrelated ethnic Han Chinese. All data were collected by trained clinical research staff and were subsequently double entered into computer databases. Baseline information on personal and clinical characteristics was complete for all 1206 patients involved in the study.

3.3. Candidate Genes and SNPs Selection

Six candidate genes (*IL-1A*, *IL-1B*, *IL-6*, *IL-6R*, *IL-8* and *IL-10*) involved in the pathogenesis of inflammation and CAD were selected a priori based on previous transcription profiling in humans, pathway analysis, association studies reported in the literature, and expert opinion [18,19]. Seven typical SNPs, *IL-1A* rs1800587 C/T, *IL-1B* rs16944 G/A, *IL-6* rs1800796 C/G, *IL-6R* rs7529229 T/C, *IL-8* rs4073 T/A, *IL-10* rs1800872 A/C, and *IL-10* rs1800896 A/G, were subsequently selected in these candidate genes, based on function importance [20,21].

3.4. Isolation of DNA and Genotyping by Matrix-Assisted Laser Desorption/Ionization Time-of-Flight Mass Spectrometry (MALDI-TOF MS) and Ligation Detection Reaction (LDR)

Blood samples were collected using vacutainers and transferred to test tubes containing ethylenediamine tetra-acetic acid (EDTA). Genomic DNA was isolated from the lymphocytes of whole blood samples using the Wizard Genomic DNA Purification Kit (Promega, Madison, WI, USA). For quality control, all sample DNAs were conducted polymerase chain reaction (PCR), analyzed on a 3% agarose gel and visualized by ethidium bromide staining. *IL-1A* rs1800587 C/T, *IL-1B* rs16944 G/A, *IL-6* rs1800796 C/G, *IL-10* rs1800872 A/C and *IL-10* rs1800896 A/G genotyping was done by MALDI-TOF MS support from CapitalBio Corporation (Beijing, China), using the MassARRAY system (Sequenom, San Diego, CA, USA) as previously described [22]. LMCAD and MPCAD at a proportion of ≈1:2 were assayed. Completed genotyping reactions were spotted onto a 384-well spectroCHIP (Sequenom) using a MassARRAY Nanodispenser (Sequenom), and analyzed by MALDI-TOF-MS (Sequenom, San Diego, CA, USA). Genotype calling was done in real time with MassARRAY RT software (version 3.1; Sequenom), and analyzed using MassARRAY Typer software (version 4.0; Sequenom). For *IL-6R* rs7529229 T/C and *IL-8* rs4073 T/A, genotyping study was performed using the LDR method [23,24], with technical support from Shanghai Biowing Applied Biotechnology Company (Shanghai, China).

3.5. Statistical Analyses

Differences in demographics, variables, and genotypes of the seven SNPs were evaluated using a chi-squared test. The associations between the seven SNPs and risk of CAD phenotype were estimated by computing odds ratios (ORs) and 95% confidence intervals (CIs) using logistic regression analyses, and by using crude ORs and adjusted ORs. Statistical differences with $p < 0.05$ were considered significant, and all statistical analyses were done with SAS software (version 9.1.3; SAS Institute, Cary, NC, USA).

4. Conclusions

In this study, we found that the *IL-6R* rs7529229 T/C polymorphism may increase the risk of LMCAD compared with MPCAD. The CC genotype is associated with higher risk of LMCAD phenotype. None of the other six polymorphisms showed any overall predisposition to LMCAD phenotype susceptibility. To the best of our knowledge, this study provides the first evidence linking variation in the *IL-6R* gene and LMCAD phenotype risk in a Chinese population.

This finding is supported by studies displaying a higher heritability of LMCAD. Capodanno *et al.* [25] investigated the epidemiology and the clinical impact of different anatomical phenotypes of the LM coronary artery, and Iwasaki *et al.* [26] investigated the distribution of coronary atherosclerosis in patients with CAD. Their findings suggested that LM phenotypes are more likely to present with atherosclerotic disease and significant stenosis and are particularly heritable [25,26].

Our results showed that *IL-6R* rs7529229 CC genotype as well as C allele was more prevalent in the LMCAD group than in the MPCAD control group. *IL-6R* rs7529229 T/C polymorphism was associated with increased risk of LMCAD, which may lead to LMCAD by enhancing inflammation. However, it was not found to be statistically significant in other six candidate gene variants.

IL-6 is an important pleiotropic cytokine that has a broad range of humoral and cellular immune properties relating to inflammation, tissue injury and contributes to the clinical evolution of CAD [27]. IL-6 exerts its biological activities through the IL-6R. The *IL-6R* gene is located on human chromosome 1q21, a region that previous studies have reported to be linked to metabolic syndrome, type 2 diabetes and atrial fibrillation [28–30]. The human *IL-6R* gene is highly polymorphic and there is considerable variation in its expression between individuals. A number of SNPs have been reported and genetic variants in *IL-6R* gene are associated with several different kinds of diseases, including CAD [31]. *IL-6R* rs7529229, in which the polymorphism is localized to a functional domain of the receptor protein, is a T/C variation in the *IL-6R* gene (intronic) on human chromosome 1. In 40 studies including up to 133449 individuals, mendelian randomization analyses revealed that *IL-6R* rs7529229 T/C marking a non-synonymous *IL-6R* variant (rs8192284; p.Asp358Ala), was associated with increased circulating log IL-6 concentration as well as reduced C-reactive protein and fibrinogen concentrations [32]. This suggests that *IL-6R* rs7529229 variant is strongly associated with CAD and cardiovascular events.

In our study, Using Power and Sample Size Calculation (PS, version 3.0, 2009), considering *IL-6R* rs7529229 T/C mutant alleles in the LMCAD group, OR, LMCAD samples and MPCAD samples, the power of our analysis (α = 0.05) was 0.857 in 396 LMCAD cases and 793 MPCAD cases with adjusted OR = 1.46. This suggests that SNP of *IL-6R* gene seems to also have a causal role in development of LMCAD. Thus, the findings indicate that *IL-6R* rs7529229 polymorphism may have potential importance in screening individuals at high risk for developing LMCAD, and targeting of IL-6R could provide a novel therapeutic approach for the prevention of LMCAD.

Several limitations of the present study need to be addressed. First, this was a hospital-based study; selection bias was unavoidable; Second, the polymorphisms we investigated, based on their functional considerations, may not offer a comprehensive view of the genetic variability. Further fine-mapping analysis of *IL* genes or high-density whole genome genetic analyses evaluating different CAD phenotypes might give further insights to the pathophysiologic mechanisms underlying LMCAD; Third, a single case–case study is not sufficient to fully interpret the relationship between *IL-6R* rs7529229 T/C polymorphism and susceptibility to LMCAD because of the relatively small number of patients with LMCAD. Replication studies with larger numbers of subjects are necessary to confirm our findings; Finally, we did not evaluate plasma IL-6 levels and the function of *IL-6R* rs7529229 T/C, which restricted our analyses.

Despite the limitations, the present study provided strong evidence that *IL-6R* rs7529229 T/C functional polymorphism may contribute to the risk of LMCAD. However, our results were obtained from a moderate-sized sample, and therefore this is a preliminary conclusion. Validation by a larger study from a more diverse ethnic population is needed to confirm these findings.

Acknowledgments

The authors thank that the support from International S & T Cooperation Program (2010DFB33140), Program for New Century Excellent Talents in University, and the Key Project in the National Science & Technology Pillar Program during the 12th 5-Year Plan Period (2011BAI11B21, 2011BAI11B02).

References

1. Roger, V.L.; Go, A.S.; Lloyd-Jones, D.M.; Benjamin, E.J.; Berry, J.D.; Borden, W.B.; Bravata, D.M.; Dai, S.; Ford, E.S.; Fox, C.S.; *et al.* Heart disease and stroke statistics—2012 update: A report from the American Heart Association. *Circulation* **2012**, *125*, e2–e220.
2. He, J.; Gu, D.; Wu, X.; Reynolds, K.; Duan, X.; Yao, C.; Wang, J.; Chen, C.S.; Chen, J.; Wildman, R.P.; *et al.* Major causes of death among men and women in China. *N. Engl. J. Med.* **2005**, *353*, 1124–1134.
3. Fajadet, J.; Chieffo, A. Current management of left main coronary artery disease. *Eur. Heart J.* **2012**, *33*, 36–50.

4. Fischer, M.; Mayer, B.; Baessler, A.; Riegger, G.; Erdmann, J.; Hengstenberg, C.; Schunkert, H. Familial aggregation of left main coronary artery disease and future risk of coronary events in asymptomatic siblings of affected patients. *Eur. Heart J.* **2007**, *28*, 2432–2437.
5. Fischer, M.; Broeckel, U.; Holmer, S.; Baessler, A.; Hengstenberg, C.; Mayer, B.; Erdmann, J.; Klein, G.; Riegger, G.; Jacob, H.J.; *et al.* Distinct heritable patterns of angiographic coronary artery disease in families with myocardial infarction. *Circulation* **2005**, *111*, 855–862.
6. Kolovou, G.; Vasiliadis, I.; Kolovou, V.; Karakosta, A.; Mavrogeni, S.; Papadopoulou, E.; Papamentzelopoulos, S.; Giannakopoulou, V.; Marvaki, A.; Degiannis, D.; *et al.* The role of common variants of the cholesteryl ester transfer protein gene in left main coronary artery disease. *Lipids Health Dis.* **2011**, *10*, 156.
7. Ross, R. Atherosclerosis—An inflammatory disease. *N. Engl. J. Med.* **1999**, *340*, 115–126.
8. Hansson, G.K. Inflammation, atherosclerosis, and coronary artery disease. *N. Engl. J. Med.* **2005**, *352*, 1685–1695.
9. Dinarello, C.A. The interleukin-1 family: 10 years of discovery. *FASEB J.* **1994**, *8*, 1314–1325.
10. Boekholdt, S.M.; Stroes, E.S. The interleukin-6 pathway and atherosclerosis. *Lancet* **2012**, *379*, 1176–1178.
11. Mihara, M.; Hashizume, M.; Yoshida, H.; Suzuki, M.; Shiina, M. IL-6/IL-6 receptor system and its role in physiological and pathological conditions. *Clin. Sci.* **2012**, *122*, 143–159.
12. Vogiatzi, K.; Apostolakis, S.; Voudris, V.; Thomopoulou, S.; Kochiadakis, G.E.; Spandidos, D.A. Interleukin 8 and susceptibility to coronary artery disease: A population genetics perspective. *J. Clin. Immunol.* **2008**, *28*, 329–335.
13. Fichtlscherer, S.; Breuer, S.; Heeschen, C.; Dimmeler, S.; Zeiher, A.M. Interleukin-10 serum levels and systemic endothelial vasoreactivity in patients with coronary artery disease. *J. Am. Coll. Cardiol.* **2004**, *44*, 44–49.
14. Harismendy, O.; Notani, D.; Song, X.; Rahim, N.G.; Tanasa, B.; Heintzman, N.; Ren, B.; Fu, X.D.; Topol, E.J.; Rosenfeld, M.G.; *et al.* 9p21 DNA variants associated with coronary artery disease impair interferon-gamma signalling response. *Nature* **2011**, *470*, 264–268.
15. McPherson, R.; Davies, R.W. Inflammation and coronary artery disease: insights from genetic studies. *Can. J. Cardiol.* **2012**, *28*, 662–666.
16. Deloukas, P.; Kanoni, S.; Willenborg, C.; Farrall, M.; Assimes, T.L.; Thompson, J.R.; Ingelsson, E.; Saleheen, D.; Erdmann, J.; Goldstein, B.A.; *et al.* Large-scale association analysis identifies new risk loci for coronary artery disease. *Nat. Genet.* **2013**, *45*, 25–33.
17. Patel, M.R.; Dehmer, G.J.; Hirshfeld, J.W.; Smith, P.K.; Spertus, J.A. ACCF/SCAI/STS/AATS/AHA/ASNC/HFSA/SCCT 2012 Appropriate use criteria for coronary revascularization focused update: A report of the American College of Cardiology Foundation Appropriate Use Criteria Task Force, Society for Cardiovascular Angiography and Interventions, Society of Thoracic Surgeons, American Association for Thoracic Surgery, American Heart Association, American Society of Nuclear Cardiology, and the Society of Cardiovascular Computed Tomography. *J. Am. Coll. Cardiol.* **2012**, *59*, 857–881.

18. Tomic, V.; Russwurm, S.; Moller, E.; Claus, R.A.; Blaess, M.; Brunkhorst, F.; Bruegel, M.; Bode, K.; Bloos, F.; Wippermann, J.; *et al.* Transcriptomic and proteomic patterns of systemic inflammation in on-pump and off-pump coronary artery bypass grafting. *Circulation* **2005**, *112*, 2912–2920.
19. Calvano, S.E.; Xiao, W.; Richards, D.R.; Felciano, R.M.; Baker, H.V.; Cho, R.J.; Chen, R.O.; Brownstein, B.H.; Cobb, J.P.; Tschoeke, S.K.; *et al.* A network-based analysis of systemic inflammation in humans. *Nature* **2005**, *437*, 1032–1037.
20. Podgoreanu, M.V.; White, W.D.; Morris, R.W.; Mathew, J.P.; Stafford-Smith, M.; Welsby, I.J.; Grocott, H.P.; Milano, C.A.; Newman, M.F.; Schwinn, D.A. Inflammatory gene polymorphisms and risk of postoperative myocardial infarction after cardiac surgery. *Circulation* **2006**, *114*, I275–I281.
21. Yamada, Y.; Izawa, H.; Ichihara, S.; Takatsu, F.; Ishihara, H.; Hirayama, H.; Sone, T.; Tanaka, M.; Yokota, M. Prediction of the risk of myocardial infarction from polymorphisms in candidate genes. *N. Engl. J. Med.* **2002**, *347*, 1916–1923.
22. Schaeffeler, E.; Zanger, U.M.; Eichelbaum, M.; Asante-Poku, S.; Shin, J.G.; Schwab, M. Highly multiplexed genotyping of thiopurine *S*-methyltransferase variants using MALD-TOF mass spectrometry: Reliable genotyping in different ethnic groups. *Clin. Chem.* **2008**, *54*, 1637–1647.
23. Chen, Z.J.; Zhao, H.; He, L.; Shi, Y.; Qin, Y.; Shi, Y.; Li, Z.; You, L.; Zhao, J.; Liu, J.; *et al.* Genome-wide association study identifies susceptibility loci for polycystic ovary syndrome on chromosome 2p16.3, 2p21 and 9q33.3. *Nat. Genet.* **2011**, *43*, 55–59.
24. Yi, P.; Chen, Z.; Zhao, Y.; Guo, J.; Fu, H.; Zhou, Y.; Yu, L.; Li, L. PCR/LDR/capillary electrophoresis for detection of single-nucleotide differences between fetal and maternal DNA in maternal plasma. *Prenat. Diagn.* **2009**, *29*, 217–222.
25. Capodanno, D.; Di, S.M.E.; Seminara, D.; Caggegi, A.; Barrano, G.; Tagliareni, F.; Dipasqua, F.; Tamburino, C. Epidemiology and clinical impact of different anatomical phenotypes of the left main coronary artery. *Heart Vessels* **2011**, *26*, 138–144.
26. Iwasaki, K.; Matsumoto, T.; Aono, H.; Furukawa, H.; Nagamachi, K.; Samukawa, M. Distribution of coronary atherosclerosis in patients with coronary artery disease. *Heart Vessels* **2010**, *25*, 14–18.
27. Papanicolaou, D.A.; Wilder, R.L.; Manolagas, S.C.; Chrousos, G.P. The pathophysiologic roles of interleukin-6 in human disease. *Ann. Intern. Med.* **1998**, *128*, 127–137.
28. Jiang, C.Q.; Lam, T.H.; Liu, B.; Lin, J.M.; Yue, X.J.; Jin, Y.L.; Cheung, B.M.; Thomas, G.N. Interleukin-6 receptor gene polymorphism modulates interleukin-6 levels and the metabolic syndrome: GBCS-CVD. *Obesity* **2010**, *18*, 1969–1974.
29. Hamid, Y.H.; Urhammer, S.A.; Jensen, D.P.; Glumer, C.; Borch-Johnsen, K.; Jorgensen, T.; Hansen, T.; Pedersen, O. Variation in the interleukin-6 receptor gene associates with type 2 diabetes in Danish whites. *Diabetes* **2004**, *53*, 3342–3345.
30. Schnabel, R.B.; Kerr, K.F.; Lubitz, S.A.; Alkylbekova, E.L.; Marcus, G.M.; Sinner, M.F.; Magnani, J.W.; Wolf, P.A.; Deo, R.; Lloyd-Jones, D.M.; *et al.* Large-scale candidate gene analysis in whites and African Americans identifies IL-6R polymorphism in relation to atrial fibrillation: the National Heart, Lung, and Blood Institute's Candidate Gene Association Resource (CARe) project. *Circ. Cardiovasc. Genet.* **2011**, *4*, 557–564.

31. Sarwar, N.; Butterworth, A.S.; Freitag, D.F.; Gregson, J.; Willeit, P.; Gorman, D.N.; Gao, P.; Saleheen, D.; Rendon, A.; Nelson, C.P.; *et al.* Interleukin-6 receptor pathways in coronary heart disease: A collaborative meta-analysis of 82 studies. *Lancet* **2012**, *379*, 1205–1213.
32. Hingorani, A.D.; Casas, J.P. The interleukin-6 receptor as a target for prevention of coronary heart disease: A mendelian randomisation analysis. *Lancet* **2012**, *379*, 1214–1224.

15

Legionella dumoffii Utilizes Exogenous Choline for Phosphatidylcholine Synthesis

Marta Palusinska-Szysz [1,*], Agnieszka Szuster-Ciesielska [2], Magdalena Kania [3], Monika Janczarek [1], Elżbieta Chmiel [1] and Witold Danikiewicz [3]

1. Department of Genetics and Microbiology, Institute of Microbiology and Biotechnology, Maria Curie-Sklodowska University, Akademicka 19 St., 20-033 Lublin, Poland; E-Mails: mon.jan@poczta.umcs.lublin.pl (M.J.); elawisniewska87@gmail.com (E.C.)
2. Department of Virology and Immunology, Institute of Microbiology and Biotechnology, Maria Curie-Sklodowska University, Akademicka 19 St., 20-033 Lublin, Poland; E-Mail: aszusterciesielska@gmail.com
3. Mass Spectrometry Group, Institute of Organic Chemistry Polish Academy of Sciences, Kasprzaka 44/52 St., 01-224 Warsaw, Poland; E-Mails: magdalena.kania@icho.edu.pl (M.K.); witold.danikiewicz@icho.edu.pl (W.D.)

* Author to whom correspondence should be addressed; E-Mail: marta.szysz@poczta.umcs.lublin.pl

---

Abstract: Phosphatidycholine (PC) is the major membrane-forming phospholipid in eukaryotes but it has been found in only a limited number of prokaryotes. Bacteria synthesize PC via the phospholipid *N*-methylation pathway (Pmt) or via the phosphatidylcholine synthase pathway (Pcs) or both. Here, we demonstrated that *Legionella dumoffii* has the ability to utilize exogenous choline for phosphatidylcholine (PC) synthesis when bacteria grow in the presence of choline. The Pcs seems to be a primary pathway for synthesis of this phospholipid in *L. dumoffii*. Structurally different PC species were distributed in the outer and inner membranes. As shown by the LC/ESI-MS analyses, PC15:0/15:0, PC16:0/15:0, and PC17:0/17:1 were identified in the outer membrane and PC14:0/16:0, PC16:0/17:1, and PC20:0/15:0 in the inner membrane. *L. dumoffii pcsA* gene encoding phosphatidylcholine synthase revealed the highest sequence identity to *pcsA* of *L. bozemanae* (82%) and *L. longbeachae* (81%) and lower identity to *pcsA* of *L. drancourtii* (78%) and *L. pneumophila* (71%). The level of TNF-α in THP1-differentiated cells induced by live and temperature-killed *L. dumoffii* cultured on a medium supplemented with choline was assessed. Live *L. dumoffii* bacteria cultured on the choline-supplemented medium induced TNF-α three-fold less efficiently than cells grown on the non-supplemented medium. There is an evident effect of PC modification, which impairs the macrophage inflammatory response.

Keywords: LC/ESI-MS; MALDI-TOF; *Legionella dumoffii*; phosphatidylcholine; TNF-α

1. Introduction

Legionella are Gram-negative bacilli that are highly successful in colonizing natural and artificial aquatic environments. Dissemination in the environment is facilitated by their characteristic biphasic lifestyle; *Legionella* may adapt to distinct intracellular and aquatic environments by alternating between a "replicative" and a "virulent" form in response to growth conditions. In water systems, *Legionella* infects and replicates within protozoa, colonize surfaces, and grow in biofilms [1]. Bacteria enclosed in water-air aerosol are inhaled into the lower respiratory tract and subsequently engulfed by enteric pulmonary macrophages. The capability of *Legionella* of intracellular proliferation in immune cells designed to kill bacteria and using them as their host cell is crucial for development of pneumonia known as Legionnaires' disease. Currently, the family *Legionellaceae* is composed of 58 species isolated from environmental sources, but 21 of them have been isolated from humans [2,3]. Among *Legionella* species that cause human pneumonia, *L. pneumophila* is the most common causative agent, while *L. dumoffii* is the fourth [4]. Pneumonia caused by *L. dumoffii* is rapidly progressive and fulminant owing to the ability of this bacterium to invade and proliferate in human alveolar epithelial cells [5]. The disease is often fatal, especially in immunocompromised patients. *L. dumoffii* has been isolated from pericarditis, prosthetic valve endocarditis, and septic arthritis, which indicates that *L. dumoffii* is responsible for extrapulmonary infections [6,7].

The way of bacterial penetration into the host cell and factors indispensable for settling the specific microniche, *i.e.*, the digestive vacuole, are dependent on virulence factors released from the cell, unique properties of surface components, and the ability to utilize host metabolites. Crucially for the biogenesis and maintenance of the bacterial replicative vacuole, *L. pneumophila* uses a type IV secretion system (Dot/Icm) to deliver a large number of effector proteins to the host cell [8]. The coordinate actions of the bacterial effectors allow *Legionella* to subvert innate immune response and evade host destruction. Moreover, the components of the cell envelope: proteins, peptidoglycan, lipopolysaccharide (LPS), and phospholipids participate in the highly specific *Legionella*-host interactions. *Legionella* phospholipids are characterized by a high phosphatidylcholine (PC) content, which is untypical of bacteria and specific for a narrow group of pathogenic and symbiotic microbes whose life cycle is strictly associated with eukaryotic cells. PC is a major component of eukaryotic cell membranes and plays a significant role in signal transduction. The high PC content in intracellular membranes of pathogens, such as *Legionella*, makes the cells of the microbes similar to the host cells. Bacteria synthesize PC via two different routes, *i.e.*, the phospholipid *N*-methylation (Pmt) or the phosphatidylcholine synthase (Pcs) pathway. In the Pmt pathway, phosphatidylethanolamine is methylated three times to yield PC in reactions catalyzed by one or several phospholipid *N*-methyltransferases (PMTs). In the Pcs pathway, choline is condensed directly with CDP-diacylglyceride to form PC in a reaction catalyzed by a bacterium-specific Pcs enzyme [9]. Since choline is not a biosynthetic product of prokaryotes, the Pcs pathway is probably a direct sensor of environmental conditions, using choline availability as an indicator of the status of the location in which the bacterium is found. It has been shown that *L. pneumophila* and *L. bozemanae* are capable of utilisation of exogenous choline for PC synthesis [10,11]. Apart from the structural role, the exact function of PC in *Legionella* cells remains unexplained. However, PC-deficient mutants of *L. pneumophila* exhibited attenuated virulence and increased susceptibility to macrophage-mediated killing. These defects were attributed to reduced bacterial binding to macrophages and a poorly functioning Dot/Icm system. In the process of binding to macrophages, *L. pneumophila* uses the platelet-activating factor receptor (PAF receptor), which harbors the same glycerophosphocholine head group as PC. Due to this structural similarity, PC is required for efficient binding of *L. pneumophila* to macrophages via the PAF receptor [12].

In response to infection caused by *Legionella*, macrophages produce inflammatory cytokines, such as interleukin 6 (IL-6), interleukin 1α (IL-1α), interleukin 1β (IL-1β), interleukin 12 (IL-12), interferon γ (INF γ), and tumor necrosis factor α (TNF-α) [13]. Among these cytokines, TNF-α appears to be pivotal for activation of phagocytes and resolution of pneumonic infection [14,15]. Treatment of rat alveolar macrophages with TNF-α resulted in decreased intracellular growth of *L. pneumophila* [16]. In turn, inhibition of endogenous TNF-α activity via TNF-α-neutralizing antibodies resulted in enhanced growth of *L. pneumophila* in the mouse lung [17]. However, little is known about how the pathogen PC influences the induction of proinflammatory cytokines in the host.

The aim of our study was to investigate the ability of *L. dumoffii* to utilize exogenous choline for PC synthesis using Matrix-Assisted Laser-Desorption/Ionization (MALDI)-Time of Flight (TOF) MALDI/TOF and Liquid Chromatography Coupled with the Mass Spectrometry Technique Using the Electrospray Ionization Technique (LC/ESI-MS) techniques. We also wanted to determine whether the bacteria use the Pcs pathway for PC synthesis by identification of a *pcsA* gene encoding phosphatidylcholine synthase. Next, the correlations between the PC species content in the *L. dumoffii* membranes and the level of TNF-α produced by human macrophages were investigated.

2. Results

2.1. Utilization of Exogenous Choline by L. dumoffii

In order to study the ability of *L. dumoffii* to use exogenous choline for PC synthesis, the bacteria were grown on BCYE medium supplemented with deuterium-labeled choline. Lipids isolated from the cells were analyzed using MALDI-TOF mass spectrometry with a CHCA matrix (in the reflectron mode). The positive ionization MALDI-TOF spectrum contained a cluster of molecular ions at *m/z*: 692.52–785.65 corresponding to PC species (Figure 1). Within the cluster, protonated ions were identified at *m/z* 692.52 PC[29:0 + H]$^+$, 706.53 PC[30:0 + H]$^+$, 720.55 PC[31:0 + H]$^+$, 734.55 PC[32:0 + H]$^+$, 746.56 PC[33:1 + H]$^+$, 762.53 PC[34:0 + H]$^+$, 776.55 PC[35:0 + H]$^+$, and corresponding ions with mass higher by 9 at *m/z* 701.50 d_9-PC[29:0 + H]$^+$, 715.58 d_9-PC[30:0 + H]$^+$, 729.60 d_9-PC[31:0 + H]$^+$, 743.58 d_9-PC[32:0 + H]$^+$, 755.60 d_9-PC[33:1 + H]$^+$, 771.55 d_9-PC[34:0 + H]$^+$, 785.65 d_9-PC[35:0 + H]$^+$. Deuterium-labeled PC[31:0 + H]$^+$ was observed as the dominant constituent of the extract. The presence of labeled PCs in the MALDI-TOF spectrum showed that *L. dumoffii* was able to utilize exogenous choline for PC synthesis.

Figure 1. Positive ion mode MALDI-TOF spectrum of the PCs of *L. dumoffii* cultured on the medium with labeled choline: m/z = 692.52–785.65. All peaks were marked according to their m/z ratio.

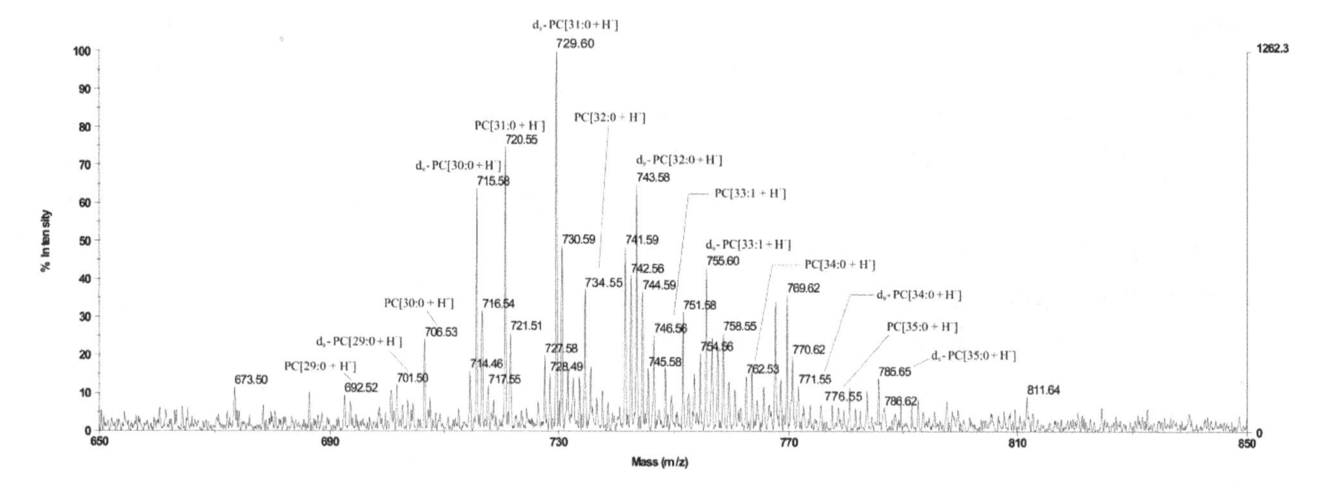

2.2. Membrane Localization of PC

2.2.1. Membrane Fractionation

To determine PC localization in the *L. dumoffii* membranes, separation of outer and inner membranes from bacteria cultured with and without labeled choline was performed by sucrose density gradient ultracentrifugation. The protein concentration in the individual fractions showed two major peaks with maxima in fractions 20–22, 31, and 32 for the choline non-supplemented bacteria. In the case of choline-supplemented bacteria, the maxima were detected in fractions 19–23 and 33–35 (Figure 2).

Figure 2. Separation of *L. dumoffii* inner (IM) and outer membrane (OM) by discontinuous sucrose density gradient centrifugation. Fractions of 1 mL were collected from the top of the gradient and assayed for the presence of protein (μg/μL). Black line—for bacteria cultured on medium non-supplemented with choline; Blue line—for bacteria cultured on medium supplemented with choline.

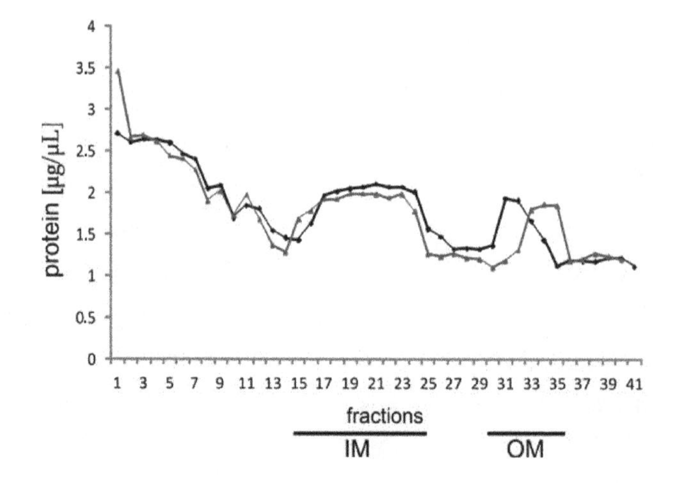

To assign the peaks to a membrane compartment, NADH oxidase and esterase activities (as characteristic for the inner and outer membrane, respectively) were determined. NADH oxidase activity was concentrated in pooled fractions corresponding to the inner membrane (fractions 16–24 for choline non-supplemented bacteria and 15–24 for bacteria cultured with choline). NADH oxidase activity was 228 μmol $min^{-1} \cdot mL^{-1}$. Esterase activity was concentrated in pooled fractions represented by the outer membrane (30–34 for bacteria cultured without choline and 32–35 for choline-supplemented bacteria). The NADH oxidase and esterase activities detected confirmed the efficiency of fractionation.

2.2.2. Localization of PC in the Outer and Inner Membranes by LC/ESI-MS Analysis

In the *L. dumoffii* membranes (inner and outer), unlabeled and labeled PC species were identified by the LC/ESI-MS technique using the precursor ion mode (PI) and neutral loss scan (NL) in the positive ion mode. The characteristic fragment of PC compounds at *m/z* 184 corresponding to the polar head group ($C_5H_{15}NPO_4^+$) was employed to determine unlabeled PC using the PI mode, while d_9-PC compounds were identified by a diagnostic ion of the head group at *m/z* 193 (Figure 3).

Figure 3. Chromatogram MS obtained with the Precursor Ion Mode for unlabeled (blue; *m/z* 184) and labeled (red; *m/z* 193) PC molecular species, recorded under NP LC/MS conditions (phase A—hexane/isopropanol (3:2, *v/v*) and phase B—an isopropanol/hexane/5 mM aqua solution of ammonium acetate (38:56:5, *v/v/v*). The following elution program was employed: from 53% B to 80% B for 23 min, 80% B maintained for 4 min, 80% B to 100% B for 9 min, and 100% B maintained for 14 min. The flow rate was 1 mL/min) in the positive ion MS mode; the lipid extract was isolated from the inner cell membrane of *L. dumoffii* cultured on labeled choline.

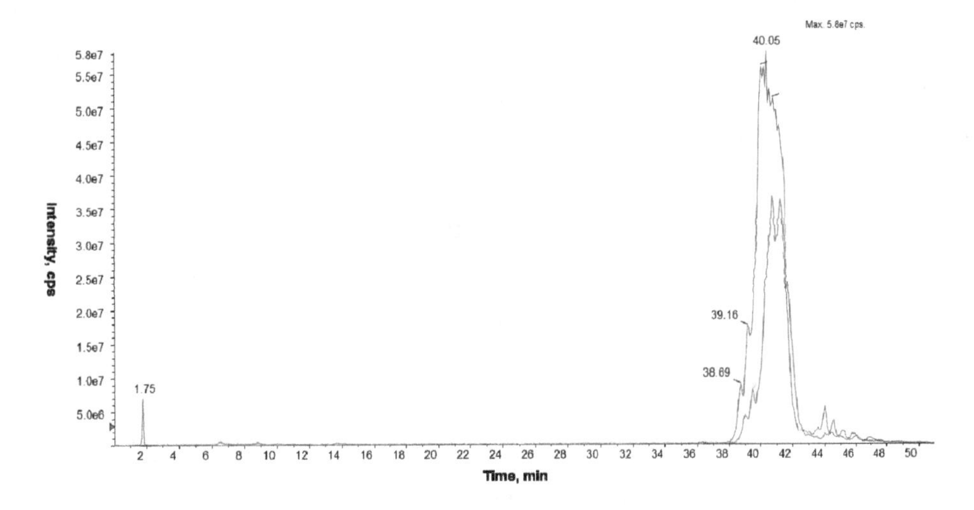

In the MS spectra of the analyzed PC compounds, $[M + H]^+$ and $[M + Na]^+$ ions were observed in the positive ion mode while in the negative ion mode—peaks corresponding to $[M + CH_3COO]^-$ adducts.

Fragmentation of the PC standard (1,2-dipalmitoyl-sn-glycero-3-phosphocholine; Figure 4) showed that the fragmentation of the sodiated PC molecule gave more information about the structure of the investigated compound. In the CID spectrum of the $[M + Na]^+$ ion of the PC standard peaks at *m/z* 441.0, 478.2, and 500.2 reflected the presence of palmitic acid in the PC molecule.

Figure 4. CID spectrum of the PC standard (1,2-dipalmitoyl-sn-glycero-3-phosphocholine) of protonated (top) and sodiated (below) molecules, recorded under NP LC/MS conditions (phase A—hexane/isopropanol (3:2, *v*/*v*) and phase B—an isopropanol/hexane/5 mM aqua solution of ammonium acetate (38:56:5, *v*/*v*/*v*). The following elution program was employed: from 53% B to 80% B for 23 min, 80% B maintained for 4 min, 80% B to 100% B for 9 min and 100% B maintained for 14 min. The flow rate was 1 mL/min) in the positive ion MS mode.

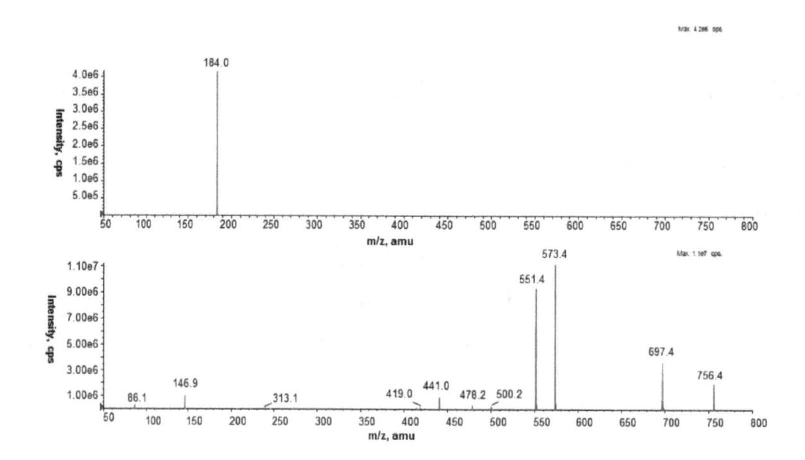

Next, neutral loss scanning was applied as an alternative method for the precursor ion mode to identify sodiated PC species in the mixture of lipids isolated from *L. dumoffii* membranes. PC compounds were found in the positive ion mode by the loss of the polar head group with a molecular weight at 183 Da (unlabeled) and 192 Da (d_9-labeled). According to Hsu and Turk, this phenomenon occurs in two steps through elimination of trimethylamine (59 Da), leading to the loss of $C_5H_{14}NPO_4$ (183 Da) from the PC structure [18]. Using the Precursor Ion and Neutral Loss scanning method, numerous PC species were found in the lipid extract obtained from *L. dumoffii* cell membranes. The PI mode seems to be more sensitive than the NL scanning technique for both unlabeled and labeled PC species. On the other hand, the NL method facilitated identification of sodiated PC molecular species in the mixture, which gave more informative CID spectra in the positive ion mode in the fragmentation process.

The presence of unlabeled and labeled PC molecules in *L. dumoffii* cell membranes was also confirmed by the CID spectra of sodiated PC ions obtained in the positive ion mode. For bacteria cultured on the medium without labeled choline, the major PCs with the following composition of fatty acids: 16:0/15:0, 17:0/15:0, and 16:0/17:1 (or cyclic 17:0, indistinguishable under the present mass spectrometry conditions) were determined in the inner membrane on the basis of the fragmentation spectra of peaks at *m*/*z* 742.4, 756.4, and 768.4 (Figure 5a). In the outer membrane, the major species of PC were diacyl 15:0/15:0, 14:0/16:0 (*m*/*z* 728.4), 16:0/17:1 (or cyclic 17:0) (*m*/*z* 768.8), and 17:0/17:1 (or cyclic 17:0) (*m*/*z* 782.8) (Figure 5b). In the CID spectrum of 16:0/15:0 PC shown in Figure 5a, loss of trimethylamine (59 Da) was observed. The peaks at *m*/*z* 559.4 and 537.4 correspond to the elimination of the PC head group giving ions $[M + Na - 183]^+$ and $[M + Na - 205]^+$, respectively. The fragment ions reflecting the presence of pentadecanoic (15:0) and palmitic (16:0) acids in the PC structure were found at *m*/*z* 405.2, 427.2, 464.2, 478.2, and 500.3. The fatty acids were eliminated as neutral molecules (*m*/*z* 427.2) or as sodium salts (*m*/*z* 405.2, 464.2, 478.2) from the $[M + Na]^+$ and $[M + Na - N(CH_3)_3]^+$ ions in the fragmentation pathway of PC molecular species in the positive ion mode.

Figure 5. (**a**) CID spectrum of unlabeled 16:0/15:0 PC molecular species identified in the lipid mixture isolated from the inner membrane (*L. dumoffii* cultivated on the medium without labeled choline), recorded under NP LC/MS conditions (phase A—hexane/isopropanol (3:2, *v*/*v*) and phase B—an isopropanol/hexane/5 mM aqua solution of ammonium acetate (38:56:5, *v*/*v*/*v*). The following elution program was employed: from 53% B to 80% B for 23 min, 80% B maintained for 4 min, 80% B to 100% B for 9 min and 100% B maintained for 14 min. The flow rate was 1 mL/min) in the positive ion MS mode; (**b**) CID spectrum of unlabeled PC molecular species with FA combination: 14:0/16:0 and 15:0/15:0 identified in the lipid mixture isolated from the outer membrane (*L. dumoffii* cultivated on the medium without labeled choline), recorded under NP LC/MS conditions (phase A—hexane/isopropanol (3:2, *v*/*v*) and phase B—an isopropanol/hexane/5 mM aqua solution of ammonium acetate (38:56:5, *v*/*v*/*v*). The following elution program was employed: from 53% B to 80% B for 23 min, 80% B maintained for 4 min, 80% B to 100% B for 9 min and 100% B maintained for 14 min. The flow rate was 1 mL/min) in the positive ion MS mode.

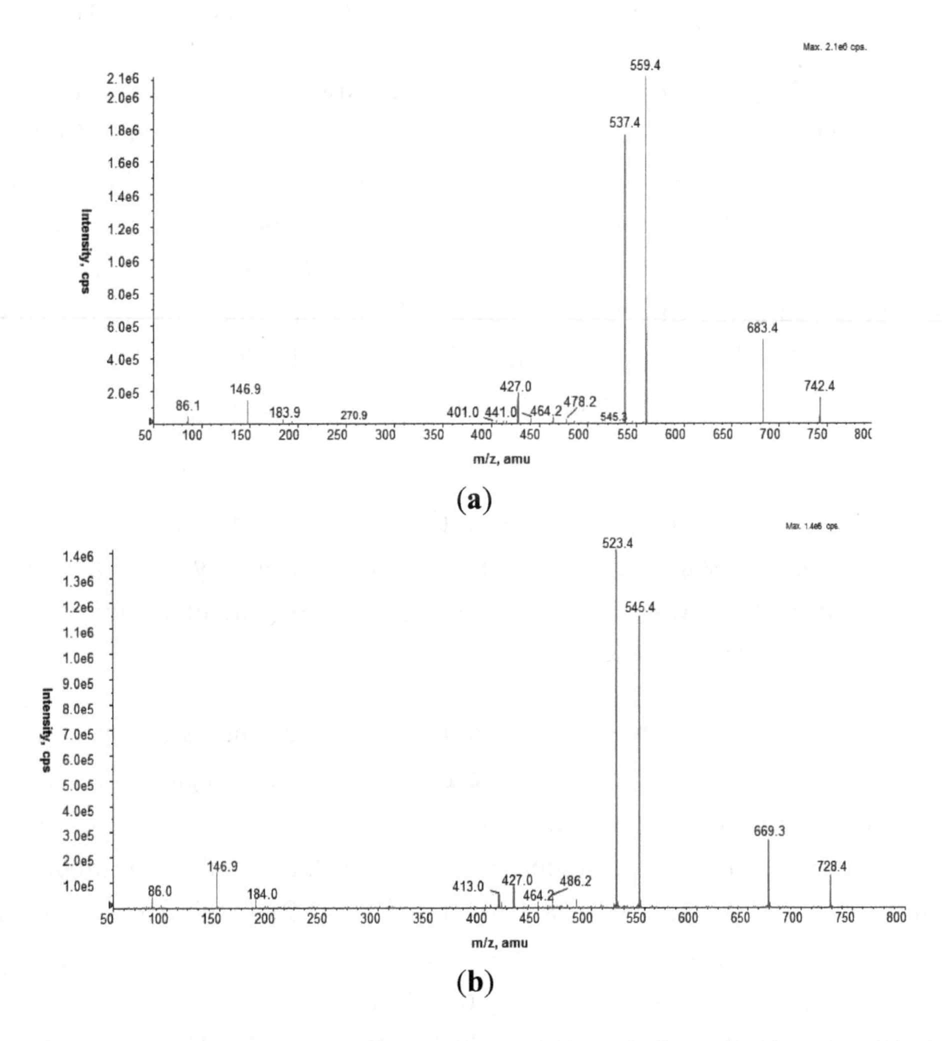

In the case of bacteria cultured on the medium with labeled choline, d_9-labeled PC molecular species were identified in the inner and outer membranes and their structures were established as shown in Table 1.

Table 1. The major d_9-labeled phosphatidycholine (PC) species determined by Liquid Chromatography Coupled with the Mass Spectrometry Technique Using the Electrospray Ionization Technique (LC/ESI-MS) in the inner and outer membrane of *L. dumoffii* cells.

m/z	PC structure, Sn-1/Sn-2 [a]	
	Inner membrane	**Outer membrane**
737.8	d_9-PC15:0/15:0	d_9-PC15:0/15:0
	d_9-PC14:0/16:0	
751.9	-	d_9-PC16:0/15:0
777.8	d_9-PC16:0/17:1	-
791.9	d_9-PC17:0/17:1	d_9-PC17:0/17:1
807.9	d_9-PC20:0/15:0	-
819.9	d_9-PC20:0/16:1	-

[a] 17:1 or cyclic 17:0, indistinguishable under the present mass spectrometry conditions.

Figure 6 presents the fragmentation spectra of individual d_9-labeled PC molecular species found in the inner and outer cell membranes recorded in the positive ion mode. In Figure 6a, the peaks at *m/z* 739.4, 615.4, and 593.4 corresponding to $[M + Na - N(CD_3)_3]^+$, $[M + Na - 192]^+$, and $[M + Na - 214]^+$ confirmed the presence of the deuterated PC head group in the PC compound. Loss of the 20:0 and 15:0 fatty acids as RCOOH and RCOONa was observed in the CID spectrum. A similar fragmentation process was also observed in the CID spectra shown in Figure 6b, where the 16:0/15:0 PC molecular species was determined. In addition to the labeled PCs, unlabeled PCs *m/z* 706.9, 720.8, 734.9, 748.7, 762.9, 776.7, and 790.7 were identified in the inner and outer membranes.

To compare the amount of the unlabeled PC phospholipid class in *L. dumoffii* cell membranes (outer and inner), the Multiple Reaction Monitoring (MRM) mass spectrometry mode was employed. The MRM mode is defined as a sensitive and a selective mass spectrometry technique. In the first step of the technique, an ion of interest was selected and subjected to the fragmentation process. Then, only predefined and characteristic fragment ions were detected.

In our studies, the MRM pairs were prepared on the basis of fragmentation spectra of peaks at *m/z* 764.9 and 778.9 corresponding to the most intense PC signals in the negative ion mode in both cell membranes. The $[M + CH_3COO]^-$ ion and the characteristic fragment ions $[M + CH_3COO - 74]^-$ and $[RCOO]^-$ of PCs obtained in the fragmentation process under the negative ion mode were chosen as MRM data (764.9/690.4, 764.9/241.2, and 778.9/704.4, 778.9/241.2). In this experiment, the MRM peak areas were compared and they showed higher values for the inner cell membranes (for the inner membrane: 764.9/690.4, peak area—1.02×10^6, 764.9/241.2—1.47×10^6; 778.9/704.4—2.08×10^6, 778.9/241.1—3.18×10^6; for the outer membrane: 764.9/690.4—4.85×10^5, 764.9/241.2—6.47×10^5, 778.9/704.4—8.57×10^5, 778.9/241.1—1.31×10^6). This indicated that the inner cell membrane seems to be richer in the PC phospholipid class.

Figure 6. (**a**) CID spectrum of labeled 20:0/15:0 PC molecular species identified in the lipid mixture isolated from the inner membrane (*L. dumoffii* cultivated on the medium with labeled choline), recorded under NP LC/MS conditions (phase A—hexane/isopropanol (3:2, *v*/*v*) and phase B—an isopropanol/hexane/5 mM aqua solution of ammonium acetate (38:56:5, *v*/*v*/*v*). The following elution program was employed: from 53% B to 80% B for 23 min, 80% B maintained for 4 min, 80% B to 100% B for 9 min and 100% B maintained for 14 min. The flow rate was 1 mL/min) in the positive ion MS mode; (**b**) CID spectrum of labeled 16:0/15:0 PC molecular species identified in the lipid mixture isolated from the outer membrane (*L. dumoffii* cultivated on the medium with labeled choline), recorded under NP LC/MS conditions (phase A—hexane/isopropanol (3:2, *v*/*v*) and phase B—an isopropanol/hexane/5 mM aqua solution of ammonium acetate (38:56:5, *v*/*v*/*v*). The following elution program was employed: from 53% B to 80% B for 23 min, 80% B maintained for 4 min, 80% B to 100% B for 9 min and 100% B maintained for 14 min. The flow rate was 1 mL/min) in the positive ion MS mode.

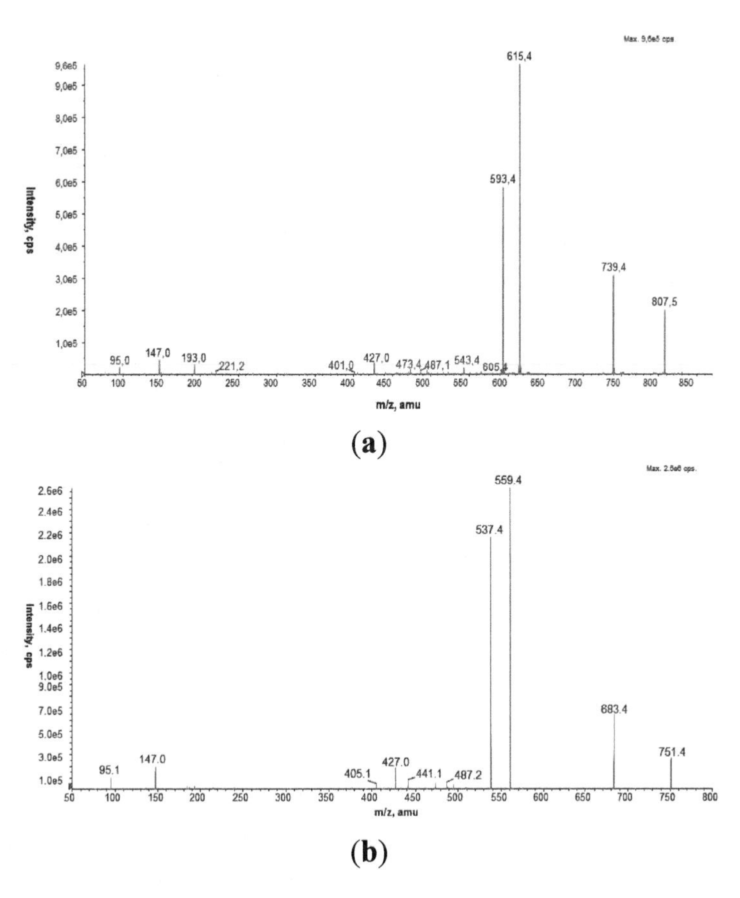

2.3. Identification of the pcsA Gene in the L. dumoffii Genome

Our biochemical analyses concerning PC synthesis in *L. dumoffii* indicated that this compound is produced in a one-step pathway, suggesting that a gene encoding phosphatidylcholine synthase is present in the genome of this bacterium. In order to identify the *pcsA* gene in *L. dumoffii*, Southern hybridization under low-stringency conditions was performed using genomic DNAs from *L. dumoffii*, *L. pneumophila*, and *E. coli* (as a negative control) digested with *Bam*HI and *Eco*RI. A DIG-labeled DNA fragment containing the *pcsA* of *L. pneumophila* was used as a probe (Figure 7). As a result, a homologue of *pcsA* was found in the *L. dumoffii* genome, although the intensity of detected signals was significantly lower in comparison to the signals obtained for *L. pneumophila*. The *pcsA* probe hybridised to a 24-kb-long *Bam*HI fragment and a 5-kb-long *Eco*RI fragment of *L. dumoffii.* In the case of *L. pneumophila*, strong positive signals were observed for the 24-kb *Bam*HI fragment and the 6-kb *Eco*RI fragment, respectively. On the contrary, only a slight unspecific signal was observed in hybridization with *E. coli* genomic DNA, confirming absence of a *pcsA* homolog in this genome (Figure 7).

Figure 7. Identification of the *pcsA* gene in the *L. dumoffii* genome by Southern hybridization using a probe containing the *pcsA* gene of *L. pneumophila*. Hybridization was performed with genomic DNAs of *L. dumoffii* (L.d.), *L. pneumophila* (L.p.), and *E. coli* (E.c.) digested with *Bam*HI and *Eco*RI as described in Materials and Methods. M—molecular size standard (λ DNA digested with *Hin*dIII and *Eco*RI).

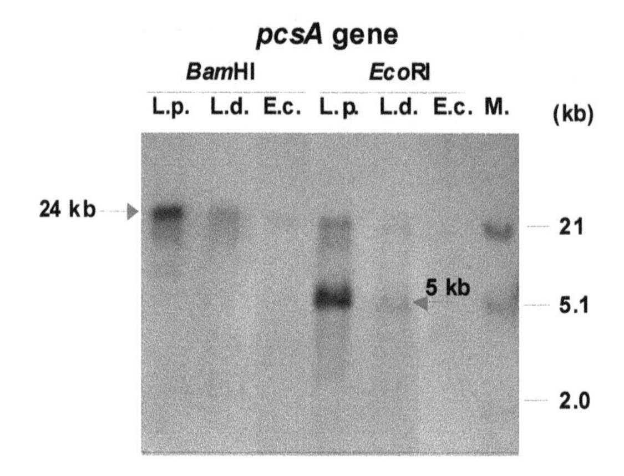

In conclusion, these results showed that the *L. dumoffii* genome contained a *pcsA* gene; however, the low intensity of the observed signal suggested a low sequence identity between *L. pneumophila pcsA* and *L. dumoffii* homolog. This observation is in agreement with literature data that described high genetic and phenotypic diversity between *L. pneumophila* and other *Legionella* species [19].

To determine the nucleotide sequence of *L. dumoffii pcsA*, a set of degenerate primers was designed based on genomic sequences of a few *L. pneumophila* strains, *L. longbeachae* D-4968 and *L. drancourtii* LLAP12. Using these primers, a 900-bp-long PCR product of *L. dumoffii* was obtained and sequenced. In the assessment of the genetic relatedness between *L. dumoffii pcsA* and this gene in other sequenced *Legionella* species, the highest sequence identity to *L. bozemanae pcsA* (82%) and *L. longbeachae* (81%) was found. *L. dumoffii pcsA* showed lower sequence identity to *pcsA* of *L. drancourtii* (78%) and *L. pneumophila* ATCC 33155 (71%). The comparison of the *pcsA* sequence indicated a high level of diversity among the *Legionella* species, especially in 5' and 3' ends of *pcsA* genes. *L. dumoffii pcsA* encodes putative 254-aa-long phosphatidylcholine synthase, whose length was identical or very similar to the other *Legionella* PcsA proteins: *L. drancourtii* (254 aa), *L. longbeachae* (253 aa), and *L. pneumophila* (255 aa). In the upstream of the *L. dumoffii pcsA* coding sequence (−135 to −86 bp), a promoter sequence of high potential activity (p = 0.99) was identified (5'-actatttttatgtttttatttttgataaaacaacaatagtttagccca-3'). In addition, 6 nt upstream of the ATG translation start codon, a potential ribosome-binding site (AGGA) was found. The PcsA of *L. dumoffii* showed the highest sequence identity to the *L. bozemanae* homolog (90% identity and 96% similarity). Lower identity of this protein was established to PcsA of *L. longbeachae* (87% identity and 94% similarity),

L. drancourtii (79% identity and 90% similarity), and *L. pneumophila* (74% identity and 84% similarity). The most conservative amino acid sequences were found in the central regions of PcsA proteins that are probably-functional domains of the enzyme, whereas their *N*- and *C*-termini are substantially more divergent.

2.4. The Cytotoxic Effect of L. dumoffii on THP-1 Cells

To determine the infectious doses of *L. dumoffii* for human macrophages (THP-1 cell line), the cytotoxic effect of different bacteria concentrations (expressed as the multiplicity of infection—MOI) was assessed with the MTT (3-(4,5 dimethyl-2-thiazolyl)-2,5-diphenyl-2H-tetrazolium bromide) method. The viability of the THP-1 differentiated cells incubated with *L. dumoffii* for 4 h at a MOI of 10, 50, 100, and 200 was found to be similar to that of the control cells (cell viability 103.05% ± 7.4%, 105.5% ± 8.0%, 101.6% ± 7.8%, 96.2% ± 7.2%, respectively). After 24-h incubation of THP-1 cells with the bacteria at a MOI of 100 and 200, a slight decrease in cell viability was observed (cell viability 99.2% ± 7.2%, 95.4% ± 7.4%, respectively). The analysis of the cytotoxic effect of *L. dumoffii* on THP-1 cells was also performed for the same doses of *L. dumoffii* cultured on choline-supplemented medium. No statistically significant differences of the toxic effect on the THP-1 cells were found between the bacteria cultured with and without choline. Based on these results and literature data [20,21], *L. dumoffii* at a MOI of 10 and 100 were chosen to study the TNF-α induction in the THP-1 differentiated cells.

2.5. TNF-α Induction by L. dumoffii in the THP-1 Differentiated Cells

2.5.1. TNF-α Induction in the THP-1 Differentiated Cells by *L. dumoffii* Cultured on Choline-Supplemented and Non-Supplemented Medium

Levels of human TNF-α in the supernatants from the THP-1 experimental cultures were measured by an enzyme-linked immunosorbent assay (ELISA) after 4-h incubation of cells with live and temperature-killed *L. dumoffii* bacteria at the MOI of 10 or 100. To study the influence of choline on TNF-α production, both live and temperature-treated bacteria were obtained from BCYE medium supplemented with choline and without this compound. Regardless of the culture conditions, dose-dependent production of TNF-α was observed in all the experiments. However, statistically significant differences in obtained results were only noted when we considered the experiments with live bacteria. Bacteria cultured on the choline-supplemented medium, regardless of their concentration, induced TNF-α production at significantly lower level ($p \leq 0.05$). However, inside study groups—live bacteria cultured with or without choline supplementation—level of TNF-α depended on bacteria concentration and was significantly higher when 100 MOI bacteria were used. The temperature-killed bacteria induced TNF-α at a considerably statistically significant lower level in comparison to live bacteria and presence of choline in the bacterial culture medium did not change the level of cytokine production (Figure 8).

Figure 8. Effects of live and temperature-treated *L. dumoffii* cells cultured on the choline-supplemented medium and without choline on the ability to induce TNF-α production in THP-1 differentiated cells. Data represent the mean ± S.D. of five independent experiments. Control THP-1 differentiated cells (PMA-activated): TNF-α 22.6 ± 0.6 (pg/mL); Control non-activated THP-1 cells: TNF-α 18.32 ± 0.9 (pg/mL). # Significantly different from the lower bacteria dose, $p \leq 0.05$; * Statistically significant in comparison to bacteria cultured on the choline non-supplemented medium, $p \leq 0.05$ (ANOVA and Tukey's *post hoc* test). [a,b] Statistically significant in comparison with similar experiments preformed with temperature-treated bacteria ([a] $p \leq 0.05$, [b] $p \leq 0.01$) c (ANOVA and Tukey's *post hoc* test) —live *L. dumoffii*; —temperature-treated *L. dumoffii*; —live *L. dumoffii* cultured on choline-supplemented medium; —temperature-treated *L. dumoffii* cultured on choline-supplemented medium.

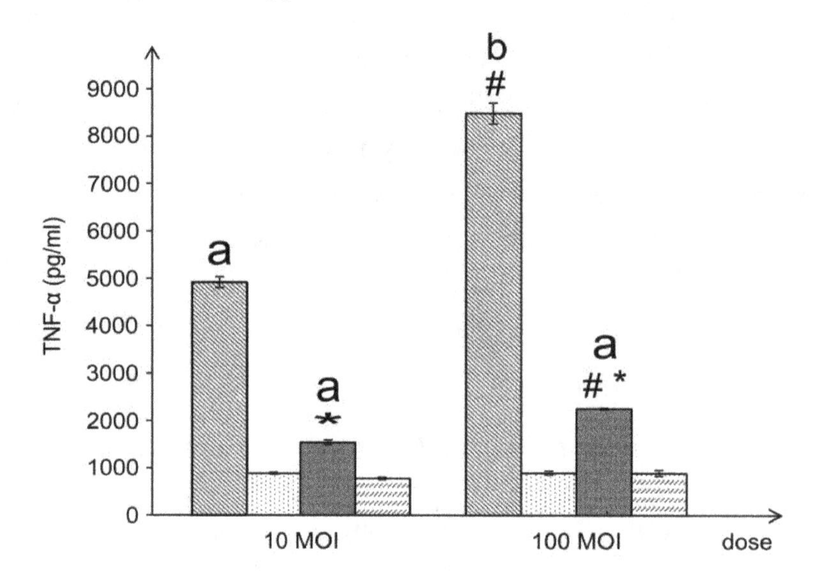

2.5.2. TNF-α Induction by the Outer and Inner Membrane of *L. dumoffii* in the THP-1 Differentiated Cells

Inner and outer membranes (10, 100, and 1000 ng/mL concentrations) isolated from *L. dumoffii* cultured with and without choline were used for TNF-α induction (Figure 9). Both the inner and outer temperature-treated membranes induced TNF-α production in each dose and in a dose-dependent manner more efficiently than the temperature non-treated membranes. A comparison of the ability of the membranes to induce TNF-α showed that the inner membrane induced TNF-α more efficiently than the outer membrane, irrespective of the doses used and temperature treatment or no treatment.

Choline supplementation influenced TNF-α production in a different manner depending on the membrane type. The inner membrane isolated from choline-supplemented bacteria induced TNF-α less efficiently than the inner membrane of bacteria cultured without choline. However, only in the presence of 100 ng/mL of these membranes, the decrease in the TNF-α production was statistically significant (Figure 9A,C). In turn, the same concentration of the outer membranes of choline-supplemented bacteria (both, non- and temperature-treated) caused opposite effect—slightly increased the level of TNF-α in comparison with the outer membranes from bacteria cultured without choline (Figure 9B,D).

Figure 9. TNF-α (pg/mL) induction in the THP-1 differentiated cells by the outer and inner membrane of *L. dumoffii* (**a**,**b**,**c**,**d**). Control THP-1 differentiated cells (PMA-activated): TNF-α 22.6 ± 0.60 (pg/mL); Control non-activated THP-1 cells: TNF-α 18.32 ± 0.90 (pg/mL). Data represent the mean ± S.D. of five independent experiments; #—Significantly different from the lower (10 ng/mL) membrane concentration, $p \leq 0.05$; ##—Significantly different from the lower (100 ng/mL) membrane concentration, $p \leq 0.05$; *—Statistically significant in comparison to bacteria cultured on the choline non-supplemented medium, $p \leq 0.05$ (ANOVA and Tukey's *post hoc* test). ■—membrane from bacteria cultured on the choline non-supplemented medium. ▒—membrane from bacteria cultured on the choline-supplemented medium. ■ —temperature-treated membrane from bacteria cultured on the choline non-supplemented medium. □—temperature-treated membrane from bacteria cultured on the choline-supplemented medium.

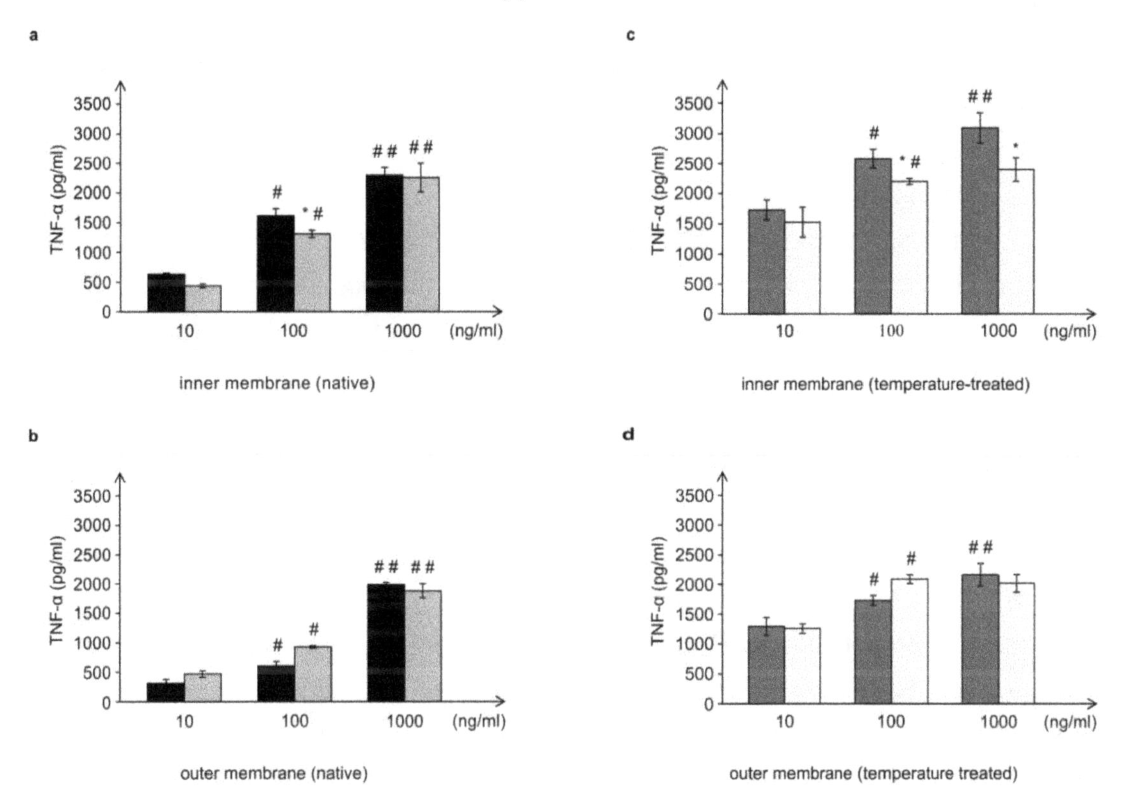

2.6. *In Vitro Infection of Differentiated THP-1 Cells with L. dumoffii*

To compare the levels of *L. dumoffii* internalization by differentiated THP1 cells, bacteria cultured on the medium with and without choline were incubated with macrophages for 2 h. Next, bacteria that were not phagocytised by macrophages were killed by gentamycin treatment. Only intracellular bacteria released after macrophage lysis were seeded onto the BCYE agar medium. The number of colonies was higher for bacteria cultured with choline before the internalization process (171.8 ± 4.2 for the choline-grown bacteria, 143.2 ± 3.8 for the non-choline grown bacteria). This investigation indicates that choline-grown bacteria undergo association by macrophages more readily than non-choline grown bacteria. However, the results were not statistically significant. *L. dumoffii* did not replicate in differentiated THP1 cells irrespective of the bacterial culture conditions (data not shown).

3. Discussion

Our previous and current investigations showed that *L. lytica*, *L. bozemanae*, and *L. dumoffii* form PC [11,22]. In this study, the ability of *L. dumoffii* to synthesize PC in a choline-dependent manner was investigated. It was evidenced that the bacteria used exogenous choline for PC synthesis via the Pcs pathway. The main PC d_9-PC[31:0 + H]$^+$, identified on the MALDI/TOF spectrum and synthesised via this pathway, was confirmed by LC/ESI-MS as PC 16:0/15:0, also identified in *L. bozemanae*. Both species exhibit blue-white fluorescence; and the presence of PC 16:0/15:0 is an additional chemotaxonomic feature that classifies both species as representatives of the genus *Fluoribacter* in the *Legionellaceae* family [23]. The Pcs pathway in *L. dumoffii* was confirmed by identification of the *pcsA* gene encoding the phosphatidylcholine synthase. All *Legionella* spp. genomes characterized so far contain the *pcsA* gene encoding phosphatidylcholine synthase, suggesting a significant function of this protein in their metabolism. Previously, we have identified the *pcsA* gene in *L. bozemanae* and found that PC is effectively produced in a one-step pathway, in which the PcsA enzyme is engaged [11]. Likewise, the *L. dumoffii* PcsA presented in this study and *L. bozemanae* PcsA exhibited the highest sequence amino acids identity to the *L. longbeachae* PcsA, which confirmed genetic relatedness of these three *Legionella* species.

The presence of unlabeled PC in the MALDI-TOF spectrum obtained from lipids of bacteria cultured on choline has indicated that *L. dumoffii* forms PC also by triple PE methylation. LC/ESI-MS analysis of PC species present in both membranes labeled with a deuterated precursor allowed us to distinguish PC species synthesized from the CDP-choline pathway and the PE methylation pathway. Our previous investigations showed that *L. bozemanae* produced PC via two independent pathways PmtA and Pcs [11]. Similarly, both PC synthesis pathways function in *L. pneumophila*. The pathway in which *Legionella* spp. utilize exogenous choline is dominant and seems to be more energetically efficient [10]. However, in the genetic experiments that were conducted (hybridization, PCR analyses using several degenerative primers and amplicon sequencing), we were unable to identify a *pmtA* homologue in the *L. dumoffii* genome (data not shown). Both PC synthesis pathways have also been reported in legume endosymbionts (*Rhizobium leguminosarum*, *Sinorhizobium meliloti*, *Bradyrhizobium japonicum*) and in the plant pathogen *Agrobacterium tumefaciens*. In human pathogens such as *Pseudomonas aeruginosa*, *Brucella melitensis*, and *Borrelia burgdorferi*, PC is produced only via the Pcs pathway [24].

Several studies have shown that the PC of bacterial membranes can be important to host-associated bacteria in pathogenesis and symbiosis. Our results have indicated that bacteria cultured on choline undergo internalization by THP1 macrophages more readily than bacteria cultured on non-choline-supplemented medium. The increase in internalization in the case of bacteria cultured on choline was not statistically significant, which may suggest a low level of participation of PC species in this process, although other papers indicate that some bacterial PC derivatives play a direct role in association with macrophages [25].

PC-deficient mutants of *B. japonicum* and *S. meliloti* were characterized by substantially reduced symbiosis with their plant hosts [26,27]. PC is indispensable for the plant pathogen *A. tumefaciens* to assembly T4SS components, important factors in formation of plant crown-gall tumors [28,29]. A *Brucella abortus pcs* mutant exhibited an altered cell envelope; therefore, it did not establish a replication niche inside the macrophages. Additionally, it showed a severe virulence defect in a murine model of infection [30,31].

A PC-deficient *P. aeruginosa* mutant exhibited the same level of sensitivity to antibiotics and antimicrobial peptides as wild strains. PC deficiency did not change the mobility and capability of biofilm formation on an abiotic surface. However, PC may have a specific role in the interaction with eukaryotic hosts, e.g., it might aid in assembly or localization of specific proteins in *P. aeruginosa* [32].

L. dumoffii cultured on choline-supplemented medium exhibited altered sensitivity to *Galleria mellonella* antimicrobial defense factors such as defensin and apoLp-III [33]. Replacement of PE with PC induced concurrent structural and functional changes in the ABC multidrug exporter of *Lactococcus brevis* [34].

The membrane fraction experiment and LC/ESI-MS analyses suggest that PC in *L. dumoffii* is present in the outer and inner membranes. In *L. pneumophila* and *P. aeruginosa*, PC was also localized in both membrane compartments [35,36]. MRM analysis showed that the amount of PC in the inner membrane of *L. dumoffii* was higher than in the outer membrane. Moreover, there were differences in the structure of the PC species present in bacteria cultured with and without choline. These differences might be important for interactions of bacteria with host cells.

Several lines of evidence indicate that PC is able to modulate the inflammatory functions of monocytic cells. Tonks *et al.* showed that PC 16:0/16:0 significantly inhibited TNF-α release from the human monocytic cell line MonoMac-6 in a dose-dependent manner. In contrast, PC 20:4/16:0 did not reduce TNF-α, which indicated that regulation of the inflammatory response was associated with the composition of fatty acids forming PC [37]. In our study, live *L. dumoffii* bacteria cultured on the choline-supplemented medium, induced three times less TNF-α than the cells grown on the non-supplemented medium. Similarly, the inner (temperature-treated and -untreated) membranes isolated from bacteria cultured on choline, applied at almost all the doses, induced a decreased level of TNF-α, in comparison to bacteria cultured without choline. Bacteria grown on choline incorporate into their inner membranes PC species with longer fatty acids than bacteria cultured without choline, which may have a significant effect on the lower induction of TNF-α level. In comparison to live bacteria, temperature-treated bacteria induced significantly lower TNF-α level. It is connected with the fact, that in live Legionella HSP, flagellin, and LPS are the main inducer of TNF-α. Temperature heating (90 °C, 20 min) causes HSP and flagellin degradation, therefore still present LPS is only inducer of this cytokine. This may be similar to the case of *L. pneumophila* LPS [38], although it is not known what the efficiency of *L. dumoffii* LPS in the induction of TNF-α is, since no such investigations have been carried out and the structure of *L. dumoffii* LPS is not known.

In the case of the 10- and 100-ng/mL concentrations of the outer membrane of live bacteria cultured on choline, TNF-α induction was higher than in the case of bacteria without choline; however, these differences were not statistically significant. It has been shown that lipopolysaccharide (LPS) of Gram-negative bacteria, a well-known TNF-α inducer located in the outer membrane, had an influence on the level of the cytokine as well. However, *L. pneumophila* LPS is about 1000 times less potent in its ability to induce pro-inflammatory cytokines (TNF-α, IL-1β, IL-6, IL-8) in Mono Mac 6 cells than the LPS of the *Enterobacteriaceae* members [21]. Cao *et al.* showed that addition of PC18:2/18:2 into the culture of Kupffer cells significantly reduced LPS-stimulated TNF-α generation [39]. We did not observe such an effect. Probably, the PC structure has a significant impact on the level of induction of this cytokine.

4. Experimental Section

4.1. Bacterial Strain and Growth Conditions

L. dumoffii strain ATCC 33279, *L. pneumophila serotype 3* ATCC 33155 were cultured on buffered charcoal-yeast extract (BCYE) agar plates, which contained a *Legionella* CYE agar base (Oxoid, Basingstoke, Hampshire, UK) supplemented with the Growth Supplement SR0110A (ACES buffer/potassium hydroxide, ferric pyrophosphate, L-cysteine HCl, α-ketoglutarate; Oxoid, Basingstoke, Hampshire, UK) for three days at 37 °C in a humid atmosphere and 5% CO_2 [40]. *L. dumoffii* were also cultivated on this medium enriched with 100 $\mu g \cdot mL^{-1}$ of choline-trimethyl-d_9 chloride (Sigma-Aldrich, Steinheim, Germany). The bacteria collected from this medium were washed three times with water by intensive vortexing and centrifugation at 8000× *g* for 10 min. Bacteria cultured with and without choline were killed by 90 °C for 20 min. The efficiency of temperature inactivation of bacteria was checked by streaking the bacteria on BCYE plates. *Escherichia coli* DH5α strain was grown in Luria-Bertani (LB) medium at 37 °C [41].

4.2. Fractionation of L. dumoffii Cultured on BCYE Medium with and without Choline

The bacteria were collected from 12 BCYE plates supplemented with labeled choline and 12 non-supplemented BCYE plates. They were washed twice in saline in order to remove the remaining medium. Cell membrane isolation was performed essentially as described by Hindahl and Iglewski [35].

The cells were washed twice with cold 10 mM HEPES (*N*-2-hydroxyethylpiperazine-*N*-2-ethanesulfonic acid; Sigma-Aldrich, Steinheim, Germany) buffer (pH 7.4) and centrifuged at 8000× *g*, 15 min in 4 °C. The cell pellets were suspended in 15 mL of 10 mM HEPES buffer containing 20% sucrose (*w*/*v*) and incubated with DNase (0.3 mg) (Sigma-Aldrich, Steinheim, Germany) and RNase (0.3 mg) (Sigma-Aldrich, Steinheim, Germany) at 37 °C for 30 min 0.8-mL suspensions were lysed by three passages through a French press (SLM-Amico Instruments, Thermo Spectronic, Rochester, NY, USA) at 18000 lb/in^2. After centrifugation for 20 min at 1000× *g*, 4 °C performed to remove cell debris and undisrupted cells, total membrane fractions were collected by centrifugation for 60 min at 100,000× *g*, 4 °C (SW 32Ti rotor, Beckman Coulter, Brea, CA, USA). Next, they were washed twice in cold 10 mM HEPES buffer by centrifugation at 100,000× *g* for 1 h, 4 °C, the pelleted membranes were suspended in 2.5 mL of 10 mM HEPES buffer and layered onto a seven-step sucrose gradient.

4.3. Isolation of the Outer and Inner Membranes

Two milliliters of each cell membrane suspension were loaded on the top of a discontinuous gradient prepared by combining sucrose solutions of the following concentrations: 6 mL 70%, 9 mL 64%, 8 mL 58%, 5 mL 52%, 4 mL 48%, 3mL 42%, 3 mL 36% (*w*/*v*) in 10 mM HEPES buffer, pH 7.5. The gradient was centrifuged at 114,000× *g*, 20 h, 4 °C (SW 32Ti rotor, Beckman Coulter, Brea, CA, USA), and 1 mL fractions were subsequently collected from the top of the centrifugation tube. The protein content was determined in the individual fractions and the fractions from the upper band of the gradient and the lower band of the gradient were collected separately. Next, the upper and lower fractions pooled separately were suspended in 10 mM HEPES buffer and centrifuged at 54,000× *g* for 1 h (MLA 80, Optima MAX-XP, Beckman Coulter, Brea, CA, USA). The membrane pellets were washed three times in cold deionized water and centrifuged at 54,000× *g* for 1 h (MLA 80, Optima MAX-XP, Beckman Coulter, Brea, CA, USA). Next, they were diluted in 250 μL of water (MQ, Millipore, Billerica, MA, USA) for further analyses, *i.e.*, enzyme assays and lipid isolation.

The efficacy of the membrane isolation procedure was confirmed by measuring the activity of NADH oxidase [42] and esterase. The concentration of protein in the fractions was determined using the Bradford method and bovine serum albumin as a standard [43].

4.4. Enzyme Assays

As marker enzymes for the inner and outer membrane, activities of NADH oxidase and esterase, respectively, were determined by spectrophotometric analysis of the absorbance decrease at 340 nm for NADH oxidase and the absorbance increase at 405 nm for esterase. For NADH oxidase, 62 μL of the reaction mixture (50 mM Tris/HCl, pH 7.5, 0.2 mM DTT, 0.12 mM NADH) (Sigma-Aldrich, Steinheim, Germany) was incubated for 15 min at 37 °C with 8 μL of each fraction in a microtitre plate. For esterase activity, 63 mg of *p*-nitrophenyl acetate (Sigma-Aldrich, Steinheim, Germany) was dissolved in 10 mL of ethanol. One milliliter of this solution was slowly added to 100 mL of distilled water. Ten microliters of each fraction were added to 90 μL of the substrate solution, incubated for 10 min at 25 °C, and absorbance at 405 nm was recorded. Fractions with the NADH oxidase and esterase activities representing the inner and outer membrane fractions were lyophilized and weighed.

4.5. Isolation of Lipids

Lipids were isolated from bacterial cells cultivated on BCYE medium enriched with labeled choline using the Bligh and Dyer (1959) method: chloroform/methanol (1:2 *v*/*v*) [44]. Lipids were analyzed with MALDI/TOF mass spectrometry.

Lipids from the inner and outer membranes prepared from bacteria grown on the choline-supplemented and non-supplemented medium were extracted using the same extraction method. The dried organic phase was then purified with a mixture of hexane/isopropanol (3:2, *v*/*v*). The extracts were dried under nitrogen before weighing and then dissolved in chloroform for further LC/ESI-MS analysis.

4.6. Matrix-Assisted Laser-Desorption/Ionization (MALDI)-Time of Flight (TOF) Mass Spectrometry

MALDI-TOF mass spectrometry analysis was performed on a Voyager-Elite instrument (PE Biosystems, Foster City, CA, USA) using delayed extraction in the reflectron mode. The dry lipid extract was dissolved in a $CHCl_3/CH_3OH$ mixture (2:1, *v*/*v*). The sample constituents mixed with 0.5 M aniline salt of α-cyano-4-hydroxycinnamic acid (CHCA) as a matrix were desorbed and ionized with a nitrogen laser at an extraction voltage of 20 kV. Angiotensin was used as an internal standard. Each spectrum was the average of about 256 laser shots.

4.7. Liquid Chromatography Coupled with the Mass Spectrometry Technique Using the Electrospray Ionization Technique (LC/ESI-MS)

The LC/ESI-MS analyses were performed on a Prominence *LC-20* (Shimadzu, Kyoto, Japan) liquid chromatograph coupled with a tandem mass spectrometer 4000 Q TRAP (Applied Biosystems Inc., Foster City, CA, USA), equipped with an electrospray (ESI) ion source (TurboIonSpray, Applied Biosystems Inc., Foster City, CA, USA) and a triple quadrupole/linear ion trap mass analyzer. The separation of the phospholipid mixture was carried out by normal phase chromatography using a 4.6 × 150 mm Zorbax SIL RX column (Agilent Technologies, Palo Alto, CA, USA). Hexane/isopropanol (3:2, *v*/*v*) was used as solvent A and an isopropanol/hexane/5 mM aqua solution of ammonium acetate (38:56:5, *v*/*v*/*v*) as solvent B. The following elution program was employed: from 53% B to 80% B for 23 min, 80% B maintained for 4 min, 80% B to 100% B for 9 min and 100% B maintained for 14 min. The flow rate was 1 mL/min. The lipid extracts were dissolved in phase A.

To identify PC species, the precursor ion mode (PI) and neutral loss scan (NL) mass spectrometry techniques were employed in the positive ion mode. The measurements were performed using an electrospray ion source (ESI) with the following parameters: ion spray voltage (IS)—5500 V, declustering potential (DP)—40 V, and entrance potential (EP)—10 V. Nitrogen was used as a curtain (CUR 20 psi), a nebulizer (GS1 50 psi), and a collision gas. The source temperature was set at 250 °C. The collision-induced dissociation (CID) spectra were obtained in the positive and the negative ion mode using collision energy of 50 eV.

The Multiple Reaction Monitoring (MRM) technique performed in the negative ion mode was employed to compare the PC amounts in the analyzed cell membranes. The most abundant PC molecular species at *m*/*z* 778.9 (16:0/15:0) and 764.9 (15:0/15:0) were chosen in this experiment. Based on the fragmentation spectra of the selected peaks in the negative ion mode, the following MRM pairs were used: 764.9/690.4, 764.9/241.2, and 778.9/704.4, 778.9/241.2.

Chemicals for the LC/MS system (hexane, isopropanol, acetonitrile LC grade) were obtained from Merck (Darmstadt, Germany), and ammonium acetate from Sigma-Aldrich Fluka (Steinheim, Germany).

4.8. DNA Methods, PCR Amplification, and Southern Hybridization

Restriction enzyme digestion, agarose gel electrophoresis, DNA labeling, and Southern hybridization were used for genomic DNA isolation [41]. To amplify the DNA fragment containing *pcsA* of *L. pneumophila* serotype 3 (ATCC 33155), the PCR reaction was performed using 100 ng of genomic DNA, REDTaq ReadyMix (Sigma-Aldrich, Steinheim, Germany) and 0.2 µM of each forward (pcsF: 5'-CTCTAGGATCCGTAATGAATCCAATAAA-3') and reverse (pcsR: 5'-CATAAATTGG ATCCAAACTCAATCTTTATTAT-3') primers in a 50-µL final volume. The PCR reaction was performed with the following temperature profile: initial denaturation at 94 °C for 4 min, 30 cycles of denaturation 94 °C for 1 min, annealing at 48 °C for 40 s, extension at 72 °C for 60 s, and final extension at 72 °C for 4 min. This PCR fragment was labeled using the non-radioactive DIG DNA Labeling and Detection kit according to the manufacturer's instruction (Roche Applied Science, Penzberg, Germany). The 800-bp-long DIG-labeled amplicon with the *pcsA* gene was used as a probe in the Southern hybridization. Additionally, DIG-labeled DNA of phage λ digested with *Eco*RI and *Hin*dIII restriction enzymes were used as a molecular size marker. Genomic DNA from *L. pneumophila* serotype 3, *L. dumoffii*, and *E. coli* DH5α (used as a negative control) were digested with *Bam*HI and *Eco*RI enzymes, separated by 0.7% agarose gel electrophoresis, and blotted. In addition, λ DNA digested with both *Eco*RI and *Hin*dIII enzymes was used as molecular markers. The hybridization experiments were performed at reduced-stringency conditions at 37 °C using 20% formamide in pre-hybridization and hybridization solutions as described previously [11].

To amplify and sequence the *L. dumoffii pcsA* gene, a set of degenerate primers has been designed based on the genomic sequences of *L. pneumophila* strain Philadelphia 1, *L. longbeachae* D-4968, and *L. drancourtii* LLAP12. Primers pcsLd-F (5'-ACTTTTKATWATYGATRMTATTTT-3') and pcsLd-R (5'-TAATCATWAAADABYCAAAGTCTAT-3') allowed amplification of the longest 900-bp fragment containing *pcsA* gene. The PCR reactions were performed with the temperature profile described above, except the annealing temperature that was decreased to 43 °C. The amplified DNA fragment was purified on the columns (A&A Biotechnology, Gdynia, Poland) and sequenced using the BigDye terminator cycle sequencing kit (Applied Biosystems, Inc., Foster City, CA, USA) and the ABI Prism 310 sequencer. Database searches were done with the BLAST and FASTA programs available at the National Center for Biotechnology Information (Bethesda, MD, USA) and the European Bioinformatics Institute (Hinxton, UK). The sequence of *L. dumoffii pcsA* obtained in this study has been deposited in NCBI GenBank under accession number KC197708. Promoter prediction in the upstream region of *L. dumoffii pcsA* was done using BDGP Neural Network Promoter Prediction (fruitfly.org).

4.9. THP-1 Cell Culture

The human acute monocytic leukemia *cell line* (THP-1) was obtained from the European Collection of Cell Cultures (Cat No. 88081201). Cells were cultured at (0.5–7) × 10^5 cells/mL in RPMI 1640 supplemented with 10% heat-inactivated fetal calf serum (FCS), 10 mM HEPES, 2 mM glutamine, 100 U/mL penicillin, and 100 μg/mL streptomycin. The cells cultured in tissue culture flasks (Falcon, Bedford, MA, USA) were incubated at 37 °C in a humidified atmosphere of 5% CO_2. The culture media, antibiotics, and FCS were purchased from Sigma-Aldrich (Steinheim, Germany).

4.10. THP-1 Cell Differentiation

THP-1 were seeded onto 24-well plastic plates (Nunc, Roskilde, Denmark) at a density of 5 × 10^5 cells/well in RPMI 1640 supplemented with 10% FCS, and treated with a final concentration of 50 ng/mL phorbol 12-myristate 13-acetate (PMA) (Sigma-Aldrich, Steinheim, Germany) for three days to induce maturation toward adherent macrophage-like cells. Subsequently, unattached cells were removed and after three-time washing adherent THP-1 cells were cultured in medium without PMA for three consecutive days with daily fresh medium change.

4.11. THP-1 Cell Viability Assay

The viability of THP-1 cells exposed to *L. dumoffii* bacteria was determined by the MTT assay, in which the yellow tetrazolium salt 3-(4,5 dimethyl-2-thiazolyl)-2,5-diphenyl-2H-tetrazolium bromide (MTT) is metabolized by viable cells to purple formazan crystals. The cells at a density of 5×10^4 cells/well were seeded onto 96-well plates (Nunc, Roskilde, Denmark) and cultured in 10% RPMI 1640 for 24 h. Next, the medium was replaced with a fresh one with addition of 2% FCS and different concentrations of *L. dumoffii* at a MOI of 10, 50, 100, and 200. The treated cells were maintained in a humidified CO_2-incubator at 37 °C for 4 and 24 h. Subsequently, MTT solution (25 μL of 5 mg/mL in PBS) (Sigma-Aldrich, Steinheim, Germany) was added to each well. The cells were incubated at 37 °C for 3 h, and 100 μL of SDS in 0.01 M HCl was added to dissolve the formazan crystals during overnight incubation. The controls included native (non-treated) cells and the medium alone. The spectrophotometric absorbance was measured at 570 nm wavelength using a VICTOR X4 Multilabel Plate Reader (Perkin Elmer, Waltham, MA, USA). The data are presented as percentage of control cell viability.

4.12. In Vitro Infection of Differentiated THP-1 with L. dumoffii—Control of Internalization and Intracellular Growth

Differentiated THP-1 cells (as described in 4.10.) were infected with 10 MOI of live *L. dumoffii* cultured on choline-supplemented and non-supplemented medium. After incubation for 2 h at 37 °C, 5% CO_2, nonphagocytized bacteria were killed by the addition of 100 mg of gentamicin/mL for 1 h. Next, supernatants were removed and the macrophages were washed three-times with PBS. 1 mL of sterile distilled water (for bacterial internalization study) or 1 mL of RPMI 1640 supplemented with 10% FCS (without antibiotics) was added and the macrophages were incubated for 24, 48, and 72 h (intracellular bacterial growth study).

4.12.1. Bacteria Internalization Assay

Cells suspended in 1 mL of sterile distilled water were disrupted by aspiration through a 25-gauge needle and then series of 10-fold dilutions were made. Subsequently, 0.1 mL of each dilution was inoculated onto BCYE agar and colonies of culturable *L. dumoffii* were counted after three days of incubation at 37 °C, 5% CO_2.

4.12.2. Intracellular Bacteria Growth Assay

After 24, 48, and 72 h of cell culture incubation, supernatants were collected into sterile tubes, centrifuged (8000 rpm/min for 10 min), and washed with sterile distilled water. One milliliter of sterile distilled water was added to THP-1 cells, which were disrupted as described above and pooled with centrifuged pellet of the respective supernatant. 0.1 mL of series of 10-fold dilutions of each sample were inoculated onto BCYE agar and incubated as above.

Formation of colonies was determined in triplicate for at least two independent experiments.

4.13. Induction of TNF-α with L. dumoffii in THP-1 Cells

The differentiated macrophages were treated with live or dead *L. dumoffii* for 4 h at a MOI = 10 or 100. In another experiment, the macrophages were treated with different concentrations (10–1000 ng/mL) of *L. dumoffii* outer or inner membranes non-treated and treated with temperature (90 °C, 20 min).

Additional controls were performed: non-treated macrophages, non-activated THP-1 cells.

In all experiments, the bacteria had been previously cultured on medium with or without addition of choline (choline-trimethyl-d_9 chloride, Sigma-Aldrich, Steinheim, Germany). After incubation for 4 h at 37 °C, 5% CO_2, cell culture supernatants were collected and frozen immediately at −80 °C for further TNF-α determination. TNF-α level was measured by the ELISA method using a commercial kit from R&D Systems according to the manufacturer's instructions (R&D Systems Inc., Minneapolis, MN, USA). All experiments were conducted in five independent replicates.

4.14. Statistics

Values are expressed as mean ± S.D. Results were statistically evaluated using two-way ANOVA and Tukey's *post hoc* tests (Statistica software ver. 6.0, StatSoft Inc., Tulsa, OK, USA, 2001). p values of ≤0.05 were considered significant.

5. Conclusions

We have demonstrated that *L. dumoffii* has an ability to utilize extracellular choline in the Pcs pathway, which may be a dominant pathway, as indicated by the presence of a strong promoter for the phosphatidylcholine synthase gene. The use of the LC/ESI-MS technique allowed us to provide evidence that the PCs synthesized by *L. dumoffii* are localized in both the inner and outer membranes, but they are structurally different. This technique allowed identification of the PC species in total lipid extracts isolated from live bacteria without any chemical modification, thereby showing the physiological state of the cells. The structural differences in the PC species affect the induction of TNF-α—a key player in immuno-inflammatory response. In the presence of choline, the bacteria exhibited a reduced capability of TNF-α induction. The reduction of inflammatory response caused by PC species might collaborate to evade the immune system, thus, allowing *L. dumoffii* to establish in the host. The identification of selective inhibitors of phosphatidylcholine synthase may contribute to development of specific adjuvant therapy in treatment of legionellosis.

Acknowledgments

The authors acknowledge Anna Skorupska (Department of Genetics and Microbiology, Maria Curie-Sklodowska University, Lublin, Poland) for critical reading and comments to the manuscript. This work was supported by the grant from the Ministry of Science and Higher Education No. N N303 822640.

References

1. Fields, B.S.; Benson, R.F.; Besser, R.E. *Legionella* and Legionnaires' disease: 25 years of investigation. *Clin. Microbiol. Rev.* **2002**, *15*, 506–526.
2. Euzeby, J.P. List of Bacterial Names with Standing in Nomenclature. Available online: http://www.bacterio.cict.fr/ (accessed on 3 April 2014).
3. Diederen, B.M. *Legionella* spp. and Legionnaires' disease. *J. Infect.* **2008**, *56*, 1–12.

4. Yu, V.L.; Plouffe, J.F.; Pastoris, M.C.; Stout, J.E.; Schousboe, M.; Widmer, A. Distribution of *Legionella* species and serogroups isolated by culture in patients with sporadic community-acquired legionellosis: an international collaborative survey. *J. Infect. Dis.* **2002**, *186*, 127–128.
5. Fujita, I.H.; Tsuboi, M.; Ohotsuka, I.; Sano, Y.; Murakami, H.; Akioka, H. *Legionella dumoffii* and *Legionella pneumophila* serogroup 5 isolated from 2 cases of fulminant pneumonia. *J. Jpn. Assoc. Infect. Dis.* **1989**, *63*, 778–780.
6. Tompkins, L.S.; Roessler; B.J; Redd, S.C.; Markowitz, L.E.; Cohen, M.L. *Legionella* prosthetic-valve endocarditis. *N. Engl. J. Med.* **1988**, *318*, 530–535.
7. Flendrie, M.; Jeurissen, M.; Franssen, M.; Kwa, D.; Klaassen, C.; Vos, F. Septic arthritis caused by *Legionella dumoffii* in a patient with systemic lupus erythematosus-like disease. *J. Clin. Microbiol.* **2011**, *49*, 746–749.
8. Ensminger, A.W.; Isberg, R.R. *Legionella pneumophila* Dot/Icm translocated substrates: A sum of parts. *Curr. Opin. Microbiol.* **2009**, *12*, 67–73.
9. Sohlenkamp, C.; Lopez Lara, I.M.; Geiger, O. Biosynthesis of phosphatidylcholine in bacteria. *Prog. Lipid Res.* **2003**, *42*, 115–162.
10. Martinez-Morales, F.; Schobert, M.; Lopez-Lara, I.M.; Geiger, O. Pathways for phosphatidylcholine biosynthesis in bacteria. *Microbiology* **2003**, *149*, 3461–3471.
11. Palusinska-Szysz, M.; Janczarek, M.; Kalitynski, R.; Dawidowicz, A.L.; Russa, R. *Legionella bozemanae* synthesizes phosphatidylcholine from exogenous choline. *Microbiol. Res.* **2011**, *166*, 87–98.
12. Conover, G.M.; Martinez-Morales, F.; Heidtman, M.I.; Luo, Z.Q.; Tang, M.; Chen, C.; Geiger, O.; Isberg, R.R. Phosphatidylcholine synthesis required for optimal function of *Legionella pneumophila* virulence determinants. *Cell. Microbiol.* **2008**, *10*, 514–528.
13. McHugh, S.L.; Yamamoto, Y.; Klein, T.W.; Friedman, H. Murine macrophages differentially produce proinflammatory cytokines after infection with virulent *vs.* avirulent *Legionella pneumophila*. *J. Leukoc. Biol.* **2000**, *67*, 863–868.
14. Blanchard, D.K.; Djeu, J.Y.; Klein, T.W.; Friedman, H.; Stewart, W.E. Protective effects of tumor necrosis factor in experimental *Legionella pneumophila* infections of mice via activation of PMN function. *J. Leukoc. Biol.* **1988**, *43*, 429–435.
15. Skerrett S.J.; Bagby, G.J.; Schmidt, R.A.; Nelson, S. Antibody-mediated depletion of tumor necrosis factor-alpha impairs pulmonary host defenses to *Legionella pneumophila*. *J. Infect. Dis.* **1997**, *176*, 1019–1028.
16. Skerrett, S.J.; Martin, T.R. Roles for tumour necrosis factor alpha and nitric oxide in resistance of rat alveolar macrophages to *Legionella pneumophila*. *Infect. Immun.* **1996**, *64*, 3236–3243.
17. Brieland, J.; Remick, D.G.; Freeman, P.T.; Hurley, M.; Fantone, J.C.; Engleberg, N.C. *In vivo* regulation of replicative *Legionella pneumophila* lung infection by endogenous tumor necrosis factor alpha and nitric oxide. *Infect. Immun.* **1995**, *63*, 3253–3258.
18. Hsu, F.F.; Turk, J. Electrospray Ionization/Tandem Quadrupole Mass Spectrometric studies on phosphatidylcholines: The fragmentation processes. *J. Am. Soc. Mass Spectrom.* **2003**, *14*, 352–363.

19. Cazalet, C.; Rusniok, C.; Bruggemann, H.; Zidane, N.; Maginer, A. Evidence in the *Legionella pneumophila* genome for exploitation of host cell functions and high genome plasticity. *Nat. Genet.* **2004**, *36*, 1165–1173.
20. Morinaga, Y.; Yanagihara, K.; Nakamura, S.; Hasegawa, H.; Seki, M.; Izumikawa, K.; Kakeya, H.; Yamamoto, Y.; Yamada, Y.; Kohno, S.; *et al. Legionella pneumophila* induces cathepsin B-dependent necrotic cell death with releasing high mobility group box1 in macrophages. *Respir. Res.* **2010**, *11*, 158–166.
21. Franco, I.S.; Shohdy, N.; Shuman, H. The *Legionella pneumophila* effector VipA is an actin nucleator that alters host cell organelle trafficking. *PLoS Pathogens* **2012**, *8*, e1002546.
22. Palusinska-Szysz, M.; Kalitynski, R.; Russa, R.; Dawidowicz, A.L.; Drożański, W.J. Cellular envelope phospholipids from *Legionella lytica. FEMS Microbiol. Lett.* **2008**, *283*, 239–246.
23. Garrity, G.M.; Brown, A.; Vickers, R.M. *Tatlockia* and *Fluoribacter*: Two new genera of organisms resembling *Legionella pneumophila. Int. J. Syst. Bacteriol.* **1980**, *30*, 609–614.
24. Geiger, O.; Lopez-Lara, I.M.; Sohlenkamp, C. Phosphatidylcholine biosynthesis and function in bacteria. *Biochim. Biophys. Acta* **2013**, *183*, 503–513.
25. Lopez-Lara, I.M.; Geiger, O. Novel pathway for phosphatidylcholine biosynthesis in bacteria associated with eukaryotes. *J. Biotechnol.* **2001**, *91*, 211–221.
26. Minder, A.C.; de Rudder, K.E.; Narberhaus, F.; Fischer, H.M.; Hennecke, H. Phosphatidylcholine levels in *Bradyrhizobium japonicum* membranes are critical for an efficient symbiosis with the soybean host plant. *Mol. Microbiol.* **2001**, *39*, 1186–1198.
27. De Rudder, K.E.; Lopez-Lara, I.M.; Geiger, O. Inactivation of the gene for phospholipid *N*-methyltransferase in *Sinorhizobium meliloti*: Phosphatidylcholine is required for normal growth. *Mol. Microbiol.* **2000**, *37*, 763–772.
28. Wessel, M.; Klusener, S.; Godeke, J.; Fritz, C.; Hacker, S.; Narberhaus, F. Virulence of *Agrobacterium tumefaciens* requires phosphatidylcholine in the bacterial membrane. *Mol. Microbiol.* **2006**, *62*, 906–915.
29. Klüsener, S.; Aktas, M.; Thormann, K.M.; Wessel, M.; Narberhaus, F. Expression and physiological relevance of *Agrobacterium tumefaciens* phosphatidylcholine biosynthesis genes. *J. Bacteriol.* **2009**, *191*, 365–374.
30. Comerci, D.J.; Altabe, S.; de Mendoza D.; Ugalde, R.A. *Brucella abortus* synthesizes phosphatidylcholine from choline provided by the host. *J. Bacteriol.* **2006**, *188*, 1929–1934.
31. Conde-Alvarez, R.; Grillo, M.J.; Salcedo, S.P.; de Miguel M.J., Fugier, E. Synthesis of phosphatidylcholine, a typical eukaryotic phospholipid, is necessary for full virulence of the intracellular bacterial parasite *Brucella abortus. Cell. Microbiol.* **2006**, *8*, 1322–1335.
32. Malek, A.A.; Wargo, M.J.; Hogan, D.A. Absence of membrane phosphatidylcholine does not affect virulence and stress tolerance phenotypes in the opportunistic pathogen *Pseudomonas aeruginosa. PLoS One* **2012**, *7*, e30829.
33. Palusinska-Szysz, M.; Zdybicka-Barabas, A.; Pawlikowska-Pawlęga, B.; Mak, P.; Cytryńska, M. Anti-*Legionella dumoffii* activity of *Galleria mellonella* defensin and apolipophorin III. *Int. J. Mol. Sci.* **2012** *13*, 17048–17064.

34. Gustot, A.; Smriti Ruysschaert, J.M.; McHaourab, H.; Govaerts, C. Lipid composition regulates the orientation of transmembrane helices in HorA, an ABC multidrug transporter. *J. Biol. Chem.* **2010**, *285*, 14144–14151.
35. Hindahl, M.S.; Iglewski, B.H. Isolation and characterization of the *Legionella pneumophila* outer membrane. *J. Bacteriol.* **1984**, *159*, 107–113.
36. Wilderman, P.J.; Vasil, A.I.; Martin, W.E.; Murphy, R.C.; Vasil, M.L. *Pseudomonas aeruginosa* synthesizes phosphatidylcholine by use of the phosphatidylcholine synthase pathway. *J. Bacteriol.* **2002**, *184*, 4792–4799.
37. Tonks, A.R.; Morris, H.K.; Price, A.J.; Thomas, A.W., Jones, K.P.; Jackson, S.K. Dipalmitoylphosphatidylcholine modulates inflammatory functions of monocytic cells independently of mitogen activated protein kinase. *Clin. Exp. Immunol.* **2001**, *124*, 86–94.
38. Arata, S.; Newton, C.; Klein, T.W.; Yamamoto, Y.; Friedman, H. *Legionella pneumophila* induced tumor necrosis factor production in permissive versus nonpermissive macrophages. *Proc. Soc. Exp. Biol. Med.* **1993**, *203*, 26–29.
39. Cao, Q.; Mak, K.M.; Lieber, C.S. Dilinoleoylphosphatidylcholine decreases acetaldehyde-induced TNF-alpha generation in Kupffer cells of ethanol-fed rats. *Biochem. Biophys. Res. Commun.* **2002**, *6*, 459–464.
40. Edelstein, P.H. Improved semiselective medium for isolation of *Legionella pneumophila* from contaminated clinical and environmental specimens. *J. Clin. Microbiol.* **1981**, *14*, 298–303.
41. Sambrook, J.; Fritsch, E.F.; Maniatis, T. *Molecular Cloning: A Laboratory Manual*, 2nd ed.; Cold Spring Harbor Laboratory Press: Cold Spring Harbor, NY, USA, 1989.
42. Osborn, M.J.; Gander, J.E.; Parisi, E.; Garson, J. Mechanism of assembly of the outer membrane of *Salmonella typhimurium*. *J. Biol. Chem.* **1972**, *247*, 3962–3972.
43. Bradford, M.M. A rapid and sensitive method for the quantitation of microgram quantities of protein utilizing the principle of protein-dye binding. *Anal. Biochem.* **1976**, *72*, 248–254.
44. Bligh, E.G.; Dyer, J.W. A rapid method of total lipid extraction and purification. *Can. J. Biochem. Physiol.* **1959**, *37*, 911–917.

16

Radiation-Induced Changes in Serum Lipidome of Head and Neck Cancer Patients

Karol Jelonek [1,*], Monika Pietrowska [1], Malgorzata Ros [1,2], Adam Zagdanski [3,5], Agnieszka Suchwalko [4,5], Joanna Polanska [6], Michal Marczyk [6], Tomasz Rutkowski [1], Krzysztof Skladowski [1], Malcolm R. Clench [7] and Piotr Widlak [1]

[1] Center for Translational Research and Molecular Biology of Cancer, Maria Sklodowska-Curie Memorial Cancer Center and Institute of Oncology, Gliwice Branch, Wybrzeze Armii Krajowej 15, 44-100 Gliwice, Poland; E-Mails: m_pietrowska@io.gliwice.pl (M.P.); ros.malgorzata@gmail.com (M.R.); tomr22@tlen.pl (T.R.); skladowski@io.gliwice.pl (K.S.); widlak@io.gliwice.pl (P.W.)

[2] Polish-Japanese Institute of Information Technologies, Koszykowa 86, 02-008 Warszawa, Poland

[3] Institute of Mathematics and Computer Science, Wroclaw University of Technology, Janiszewskiego 14a, 50-370 Wroclaw, Poland; E-Mail: Adam.Zagdanski@pwr.wroc.pl

[4] Institute of Biomedical Engineering and Instrumentation, Wroclaw University of Technology, Wybrzeze Wyspianskiego 27, 50-370 Wroclaw, Poland; E-Mail: agnieszka.suchwalko@pwr.wroc.pl

[5] MedicWave AB, Kristian IV:s vag 3, 302 50 Halmstad, Sweden

[6] Faculty of Automatic Control, Electronics and Computer Science, Silesian University of Technology, Akademicka 16, 44-100 Gliwice, Poland; E-Mails: Joanna.Polanska@polsl.pl (J.P.); Michal.Marczyk@polsl.pl (M.M.)

[7] Biomedical Research Centre, Sheffield Hallam University, Sheffield S1 1WB, UK; E-Mail: M.R.Clench@shu.ac.uk

* Author to whom correspondence should be addressed; E-Mail: kjelonek@io.gliwice.pl

---

Abstract: Cancer radiotherapy (RT) induces response of the whole patient's body that could be detected at the blood level. We aimed to identify changes induced in serum lipidome during RT and characterize their association with doses and volumes of irradiated tissue. Sixty-six patients treated with conformal RT because of head and neck cancer were enrolled in the study. Blood samples were collected before, during and about one month after the end of RT. Lipid extracts were analyzed using MALDI-oa-ToF mass spectrometry in positive ionization mode. The major changes were observed when pre-treatment and within-treatment samples were compared. Levels of several identified phosphatidylcholines, including (PC34), (PC36) and (PC38) variants, and lysophosphatidylcholines, including (LPC16) and (LPC18) variants, were first significantly decreased and then increased in post-treatment samples. Intensities of changes were correlated with doses of radiation received by patients. Of note, such correlations were more frequent when low-to-medium doses of radiation delivered during conformal RT to large volumes of normal tissues were analyzed. Additionally, some radiation-induced changes in serum lipidome were associated with toxicity of the treatment. Obtained results indicated the involvement of choline-related signaling and potential biological importance of exposure to clinically low/medium doses of radiation in patient's body response to radiation.

Keywords: dose-volume effect; intensity-modulated radiation therapy; mass spectrometry; radiation toxicity; serum lipidome

---

1. Introduction

Metabolomics, an emerging field of the "omics" sciences, has a capacity to deliver essential information about small biomolecules (<1 kDa) that are end-products of all cellular processes. Lipidomics, which deals with dynamic changes of cellular lipids and their derivatives, is one of the most complex areas of metabolomics [1]. More than 500 different lipid species was reported to be present in human plasma specimens [2]. The most abundant category of lipids are glycerophospholipids (phospholipids; PLs). PLs are both key components of biological membranes and important players in different cellular mechanisms [3,4]. Derivatives of PLs are important signaling molecules involved in regulation of proliferation and apoptosis [5,6]. Of note, metabolism of phosphatidylcholines (PCs) and other PLs is significantly disturbed in cancer cells, hence elevated serum levels of their precursors (e.g., choline) and derivatives (e.g., lysophosphatidylcholines, LPCs) are promising cancer markers [7]. Changes in level of choline-containing lipids were observed in malignant tumors during anti-cancer therapy [8]. Metabolism and blood levels of PLs changed also after exposure to ionizing radiation [9,10]. Although such effects have only been studied in animal models until now, they indicated applicability of serum phospholipid profiles in assessment of radiation exposure.

Radiotherapy (RT), either alone or in combination with chemotherapy, is an effective treatment of different types of cancer allowing preservation of structure and function of a target organ. The effects of ionizing radiation concerns damage induced not only in cancer cells, but also in adjacent normal tissue. Conformal methods of radiotherapy, like intensity-modulated radiation therapy (IMRT), were developed to allow precise delivery of high radiation doses to a tumor volume, minimizing the dose delivered to surrounding normal tissues [11]. This technique is being used most extensively in treatment of tumors located near critical structures, such as head and neck cancers [12]. IMRT is accomplished by application of many non-coplanar radiation fields that markedly extends the volume of normal tissues being exposed to low doses of radiation, for which biological relevance is not clear at the moment [13]. Radiation-induced damage of normal tissues could lead to acute and/or late injury reactions, which in extreme cases might significantly affect patient's comfort and effectiveness of the treatment. For this reason planning and monitoring of radiotherapy would be greatly facilitated if molecular markers of individual response to radiation were available in the clinical practice. In addition, molecular markers of exposure to ionizing radiation would have a great applicability in the epidemiology field and for exposure assessment after radiation accidents [14].

Local irradiation during cancer radiotherapy induces patient's whole body response that could be detected at the level of blood components. Markers of human exposure to ionizing radiation have been searched in blood cells using different genetic and genomics approaches [15–18]. Mass spectrometry-based proteomics approaches have been also explored, which allowed identifying of radiotherapy-related changes in serum proteome of cancer patients [19,20]. More recently, it has been shown that IMRT-induced changes in the low-molecular-weight fraction of serum proteome of head and neck cancer patients were affected by clinically irrelevant doses of radiation delivered to large volumes of normal tissues [21]. Here we aimed to extend the analysis of radiotherapy-related changes and radiation dose-effects on the lipid component of serum. MALDI-oa-ToF profiling was applied for the first time to search for radiation-induced changes in human serum lipidome. The positive mode of MALDI ionization was selected in order to favor the analysis of choline-based compounds and other phospholipids, which already have been proposed as potential markers of the response to radiation and anti-cancer treatment [8,10].

2. Results

2.1. Exposure to Radiation during Radiotherapy Induced Changes in the Serum Lipidome Profiles

In the analyzed mass range 350–900 Da 842 spectral components (*i.e.*, lipid species with their isotope variants) common for all mass profiles were detected (an average mass profile is presented on Figure 1A). In order to find radiotherapy-related changes individual differential spectra were computed paired with respect to consecutive time points (*i.e.*, changes AΔB, BΔC and AΔC), and then the statistical significance of differences in component's abundances was estimated (Figure 1B shows resulting differential spectra). Several spectral components changed their abundances significantly between compared time points (FDR < 5% was selected as a statistical significance threshold), which are listed in Table 1 (complete data regarding all registered components are presented in Supplementary Table S1). We observed that major changes occurred between pre-treatment and within-treatment samples (the AΔB change), where 27 spectral components (lipid species) changed their abundance with high level of statistical significance (FDR < 5%). When within-treatment samples were compared with post-treatment samples (the BΔC change), 14 spectral components showed significantly changed abundance. However, abundances of only three spectral components remained different at high level of statistical significance when pre-treatment and post-treatment samples were compared (the AΔC change). Of note, we observed that seven spectral components significantly differentiated samples B from both samples A and samples C (registered *m*/*z* values = 520.36, 522.39, 603.68, 749.51, 760.63, 786.64 and 788.65 Da). Moreover, one spectral component (*m*/*z* value = 751.47 Da) differentiated samples A both from samples B and samples C. Half of the differentiating components were identified with respect to their lipid class (see Table 1), almost all of them being phospholipids containing the choline "head": phosphatidylcholines (PC; 10 compounds), lysophosphatidylcholines (LPC; 4 compounds) and sphingomyelines (SM; 2 compounds).

Table 1. Spectral components that changed abundances significantly between analyzed time points. Shown here is the registered *m/z* value, significant change in abundance, real pattern of changes, cluster number (hypothetical pattern of changes) and identification (lipid class and length of fatty acyl chains) of analyzed spectral components (*i.e.*, lipid species); components of isotopic envelope were excluded from analysis.

Ion mass[*m/z*]	Significant change (FDR < 5%)	Pattern of changes	Cluster number	Lipid class identification
373.08	AΔB	A < B > C	#4	not assigned
496.36	AΔB	A > B < C	#2	**LPC(16:0) + H^+**
520.36	AΔB;BΔC	A > B < C	#2	**LPC(18:2) + H^+**
522.39	AΔB;BΔC	A > B < C	#2	**LPC(18:1) + H^+**
524.38	AΔB	A > B < C	#2	**LPC(18:0) + H^+**
543.39	AΔB	A > B = C	#2	not assigned
560.28	AΔB	A > B = C	#1	**[Vitamin D3 adduct] + H^+**
564.64	BΔC	A = B > C	#4	**Cer(36:2) + H^+**
587.33	AΔB	A < B > C	#4	not assigned
601.12	AΔB	A < B > C	#4	not assigned
603.68	AΔB;BΔC	A < B < C	#6	not assigned
644.11	AΔB	A < B > C	#4	not assigned
703.58	BΔC	A = B < C	#6	**SM(34:1) + H^+**
721.49	BΔC	A > B < C	#2	not assigned
726.53	AΔC	A = B < C	#6	not assigned
730.62	AΔB	A < B = C	#3	**PC(32:2) + H^+**
732.47	BΔC	A > B < C	#2	not assigned
732.63	AΔB	A < B = C	#3	**PC(32:1) + H^+**
749.51	AΔB;BΔC	A > B < C	#2	not assigned
751.47	AΔB;AΔC	A > B = C	#1	not assigned
751.61	AΔB	A < B = C	#4	not assigned
755.42	AΔB	A > B = C	#1	not assigned
755.63	AΔB	A < B = C	#3	**SM(38:3) + H^+**
758.61	BΔC	A > B < C	#2	**PC(34:2) + H^+**
760.63	AΔB;BΔC	A > B < C	#2	**PC(34:1) + H^+**
762.63	BΔC	A > B < C	#2	**PC(34:0) + H^+**
767.47	BΔC	A > B < C	#2	not assigned
777.33	AΔC	A = B > C	#5	not assigned
784.62	AΔB	A > B < C	#2	**PC(36:3) + H^+**
786.64	AΔB;BΔC	A > B < C	#2	**PC(36:2) + H^+**
786.94	AΔB	A > B < C	#2	not assigned
788.65	AΔB;BΔC	A > B < C	#2	**PC(36:1) + H^+**
790.65	AΔB	A > B < C	#2	**PC(36:0) + H^+**
808.62	AΔB	A > B < C	#2	**PC(38:5) + H^+**
825.58	AΔB	A > B = C	#2	not assigned
839.50	AΔB	A > B = C	#4	not assigned

Figure 1. Mass profiles of serum lipids were affected during radiotherapy. (Panel **A**): Averaged mass spectrum of serum lipids registered in the 350–900 Da range for pre-treatment samples (**A**); (Panel **B**): Averaged differential spectrum for pre-treatment and within-treatment samples (AΔB); components that changed their abundances significantly (FDR < 5%) are marked with red lines.

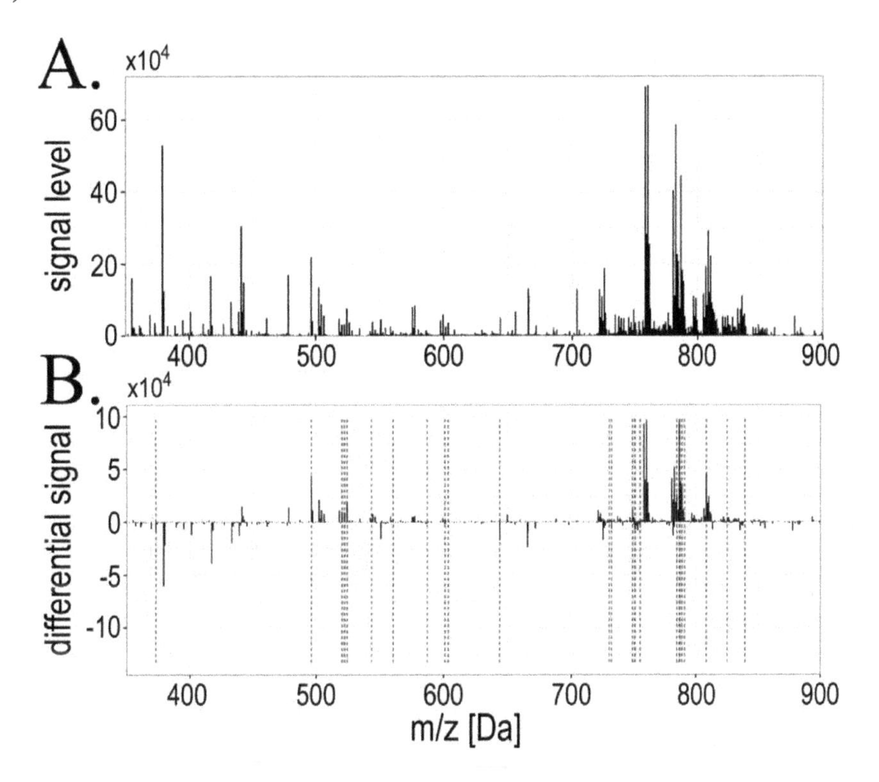

2.2. Radiotherapy-Related Changes in Lipidome Profiles Showed Different Time-Course Patterns

To identify different patterns of radiotherapy-related changes an unsupervised cluster analysis was performed. We identified six different hypothetical patterns of changes (clusters) characterized in Table 2 and depicted in Figure 2 (in case of a few spectral components where differences between time points were not statistically significant observed patterns of changes were not strictly coherent with cluster characteristics); detailed data are presented in Supplementary Table S2. Identified clusters could be further divided into three groups with two "mirrored" clusters in each, where reverse changes were observed: #1 [A > B = C] and #3 [A < B = C], #2 [A > B < C] and #4 [A < B > C], #5 [A = B > C] and #6 [A = B < C]. Of note, the second group (*i.e.*, clusters #2 and #4) where "earlier" changes (AΔB) were compensated by "later" changes (BΔC) was the most numerous (about 70% of all detected components). Furthermore, cluster #2 [A > B < C] contained the majority of differentiating components, that changed abundances significantly between consecutive time points (19 out of 36 "significant" components, see Table 1). This indicated that pattern of changes where "earlier" changes were reversed/compensated by "later" changes was the most common feature of lipidome profiles in serum from irradiated patients. Of note, the majority of "differentiating" LPCs and PCs belonged to cluster #2, and their serum levels decreased significantly during radiotherapy and then increased afterwards; these included LPC(18:2), LPC(18:1), PC(34:1), PC(36:2) and PC(36:1) (Figure 3). On the other hand PCs containing 32 carbons (32:2 and 32:1) and SM(38:3) significantly increased their levels during radiotherapy (cluster #3) (see Table 1).

Figure 2. Radiation induced changes followed different patterns. Presented are characteristics of six identified clusters of components with similar time-courses of changes; marked are average profiles for each cluster (red lines) and components that changed abundance significantly (FDR < 5%; solid black lines); other components belonging to each cluster are marked with grey lines.

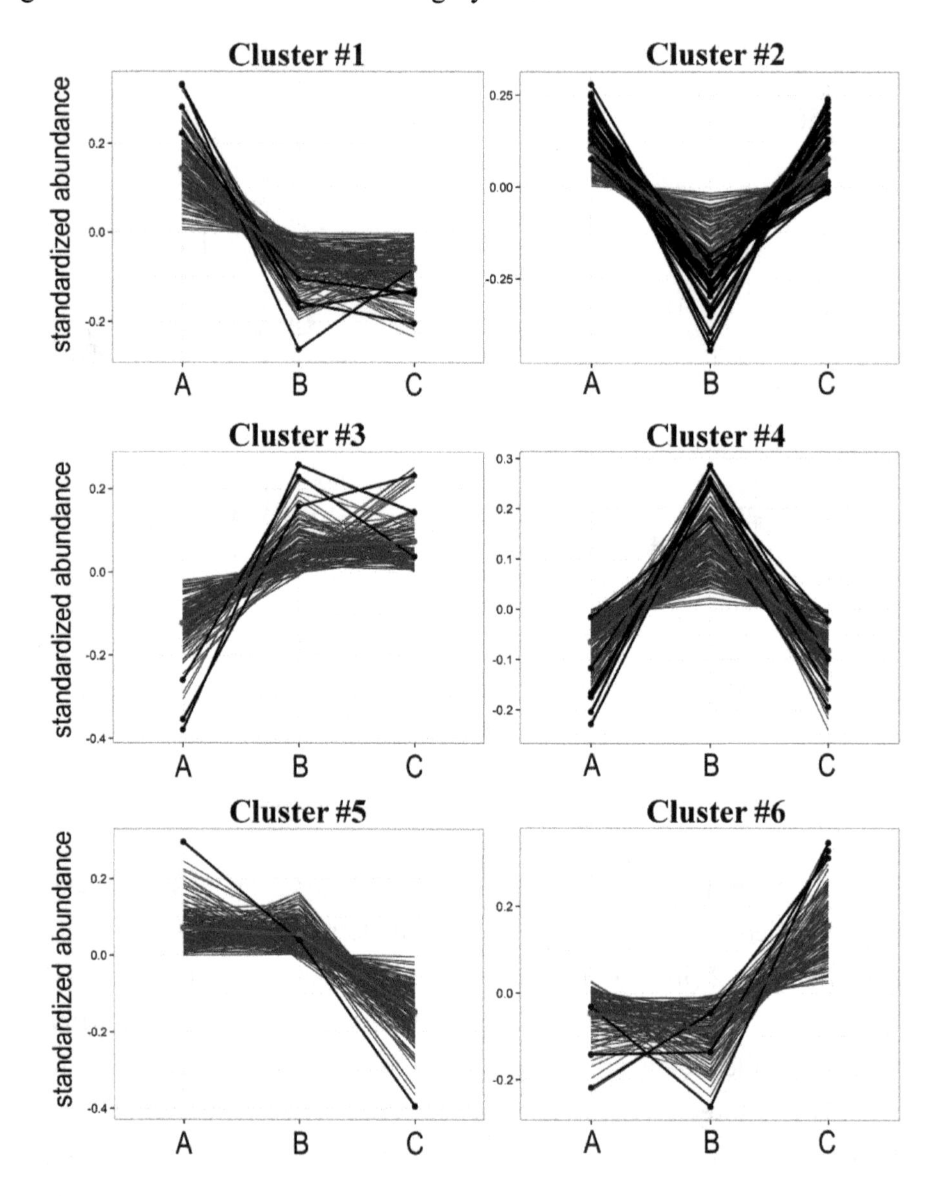

Table 2. Characteristics of identified clusters of spectral components.

Cluster	Pattern of change *	Number of components	Number of differentiating components **
#1	A > B = C	147	4
#2	A > B < C	129	19
#3	A < B = C	121	3
#4	A < B > C	170	6
#5	A = B > C	160	1
#6	A = B < C	115	3

* Pattern of change refers to the dominant characteristics of change in the specified cluster (with some not significant deviations from the pattern within the cluster); ** Components which abundances changed significantly between consecutive time points (FDR < 5%).

Figure 3. The abundance of several choline-containing phospholipids decreased markedly during radiotherapy and increased afterward. Presented are examples of lysophosphatidylcholines: LPC(18:2) [*m*/*z* = 520.36 Da] and LPC(18:1) [*m*/*z* = 522.39 Da], and phosphatidylcholines: PC(34:1) [*m*/*z* = 760.63 Da] and PC(36:1) [*m*/*z* = 788.65 Da]. Boxplots show: minimum, lower quartile, median, upper quartile and maximum values (outliers were removed from the plots for perspicuity).

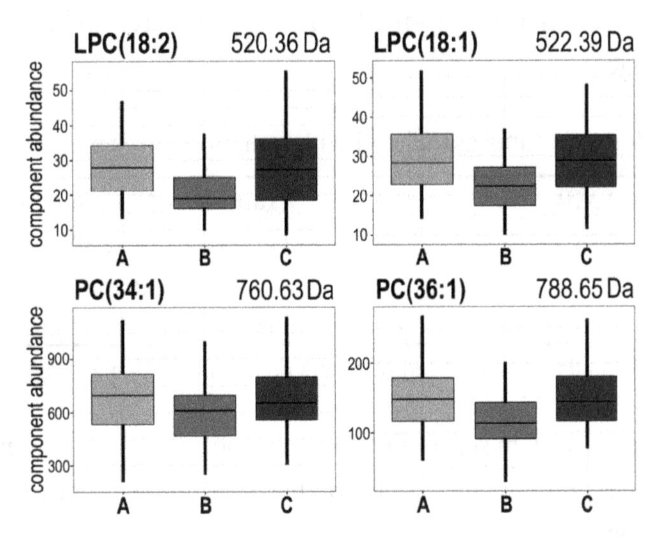

2.3. Radiotherapy-Related Changes in Serum Lipidome Were Associated with Doses of Radiation and Volumes of Irradiated Tissues

In the next step of the study we searched for association between features of serum lipidome (*i.e.*, changes in abundance of particular lipidome components) and doses of ionizing radiation received by patients (doses accumulated until a time point corresponding to the collection of sample B in case of the AΔB changes and total doses in case of the AΔC and BΔC changes). Correlations were identified between specific features of the serum lipidome and either the total (maximum) dose received by gross tumor volume (GTV), volume of the patient's body irradiated at different (smaller) doses or dose delivered to different volume of tissue. Numbers of serum lipidome components, for which changes in abundances correlated with maximum GTV doses are shown in Table 3 ($p < 0.05$ was selected as a statistical significance threshold). We found the highest number of identified correlations was observed in case of AΔC changes (60), yet clear association between maximum GTV doses and features of lipidome were detected also for the AΔB and BΔC changes (44 and 37 components, respectively). The maximum radiation doses (up to 72 Gy) were delivered only to tumor and its adjacent margins (usually 100–200 ccm), while tissue irradiated with lower doses represent much higher volumes (e.g., about 4000 ccm irradiated with 10 Gy). Hence, we searched for correlations between features of serum lipidome and volumes of tissues (including normal tissues irradiated upon IMRT treatment) irradiated with different doses (including "low" or "clinically irrelevant" doses); see Figure 4A,B. Figure 4C shows the numbers of lipidome components, which changes in abundance correlated with volume of tissue irradiated at different doses ($p = 0.05$ was selected as statistical significance threshold). We observed that association between lipidome features and dose-volume effects were the most frequent in case of larger tissue volumes irradiated with clinically low-to-medium doses (*i.e.*, less than 20 Gy in case of the AΔB change and less than 40 Gy in case of the BΔC and

AΔC changes; which corresponded to dose fractions below 1 Gy). Additionally, when reverse analysis was performed and serum lipidome features were correlated with doses delivered to a given volume similar results were obtained—majority of correlations were observed for high volumes irradiated with low doses (Figure 4D). Detailed data on correlation of serum lipidome features with volumes irradiated at a given dose or with doses delivered to a given volume are presented in Supplementary Tables S3–S5. Our results clearly indicated that radiotherapy-related changes in serum lipidome profiles depended on doses of delivered radiation, and that low-to-medium doses delivered to large volumes of normal tissue could affect observed changes.

Table 3. Serum lipidome features associated with radiation doses and acute radiation toxicity.

Change	GTV-D	AMR	Examples of components [*m/z*] *
AΔB	44	36	473.11; 514.21; 590.61; 872.42
BΔC	37	41	583.61; 669.64
AΔC	60	35	614.38; 641.33; 649.43; 655.65; 673.62; 765.64; 803.71; 886.88

* Components for which abundances correlated with both doses of radiation and radiation toxicity ($p < 0.05$).

Figure 4. Dose-volume effects in serum lipidome changes. (Panel **A**): Averaged Dose-Volume Histogram; doses corresponding to deciles of the area under curve of the histogram are marked with red lines; (Panel **B**): Correlations between volume of tissue irradiated with 13.7 Gy and changes in abundance of the m/z = 378.13 Da component in pre- and post-treatment samples (C-A; arbitrary units); Numbers of serum lipidome features correlating with tissue volumes irradiated at a given dose of radiation (Panel **C**) or doses of radiation delivered to a given tissue volume (Panel **D**). Shown here are the AΔB (black bars), BΔC (grey bars) and AΔC (empty bars) changes; p = 0.05 was selected as a statistical significance threshold (doses in parentheses refer to AΔB changes).

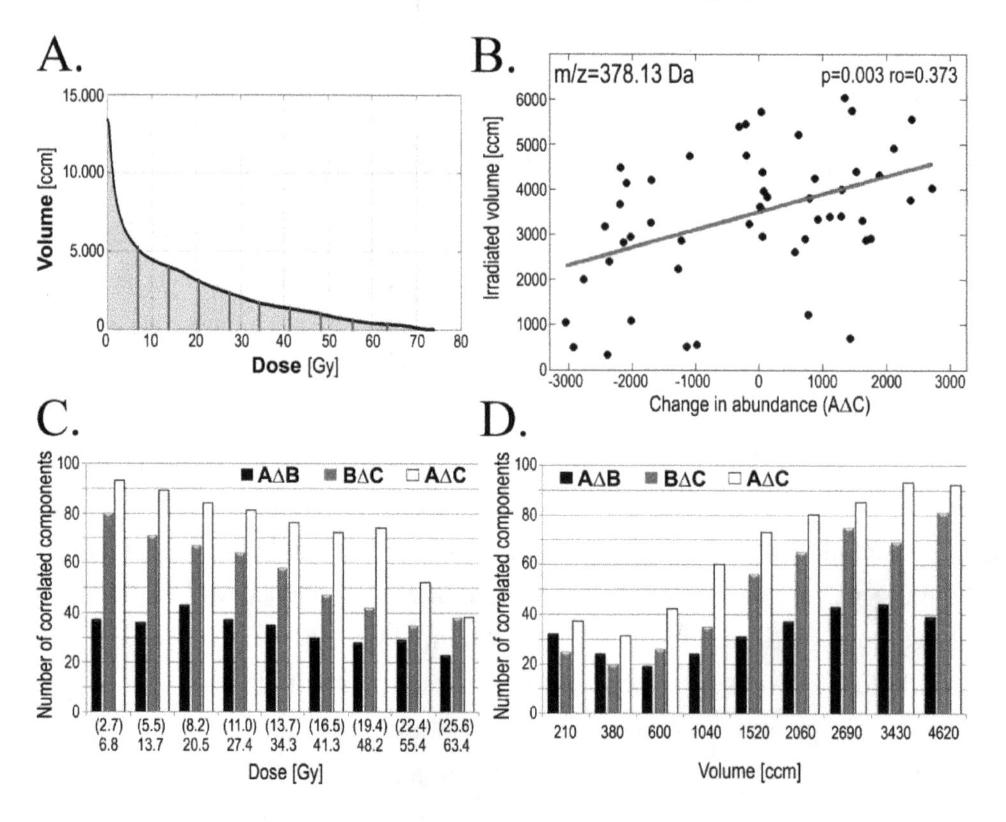

2.4. Radiotherapy-Related Changes in Serum Lipidome Were Associated with Radiation Toxicity

Finally, we searched for potential association of serum lipidome features with toxicity of the treatment. The clinically relevant response of normal tissues to toxicity of radiation was assessed using a modified Dische system [22] relying on the intensity of the acute mucosal reaction. The maximum AMR intensity correlates with both maximum GTV dose and volume of normal tissues irradiated with "intermediate" doses (about 0.8–1 Gy per fraction), which was documented in another study based on very similar group of head and neck cancer patients [21]. Here we searched for association between the early radiation toxicity and radiation-induced changes of serum lipidome features. We found correlation between changes in abundance of several lipidome components and the maximum AMR intensity: about 40 lipidome features correlated with the AMR for each of analyzed time-courses (Table 3). Of note, several serum lipidome features associated with the maximum AMR intensity also correlated with volumes of tissues irradiated at given doses of radiation (or radiation doses delivered to a given tissue volume). These features are listed in Table 3. Hence, we concluded that radiotherapy-related changes in the serum lipidome were associated with dose-related toxicity of radiation.

3. Discussion

To our knowledge, this is the first paper to analyze the response of human organism to ionizing radiation due to local cancer irradiation performed at the level of blood lipidome. The main changes in abundances of lipid serum components were observed between pre-treatment samples and samples collected during radiotherapy (the AΔB changes). Unsupervised cluster analysis revealed that major group of lipids (70% of registered spectral components) consisted of species, for which radiation-induced changes observed during radiotherapy were reversed/compensated in the post-treatment samples collected 1–2 months after the end of radiotherapy. A minor group of lipids (20% of registered spectral components) consisted of species, where radiation-induced changes detected during radiotherapy remained not reversed/compensated during the follow-up. As a consequence, only a few lipid species showed significant differences when their pre-exposure and post-exposure levels were compared (the AΔC changes). This observation indicated that in case of majority of serum lipids their return to the initial pre-exposure steady-state level was efficient enough during 1–2 months after the end of radiation treatment. Of note, when radiotherapy-related changes in serum proteome profiles were analyzed in a very similar group of patients, the major changes were observed in post-treatment samples collected 1–2 months after the end of radiotherapy (corresponding to the AΔC and BΔC changes). Such serum proteome changes apparently reflected escalation of radiation toxicity (acute mucosal reaction) and its subsequent healing during the follow-up [21]. Here, we show that radiotherapy-related changes in serum lipidome profiles are apparently "faster" compared to changes observed in the low-molecular-weight fraction of serum proteome. In fact, most radiation-induced changes in serum lipidome could be reversed within 1–2 months after completion of radiotherapy, while similar changes in serum proteome could be detected several months after the treatment. This indicated that changes in lipidome and proteome profiles observed in cancer patients treated with radiotherapy might reflect different radiation-induced mechanisms.

Lipid class identification (by MS/MS and/or annotation of registered *m/z* values at LipidMaps database) allowed annotation of the majority (85%) of lipids revealing radiation-induced changes as choline-based phospholipids. High extent of lipids referring to this type apparently resulted from both chosen conditions of serum extraction, which favored glycerophospholipids, sphingolipids, sterols and prenols [2], and positive mode of MALDI ionization, which narrowed the ionization of glycerophospholipids to neutral (zwitterionic) representatives, such as phosphatidylocholines and phosphatidylethanolamines [23]. The majority of PLs identified in this work, including different PCs and LPCs as well as SM(34:1), belong to the most abundant in their classes measured in human plasma samples [2]; SM(38:3) and Cer(36:2) are less common species. Phosphatidylocholines are the main building blocks of membrane bilayers and in plasma they are mostly located in high density lipoproteins (HDL). Decreased levels of PCs in serum of irradiated patients may be explained by their rapid turnover in stressed/damaged cells, which resulted in an increased PC's uptake from the blood. In addition to their main function as a membrane constituent, PCs have a role in signaling through the generation of LPCs (by phospholipase A_2 enzymes), SMs (by SM synthase), phosphatidic acid (PA; by phospholipase D enzymes) and/or diacylglycerols (DAG; by phospholipase C enzymes). From this point of view, significant down-regulation of major serum PCs observed during radiotherapy might be relevant for increasing capability of cell signaling pathways depending on PC-derived compounds. LPCs are reported to be the major bioactive lipid component of oxidized low density lipoproteins (LDL) and therefore mainly responsible for their pro-inflammatory functions [24]. Down-regulation of LPCs in blood was significantly correlated with activated inflammatory status in many cancer types [25]. Radiotherapy-related down-regulation of LPCs apparently indicated association between inflammatory processes and whole body response to radiation, which was previously documented at the level of serum proteome [21]. Another important class of signaling lipids derived from PCs and LPCs upon action of phospholipase D enzymes are lysophosphatidic acids (LPAs). The most prominent LPA functions include stimulation of cell proliferation, cell survival, and tumor cell invasion [26]. Down regulation of both PCs and LPCs may be therefore explained by the increased formation of LPAs. Another potential mechanism explaining down-regulation of PCs and LPCs involves the disruptive action of reactive oxygen species (ROS) appearing in high levels in irradiated tissues and causing the degradation of these lipids [27]. In contrast to PCs and LPCs, which indicated decreased levels during radiotherapy and were compensated during the follow-up, both identified sphingomyelines showed significant radiation-related up-regulation only: SM(38:3) during earlier stage of the treatment while SM(34:1) during later stage of the treatment or subsequent follow-up. New SM molecules were most probably generated from degraded PC compounds by SM synthase (this transferase utilizes a choline "head" from PC) and suitable ceramide molecules, which was coherent with observed down-regulation of Cer(36:2). SMs can be hydrolyzed back to ceramides by SMase action. This balance between sphingomyeline production and degradation is a key factor in SM-related apoptotic signaling, and generation of ceramides from SMs' degradation was reported to influence both the rate and form of cell death [28].

Although the model presented here is rather complicated and could be affected by many different processes ongoing in the patient's organism, one could expect that accumulation and subsequent healing of radiation-induced damage, such as acute mucosal reaction, would have the major influence on general therapy-related changes assessed at the level of serum lipidome. This expectation is supported by observed association of radiotherapy-related serum lipidome features with doses of radiation delivered to normal tissues and intensity of radiation-induced acute mucosal reaction. Although correlations identified here between particular lipidome components and different parameters reflecting radiation doses and toxicity possess moderate statistical power when analyzed separately, reliable conclusions could be drawn based on the general patterns of observed association. Of note, collected data indicated that low-to-medium doses delivered to large volumes of normal tissues during IMRT (considered as "therapeutically irrelevant") significantly affected whole body response observed at the level of serum lipidome. These observations are consistent with results of our earlier study, where similar association between dose-volume effects and features of the low-molecular-weight fraction of serum proteome has been observed in similar group of head and neck cancer patients [21]. The data indicated collectively, that a whole body response to the local cancer irradiation could be detected at the level of both serum proteome and lipidome. However, the majority of radiation-induced changes in abundances of serum lipids returned to pre-exposure steady-state levels within a relatively short time after the treatment, while changes in serum proteome could be detected even several months after irradiation.

4. Experimental Section

4.1. Characteristics of the Patients

Sixty-six patients with head and neck squamous cell carcinoma (HNSCC) were enrolled in this study. All of them were Caucasians (64 men) 45–82 years old (median age 63 years); 82% of them were current smokers and 86% alcohol consumers. Cancer was located mainly in larynx (45 pts.), but also in oropharynx (15 pts.) or hypopharynx (6 pts.). The primary tumor stage was scored as: T1 (21%), T2 (44%), T3 (26%) and T4 (9%); 68% of N0. All patients were subjected to IMRT using 6 MeV photons with 1.8 Gy daily fraction doses according to the continuous accelerated irradiation scheme (CAIR) [29]). Total radiation doses delivered to gross tumor volume (GTV) were in the range of 61.2–72 Gy (median 66.6 Gy). Neither surgery nor induction/concomitant chemotherapy was applied to patients enrolled in this study. Three consecutive blood samples (5 mL) were collected from each patient: pre-treatment sample A (within one week before RT; 66 pts.); within-treatment sample B (10–22 days after the start of RT, median 15 days; 66 pts.) and post-treatment sample C (23–59 days after the end of RT, median 36 days; 56 pts.). The acute mucosal reaction (AMR) was estimated using the modified Dische score system [22] every 3–5 days during the radiotherapy. The study was approved by the appropriate Ethics Committee and all participants provided informed consent indicating their conscious and voluntary participation.

4.2. Preparation of the Samples

Blood was collected into a 5 mL Vacutainer Tube (Becton Dickinson, Franklin Lakes, NJ, USA), incubated for 30 min RT to allow clotting, and then centrifuged at 1000× *g* for 10 min to remove the clot. The serum was aliquoted and stored at −70 °C until extraction. Total lipids were extracted according to modified Folch method [30]. In brief, 25 μL of serum was mixed with 350 μL of 1:1 methanol/chloroform mixture (*v*/*v*) containing antioxidants: 0.01% (*w*/*v*) 2,6-di-*tert*-butyl-4-methylphenol and 0.005% (*w*/*v*) retinol. The solution was vortex-mixed for 0.5 min and incubated for 30 min at 20 °C. Then 80 μL of water was added to the mixture, vortex-mixed for another 0.5 min and centrifuged (5 min, 10,000× *g*). Chloroform phase (the lower one) was kept and stored at −70 °C until mass spectrometry analysis (within three weeks).

4.3. MALDI Mass Spectrometry Analysis

Samples was analyzed using MALDI quadrupole/orthogonal acceleration ToF (oa-ToF) high-definition MS (HDMS) SYNAPT G2-HDMS™ system (Waters, Manchester, UK) equipped with the 355 nm Nd:YAG laser. First, 0.5 μL of sample was mixed directly on stainless steel target plate with 0.5 μL of 30 mg/mL of α-cyano-4-hydroxycinnamic acid (CHCA) matrix (Bruker Daltonics, Billerica, MA, USA) dissolved in 70% (*v*/*v*) acetonitrile containing 0.1% (*w*/*v*) trifluoroacetic acid; each sample was analyzed in triplicate (*i.e.*, placed on three different spots). Mass spectra were recorded using the positive ion mode in the 350–900 Da range with resolution of 10,000 FWHM. Spectra were calibrated with a standard solution of polyethylene glycol (PEG), and m/z scales were adjusted after acquisition using the PEG signal at m/z 701.3935 as a lock mass and centroided prior to the generation of accurate mass peak lists. Samples were spotted and analyzed in a random sequence to avoid “batch effect”.

4.4. Processing of the Mass Spectra

The initial preprocessing of spectra including alignment, detection of outlier profiles (using Dixon’s Q test), averaging of three technical replicas, additional alignment of averaged individual spectra (*i.e.*, averaged technical replicas), baseline removal and normalization of the total ion current (TIC) was performed according to procedures considering to be standard in the mass spectrometry field [31]. Preprocessed spectra were transferred to Spectrolyzer software suite (v.1.0, MedicWave AB, Halmstad, Sweden; [32]) for peak detection and binning (peak clustering) analysis. The processing steps performed in Spectrolyzer software were also consistent with the standard procedures used for spectral data processing [33,34].

4.5. Testing for Differentiating Spectral Components

For each of the 842 spectral components (spectral peaks) statistical significance of differences in abundance between different time points (*i.e.*, samples A, B and C) was estimated using appropriate tools available in R statistical software (see [35]). Individual differential spectra were computed paired with respect to consecutive time points (A–B, B–C and A–C), which resulted in 66 samples for comparison of A *vs.* B, and 56 samples for comparison A *vs.* C and B *vs.* C. To verify whether observed differences in abundances were significant, the Wilcoxon signed rank test was used (with the null hypothesis that the median value of intensities in the differential spectrum is equal to zero). To account for multiple comparisons the Storeys approach [36] that allows for FDR (false discovery rate) control was used. Statistically significant components involved also features that were identified manually as isotopes of other compounds; these components were rejected from the final list of specific components intended for identification.

4.6. Identification of Differentiating Components

Spectral components showing significant differences between analyzed time points (FDR < 5%) were analyzed by MS/MS in order to identify lipid class and length of fatty acyl chains. PCs and SM classes were recognized in MS/MS based on characteristic 184.1 Da phosphocholine fragmentation ion, while LPCs based on both 184.1 Da phosphocholine and 104.1 choline fragmentation ions Additionally, other spectral components were annotated at the LipidMaps database [37] based on their registered *m/z* values; mass tolerance 0.1 Da and no limit for category/class was applied. Compounds that were not identified experimentally (due to the low abundance of precursor or productions) were regarded as identified only if a single unique lipid record was return from the database search.

4.7. Analysis of Patterns of Changes

To investigate the general patterns of changes in abundances of spectral components between compared time points averaged "time courses" were computed based on individual time courses. Data standardization (centering and scaling separately for each of the spectral component) was performed to account for wide differences in abundance ranges observed for distinct components. Cluster analysis was performed using Partitioning Around Medoids (PAM) method, which is a classical algorithm of unsupervised analysis widely used for similar problems [38]. For a given number of clusters (k) the PAM finds k representative objects (so called medoids) that are most different from each other and assigns all the remaining objects to the most similar of the representatives. The similarity of the objects being an input for the PAM was computed based on correlation between average time courses. In order to determine the optimal k number and assess the quality of clustering results, an average Silhouette Width (SW) criterion was used [39], which revealed a six-cluster solution as the optimum.

4.8. Correlation of Component's Abundance with Radiation Parameters

Correlations between changes in abundance of spectral components and parameters reflecting absorbed doses of radiation, as well as maximal intensities of AMR, were analyzed using the Spearman's rank correlation coefficient. Total radiation dose absorbed by patient's body was estimated from the individual dose-volume histogram generated during the treatment planning. For the analysis of dose/volume effect we selected the doses corresponding to deciles of the area under the curve of the averaged dose-volume histogram (for details see [21]).

5. Conclusions

This study demonstrates for the first time the massive involvement of choline-based lipid serum components in the response of humans to ionizing radiation. Significant change in LPCs' levels suggests activation of inflammatory processes, while disturbances in levels of sphingomyelines and ceramides indicate involvement of apoptotic pathways. Additionally, correlations of lipidome changes with low and moderate radiation doses call attention to the biological relevance of "therapeutically irrelevant" doses during IMRT.

Acknowledgments

This work was supported by the European Social Fund within the INTERKADRA project UDA-POKL-04.01.01-00-014/10-00 to KJ and MR. JP was supported by the SUT, Grant BK-214/Rau-1/2013/10. PW was supported by the Polish National Science Centre, Grant 2011/01/B/NZ4/03563.

Author Contributions

KJ—Performed and interpreted experiments, identified lipid class and length of fatty acyl chains of statistically significant lipid components, prepared the final manuscript, MP—Interpreted MS results, MR—Interpreted MS results, AZ—Performed testing for differentiating spectral components, AS—Performed analysis of patterns of changes, JP—Performed analyses of correlation of component's abundance with radiation parameters, MM—Performed analyses of correlation of component's abundance with radiation parameters, TR—Collected and interpreted clinical data, KS—Collected and interpreted clinical data, MRC—Interpreted MS results, PW—Designed the project, designed and interpreted experiments, prepared the final manuscript. All authors read and approved the final manuscript.

References

1. Dennis, E.A. Lipidomics joins the omics evolution. *Proc. Natl. Acad. Sci. USA* **2009**, *106*, 2089–2090.
2. Quehenberger, O.; Armando, A.M.; Brown, A.H.; Milne, S.B.; Myers, D.S.; Merrill, A.H.; Bandyopadhyay, S.; Jones, K.N.; Kelly, S.; Shaner, R.L.; *et al.* Lipidomics reveals a remarkable diversity of lipids in human plasma. *J. Lipid Res.* **2010**, *51*, 3299–3305.
3. Marcus, A.J.; Hajjar, D.P. Vascular transcellular signaling. *J. Lipid Res.* **1993**, *34*, 2017–2031.
4. Berridge, M.J. Inositol trisphosphate and calcium signaling. *Nature* **1993**, *361*, 315–325.
5. Wright, M.M.; Howe, A.G.; Zaremberg, V. Cell membranes and apoptosis: Role of cardiolipin, phosphatidylcholine, and anticancer lipid analogues. *Biochem. Cell Biol.* **2004**, *82*, 18–26.
6. Bartke, N.; Hannun, Y.A. Bioactive sphingolipids: Metabolism and function. *J. Lipid Res.* **2009**, *50* (Suppl.), 91–96.
7. Ackerstaff, E.; Glunde, K.; Bhujwalla, Z.M. Choline phospholipid metabolism: A target in cancer cells? *J. Cell. Biochem.* **2003**, *90*, 525–533.
8. Jagannathan, N.R.; Kumar, M.; Seenu, V.; Coshic, O.; Dwivedi, S.N.; Julka, P.K.; Srivastava, A.; Rath, G.K. Evaluation of total choline from *in vivo* volume localized proton MR spectroscopy and its response to neoadjuvant chemotherapy in locally advanced breast cancer. *Br. J. Cancer* **2001**, *84*, 1016–1022.
9. Feurgard, C.; Bayle, D.; Guezingar, F.; Serougne, C.; Mazur, A.; Lutton, C.; Aigueperse, J.; Gourmelon, P.; Mathe, D. Effects of ionizing radiation (neutrons/gamma rays) on plasma lipids and lipoproteins in rats. *Radiat. Res.* **1998**, *150*, 43–51.
10. Wang, C.; Yang, J.; Nie, J. Plasma phospholipid metabolic profiling and biomarkers of rats following radiation exposure based on liquid chromatography-mass spectrometry technique. *Biomed. Chromatogr.* **2009**, *23*, 1079–1085.

11. Halperin, E.C.; Perez, C.A.; Brady, L.W. *Perez and Brady's Principles and Practice of Radiation Oncology*, 5th ed.; Wolters Kluwer Health, Lippincott Williams & Wilkins: Philadelphia, PA, USA, 2008.
12. Lee, N.; Puri, D.R.; Blanco, A.I.; Chao, K.S. Intensity-modulated radiation therapy in head and neck cancers: An update. *Head Neck* **2007**, *29*, 387–400.
13. De Neve, W.; de Gersem, W.; Madani, I. Rational use of intensity-modulated radiation therapy: The importance of clinical outcome. *Semin. Radiat. Oncol.* **2012**, *22*, 40–49.
14. Rana, S.; Kumar, R.; Sultana, S.; Sharma, R.K. Radiation-induced biomarkers for the detection and assessment of absorbed radiation doses. *J. Pharm. Bioallied Sci.* **2010**, *2*, 189–196.
15. Garaj-Vrhovac, V.; Kopjar, N. The alkaline comet assay as biomarker in assessment of DNA damage in medical personnel occupationally exposed to ionizing radiation. *Mutagenesis* **2003**, *18*, 265–271.
16. Kang, C.M.; Park, K.P.; Song, J.E.; Jeoung, D.I.; Cho, C.K.; Kim, T.H.; Bae, S.; Lee, S.J.; Lee, Y.S.; Possible biomarkers for ionizing radiation exposure in human peripheral blood lymphocytes. *Radiat. Res.* **2003**, *159*, 312–319.
17. Amundson, S.; Do, K.; Shahab, S.; Bittner, M.; Meltzer, P.; Trent, J.; Fornace, A.J. Identification of potential mRNA biomarkers in peripheral blood lymphocytes for human exposure to ionizing radiation. *Radiat. Res.* **2000**, *154*, 342–346.
18. Mah, L.J.; El-Osta, A.; Karagiannis, T.C. γH2AX: A sensitive molecular marker of DNA damage and repair. *Leukemia* **2010**, *24*, 679–686.
19. Menard, C.; Johann, D.; Lowenthal, M.; Muanza, T.; Sproull, M.; Ross, S.; Gulley, J.; Petricoin, E.; Coleman, C.N.; Camphausen, K. Discovering clinical biomarkers of ionizing radiation exposure with serum proteomic analysis. *Cancer Res.* **2006**, *66*, 1844–1850.
20. Widlak, P.; Pietrowska, M.; Wojtkiewicz, K.; Rutkowski, T.; Wygoda, A.; Marczak, L.; Marczyk, M.; Polańska, J.; Walaszczyk, A.; Domińczyk, I.; *et al.* Radiation-related changes in serum proteome profiles detected by mass spectrometry in blood of patients treated with radiotherapy due to larynx cancer. *J. Radiat. Res.* **2011**, *52*, 575–581.
21. Widlak, P.; Pietrowska, M.; Polanska, J.; Rutkowski, T.; Jelonek, K.; Kalinowska-Herok, M.; Gdowicz-Klosok, A.; Wygoda, A.; Tarnawski, R.; Skladowski, K. Radiotherapy-related changes in serum proteome patterns of head and neck cancer patients; the effect of low and medium doses of radiation delivered to large volumes of normal tissue. *J. Transl. Med.* **2013**, *11*, doi:10.1186/1479-5876-11-299.
22. Wygoda, A.; Maciejewski, B.; Skladowski, K.; Hutnik, M.; Pilecki, B.; Golen, M.; Rutkowski, T. Pattern analysis of acute mucosal reactions in patients with head and neck cancer treated with conventional and accelerated irradiation. *Int. J. Radiat. Oncol. Biol. Phys.* **2009**, *73*, 384–390.
23. Fuchs, B.; Süss, R.; Schiller, J. An update of MALDI-TOF mass spectrometry in lipid research. *Prog. Lipid Res.* **2010**, *49*, 450–475.
24. Huang, Y.H.; Schäfer-Elinder, L.; Wu, R.; Claesson, H.E.; Frostegard, J. Lysophosphatidylcholine (LPC) induces proinflammatory cytokines by a platelet-activating factor (PAF) receptor-dependent mechanism. *Clin. Exp. Immunol.* **1999**, *116*, 326–331.

25. Taylor, L.A.; Arends, J.; Hodina, A.K.; Unger, C.; Massing, U. Plasma lyso-phosphatidylcholine concentration is decreased in cancer patients with weight loss and activated inflammatory status. *Lipids Health Dis.* **2007**, *6*, 17.
26. Fang, X.; Schummer, M.; Mao, M.; Yu, S.; Tabassam, F.H.; Swaby, R.; Hasegawa, Y.; Tanyi, J.L.; LaPushin, R.; Eder, A.; *et al.* Lysophosphatidic acid is a bioactive mediator in ovarian cancer. *Biochim. Biophys. Acta* **2002**, *23*, 57–64.
27. Schiller, J.; Fuchs, B.; Arnhold, J.; Arnold, K. Contribution of reactive oxygen species to cartilage degradation in rheumatic diseases: Molecular pathways, diagnosis and potential therapeutic strategies. *Curr. Med. Chem.* **2003**, *10*, 2123–2145.
28. Green, D.R. Apoptosis and sphingomyelin hydrolysis: The flip side. *J. Cell Biol.* **2000**, *150*, 5–8.
29. Skladowski, K.; Maciejewski, B.; Golen, M.; Tarnawski, R.; Slosarek, K.; Suwinski, R.; Sygula, M.; Wygoda, A. Continuous accelerated 7-days-a-week radiotherapy for head-and-neck cancer: Long-term results of phase III clinical trial. *Int. J. Radiat. Oncol. Biol. Phys.* **2006**, *66*, 706–713.
30. Folch, J.; Lees, M.; Stanley, G.H.S. A simple method for the isolation and purification of total lipids from animal tissues. *J. Biol. Chem.* **1957**, *226*, 497–509.
31. Hilario, M.; Kalousis, A.; Pellegrini, C.; Müller, M. Processing and classification of protein mass spectra. *Mass Spectrom. Rev.* **2006**, *25*, 409–449.
32. Spectrolyzer Software Suite, MedicWave AB, Halmstad, Sweden. Available online: http://www.spectrolyzer.com/spectrolyzer/help-support/manual/ (accessed on 27 January 2014).
33. Cruz-Marcelo, A.; Guerra, R.; Vannucci, M.; Li, Y.; Lau, C.; Man, T. Comparison of algorithms for preprocessing of SELDI-TOF mass spectrometry data. *Bioinformatics* **2008**, *24*, 2129–2136.
34. Yang, C.; He, Z.; Yu, W. Comparison of public peak detection algorithms for MALDI mass spectrometry data analysis. *BMC Bioinform.* **2009**, *10*, doi:10.1186/1471-2105-10-4.
35. R Core Team. R: A language and environment for statistical computing. R Foundation for Statistical Computing. Available online: http://www.R-project.org/ (accessed on 28 January 2014).
36. Storey, J.D. A direct approach to false discovery rates. *JRSSB* **2002**, *64*, 479–498.
37. Sud, M.; Fahy, E.; Cotter, D.; Brown, A.; Dennis, E.A.; Glass, C.K.; Merrill, A.H., Jr.; Murphy, R.C.; Raetz, C.R.H.; Russell, D.W.; *et al.* LMSD: LIPID MAPS structure database. *Nucleic Acids Res.* **2006**, *35*, D527–D532.
38. Kaufman, L.; Rousseeuw, P.J. *Finding Groups in Data: An Introduction to Cluster Analysis*; John Wiley & Sons, Inc.: Hoboken, NJ, USA, 1990.
39. Handl, J.; Knowles, J.; Kell, D.B. Computational cluster validation in post-genomic data analysis. *Bioinformatics* **2005**, *21*, 3201–3212.

Permissions

All chapters in this book were first published in MDPI; hereby published with permission under the Creative Commons Attribution License or equivalent. Every chapter published in this book has been scrutinized by our experts. Their significance has been extensively debated. The topics covered herein carry significant findings which will fuel the growth of the discipline. They may even be implemented as practical applications or may be referred to as a beginning point for another development.

The contributors of this book come from diverse backgrounds, making this book a truly international effort. This book will bring forth new frontiers with its revolutionizing research information and detailed analysis of the nascent developments around the world.

We would like to thank all the contributing authors for lending their expertise to make the book truly unique. They have played a crucial role in the development of this book. Without their invaluable contributions this book wouldn't have been possible. They have made vital efforts to compile up to date information on the varied aspects of this subject to make this book a valuable addition to the collection of many professionals and students.

This book was conceptualized with the vision of imparting up-to-date information and advanced data in this field. To ensure the same, a matchless editorial board was set up. Every individual on the board went through rigorous rounds of assessment to prove their worth. After which they invested a large part of their time researching and compiling the most relevant data for our readers.

The editorial board has been involved in producing this book since its inception. They have spent rigorous hours researching and exploring the diverse topics which have resulted in the successful publishing of this book. They have passed on their knowledge of decades through this book. To expedite this challenging task, the publisher supported the team at every step. A small team of assistant editors was also appointed to further simplify the editing procedure and attain best results for the readers.

Apart from the editorial board, the designing team has also invested a significant amount of their time in understanding the subject and creating the most relevant covers. They scrutinized every image to scout for the most suitable representation of the subject and create an appropriate cover for the book.

The publishing team has been an ardent support to the editorial, designing and production team. Their endless efforts to recruit the best for this project, has resulted in the accomplishment of this book. They are a veteran in the field of academics and their pool of knowledge is as vast as their experience in printing. Their expertise and guidance has proved useful at every step. Their uncompromising quality standards have made this book an exceptional effort. Their encouragement from time to time has been an inspiration for everyone.

The publisher and the editorial board hope that this book will prove to be a valuable piece of knowledge for researchers, students, practitioners and scholars across the globe.

List of Contributors

Chia-Ming Lu
Institute of Traditional Medicine, School of Medicine, National Yang-Ming University, No. 155, Sec. 2, Li-Nong St, Beitou District, Taipei 11221, Taiwan

Lie-Chwen Lin
Institute of Traditional Medicine, School of Medicine, National Yang-Ming University, No. 155, Sec. 2, Li-Nong St, Beitou District, Taipei 11221, Taiwan
National Research Institute of Chinese Medicine, No. 155-1, Sec. 2, Li-Nong St., Beitou District, Taipei 11221, Taiwan

Tung-Hu Tsai
Institute of Traditional Medicine, School of Medicine, National Yang-Ming University, No. 155, Sec. 2, Li-Nong St, Beitou District, Taipei 11221, Taiwan
National Research Institute of Chinese Medicine, No. 155-1, Sec. 2, Li-Nong St., Beitou District, Taipei 11221, Taiwan
School of Pharmacy, College of Pharmacy, Kaohsiung Medical University, No. 100, Shih-Chuan 1st Road, Kaohsiung 80708, Taiwan
Department of Education and Research, Taipei City Hospital, No.145, Zhengzhou Rd., Datong Dist., Taipei 103, Taiwan

Artur Rydosz
Department of Electronics, AGH University of Science and Technology, Av. Mickiewicza 30, Krakow 30-059, Poland

Pei-Min Dai, Ying Wang, Lin-Liu Lu and Zhong-Qiu Liu
Department of Pharmaceutics, School of Pharmaceutical Sciences, Southern Medical University, Guangzhou 510515, Guangdong, China
International Institute for Translational Chinese Medicine, Guangzhou University of Chinese Medicine, Guangzhou 510006, Guangdong, China

Ling Ye, Shan Zeng, Zhi-Jie Zheng and Qiang Li
Department of Pharmaceutics, School of Pharmaceutical Sciences, Southern Medical University, Guangzhou 510515, Guangdong, China

Ning Liu, Ningning Sun and Xiang Gao
Central Laboratory, Jilin University Second Hospital, Changchun 130041, China

Zijian Li
Institute of Vascular Medicine, Peking University Third Hospital, Key Laboratory of Cardiovascular Molecular Biology and Regulatory Peptides, Ministry of Health, Key Laboratory of Molecular Cardiovascular Sciences, Ministry of Education and Beijing Key Laboratory of Cardiovascular Receptors Research, Beijing 100191, China

Barbora Šalovská and Martina Řezáčová
Institute of Medical Biochemistry, Faculty of Medicine in Hradec Králové, Charles University in Prague, Hradec Kralove 500 00, Czech Republic

Ivo Fabrik and Marek Link
Institute of Molecular Pathology, Faculty of Health Sciences in Hradec Králové, University of Defense in Brno, Hradec Kralove 500 01, Czech Republic

Kamila Ďurišová, Jiřina Vávrová and Aleš Tichý
Department of Radiobiology, Faculty of Health Sciences in Hradec Králové, University of Defense in Brno, Hradec Kralove 500 01, Czech Republic

Hiroki Kannen and Hisanao Hazama
Graduate School of Engineering, Osaka University, 2-1 Yamadaoka, Suita, Osaka 565-0871, Japan

Yasufumi Kaneda
Graduate School of Medicine, Osaka University, 2-2 Yamadaoka, Suita, Osaka 565-0871, Japan

Tatsuya Fujino
Graduate School of Science and Engineering, Tokyo Metropolitan University, 1-1 Minamiosawa Hachioji, Tokyo 192-0397, Japan

Kunio Awazu
Graduate School of Engineering, Osaka University, 2-1 Yamadaoka, Suita, Osaka 565-0871, Japan
Graduate School of Frontier Biosciences, Osaka University, 1-3 Yamadaoka, Suita, Osaka 565-0871, Japan
The Center for Advanced Medical Engineering and Informatics, Osaka University, 2-2 Yamadaoka, Suita, Osaka 565-0871, Japan

Armann Andaya, Nancy Villa, Weitao Jia, Christopher S. Fraser and Julie A. Leary
Department of Molecular and Cellular Biology, University of California at Davis, Davis, CA 95616, USA

Tai-Chia Chiu
Department of Applied Science, National Taitung University, 684 Section 1, Chunghua Road, Taitung 95002, Taiwan

Pieter Glibert, Liesbeth Vossaert, Katleen Van Steendam, Filip Van Nieuwerburgh, Maarten Dhaenens and Dieter Deforce
Laboratory of Pharmaceutical Biotechnology, Ghent University, 72 Harelbekestraat, B-9000 Ghent, Belgium

Stijn Lambrecht
Department of Rheumatology, Ghent University Hospital, 185 1P7 De Pintelaan, B-9000 Ghent, Belgium

Fritz Offner
Department of Hematology, Ghent University Hospital, 185 1P7 De Pintelaan, B-9000 Ghent, Belgium

Thomas Kipps
Department of Medicine, Moores Cancer Center, University of California at San Diego (UCSD), 3855 Health Sciences Drive, La Jolla, CA 92093, USA

Rongli Sun, Juan Zhang, Lihong Yin and Yuepu Pu
Key Laboratory of Environmental Medicine Engineering, Ministry of Education, School of Public Health, Southeast University, Nanjing 210009, Jiangsu, China

Ryuji Hiraguchi and Hisanao Hazama
Graduate School of Engineering, Osaka University, 2-1 Yamadaoka, Suita, Osaka 565-0871, Japan

Kenichirou Senoo and Yukinori Yahata
JEOL Ltd., 1156 Nakagamicho, Akishima, Tokyo 196-0022, Japan

Katsuyoshi Masuda
Suntory Institute for Bioorganic Research, Suntory Foundation for Life Sciences, 1-1-1 Wakayamadai, Shimamotocho, Mishimagun, Osaka 618-0024, Japan

Kunio Awazu
Graduate School of Engineering, Osaka University, 2-1 Yamadaoka, Suita, Osaka 565-0871, Japan
Graduate School of Frontier Biosciences, Osaka University, 1-1 Yamadaoka, Suita, Osaka 565-0871, Japan
The Center of Advanced Medical Engineering and Informatics, Osaka University, 2-2 Yamadaoka, Suita, Osaka 565-0871, Japan

Marko Jovanović
Department of Biotechnology, University of Rijeka, Radmile Matejčić 2, Rijeka 51000, Croatia

Richard Tyldesley-Worster
Waters Corporation, Stamford Avenue, Altrincham Road, Wilmslow SK9 4AX, UK

Gottfried Pohlentz
Institute for Hygiene, University of Muenster, Robert-Koch-Strasse 41, Muenster D-48149, Germany

Jasna Peter-Katalinić
Department of Biotechnology, University of Rijeka, Radmile Matejčić 2, Rijeka 51000, Croatia
Institute for medical Physics and Biophysics, University of Muenster, Robert-Koch-Strasse 31, Muenster D-48149, Germany

Ray K. Iles
Williamson Laboratory for Molecular Oncology, St Bartholomews Hospital, London EC1A 7BE, UK
ELK Foundation for Health Research, An Scoil Monzaird, Crieff PH7 4JT, UK
MAP Diagnostics Ltd., Ely, Cambridgeshire CB6 3FQ, UK

Laurence A. Cole
USA hCG Reference Service, Angel Fire, NM 87710, USA

Stephen A. Butler
Williamson Laboratory for Molecular Oncology, St Bartholomews Hospital, London EC1A 7BE, UK
MAP Diagnostics Ltd., Ely, Cambridgeshire CB6 3FQ, UK
USA hCG Reference Service, Angel Fire, NM 87710, USA

Feng He, Xiao Teng, Hanning Liu, Zhou Zhou, Yan Zhao, Shengshou Hu and Zhe Zheng
State Key Laboratory of Cardiovascular Disease, Fuwai Hospital, National Center for Cardiovascular Diseases, Chinese Academy of Medical Sciences and Peking Union Medical College, Beijing 100037, China

Haiyong Gu
Department of Cardiothoracic Surgery, Affiliated People's Hospital of Jiangsu University, Zhenjiang 212002, China

Marta Palusinska-Szysz, Monika Janczarek and Elżbieta Chmiel
Department of Genetics and Microbiology, Institute of Microbiology and Biotechnology, Maria Curie-Sklodowska University, Akademicka 19 St., 20-033 Lublin, Poland

Agnieszka Szuster-Ciesielska
Department of Virology and Immunology, Institute of Microbiology and Biotechnology, Maria Curie-Sklodowska University, Akademicka 19 St., 20-033 Lublin, Poland

Magdalena Kania and Witold Danikiewicz
Mass Spectrometry Group, Institute of Organic Chemistry Polish Academy of Sciences, Kasprzaka 44/52 St., 01-224 Warsaw, Poland

Karol Jelonek, Monika Pietrowska, Tomasz Rutkowski, Krzysztof Skladowski and Piotr Widlak
Center for Translational Research and Molecular Biology of Cancer, Maria Sklodowska-Curie Memorial Cancer Center and Institute of Oncology, Gliwice Branch, Wybrzeze Armii Krajowej 15, 44-100 Gliwice, Poland

Malgorzata Ros
Center for Translational Research and Molecular Biology of Cancer, Maria Sklodowska-Curie Memorial Cancer Center and Institute of Oncology, Gliwice Branch, Wybrzeze Armii Krajowej 15, 44-100 Gliwice, Poland
Polish-Japanese Institute of Information Technologies, Koszykowa 86, 02-008 Warszawa, Poland

Adam Zagdanski
Institute of Mathematics and Computer Science, Wroclaw University of Technology, Janiszewskiego 14a, 50-370 Wroclaw, Poland
MedicWave AB, Kristian IV:s vag 3, 302 50 Halmstad, Sweden

Agnieszka Suchwalko
Institute of Biomedical Engineering and Instrumentation, Wroclaw University of Technology, Wybrzeze Wyspianskiego 27, 50-370 Wroclaw, Poland
MedicWave AB, Kristian IV:s vag 3, 302 50 Halmstad, Sweden

Joanna Polanska and Michal Marczyk
Faculty of Automatic Control, Electronics and Computer Science, Silesian University of Technology, Akademicka 16, 44-100 Gliwice, Poland

Malcolm R. Clench
Biomedical Research Centre, Sheffield Hallam University, Sheffield S1 1WB, UK

Index

Printed in the USA
CPSIA information can be obtained
at www.ICGtesting.com
JSHW051413091023
49903JS00006B/404

9 781639 877362